AF732706

Organised Molecular Assemblies in the Solid State

Editorial Board

Founding Editor
Professor J. D. Dunitz, Laboratorium fur Organische Chemie, ETH-Zentrum, 8092 Zurich, Switzerland

Editors
Professor V. V. Boldyrev, Institute of Solid State Chemistry, Academy of Science, Novosibirsk 91, Russia

Professor A. Gavezzotti, Dipartimento di Chimica Strutturale e Stereochimca Inorganica, Università di Milano, via Venezian 21, 20133 Milano, Italy

Professor L. Leiserowitz, Institute of Solid State Chemistry, The Weizmann Institute of Science, Rehovot 76100, Israel

Professor J. M. McBride, Department of Chemistry, Yale University, 225 Prospect Street, PO Box 66, New Haven, Connecticut 06511, USA

Professor Y. Ohashi, Department of Chemistry, Faculty of Sciences, Tokyo Institute of Technology, 2-12-1 Ookayama, Meguro-ku, Tokyo 152, Japan

Professor J. K. Whitesell, Department of Chemistry, The University of Texas at Austin, Austin, Texas 78712-1167, USA

Books previously published in the Molecular Solid State series

Theoretical Aspects and Computer Modeling in the Molecular Solid State
Edited by Angelo Gavezzotti, University of Milan, Italy ISBN 471961876

Organised Molecular Assemblies in the Solid State

EDITED BY
JAMES K. WHITESELL
University of Texas at Austin, USA

The Molecular Solid State Volume 2

JOHN WILEY & SONS, LTD
Chichester • New York • Weinheim • Brisbane • Singapore • Toronto

Copyright © 1999 by John Wiley & Sons Ltd,
Baffins Lane, Chichester,
West Sussex PO19 1UD, England

National 01243 779777
International (+44) 1243 779777

e-mail (for orders and customer service enquiries): cs-books@wiley.co.uk.
Visit our Home Page on http://www.wiley.co.uk
or http://www.wiley.com

All Rights Reserved. No part of this publication may be reproduced, stored in a retrieval system, or transmitted, in any form or by any means, electronic, mechanical, photocopying, recording, scanning or otherwise, except under the terms of the Copyright, Designs and Patents Act, 1988 or under the terms of a licence issued by the Copyright Licensing Agency, 90 Tottenham Court Road, London, W1P 9HE, UK, without the permission in writing of the publisher.

Other Wiley Editorial Offices

John Wiley & Sons, Inc., 605 Third Avenue,
New York, NY 10158-0012, USA

WILEY-VCH Verlag GmbH, Pappelallee 3,
D-69469 Weinheim, Germany

Jacaranda Wiley Ltd, 33 Park Road, Milton,
Queensland 4064, Australia

John Wiley & Sons (Asia) Pte Ltd, 2 Clementi Loop #02-01,
Jin Xing Distripark, Singapore 129809

John Wiley & Sons (Canada) Ltd, 22 Worcester Road,
Rexdale, Ontario M9W 1L1, Canada

British Library Cataloguing in Publication Data

A catalogue record for this book is available from the British Library

ISBN 0 471 95232 X

Typeset in 10/12 pt Times by Thomson Press (India) Ltd., New Delhi
Printed and bound in Great Britain by Biddles, Guildford and King's Lynn
This book is printed on acid-free paper responsibly manufactured from sustainable forestation, for which at least two trees are planted for each one used for paper production.

Contents

Contributors

Donovan N. Chin, Department of Chemistry, Harvard University, 12 Oxford St., Cambridge, MA 02138, USA

P. H. Lakshminarasimhan, Department of Chemistry, Tulane University, New Orleans, LA 70118, USA

John C. MacDonald, Department of Chemistry, Harvard University, 12 Oxford St., Cambridge, MA 02138, USA

V. Ramamurthy, Department of Chemistry, Tulane University, New Orleans, LA 70118, USA

Rebecca J. Robbins, Department of Chemistry, Tulane University, New Orleans, LA 70118, USA

Timothy M. Swager, Department of Chemistry, Massachusetts Institute of Technology, Cambridge, MA 02139, USA

K. J. Thomas, Department of Chemistry, Tulane University, New Orleans, LA 70118, USA

Abraham Ulman, Department of Chemistry and the Herman Mark Polymer Research Institute, Polytechnic University, Six Metro Tech Center, Brooklyn, NY 11201, USA

K. Wheeler, Department of Chemistry, Delaware State University, 1200 N. Dupont Highway, Dover, DE 19901, USA

George M. Whitesides, Department of Chemistry, Harvard University, 12 Oxford St., Cambridge, MA 02138, USA

Jonathan A. Zerkowski, Department of Chemistry, Harvard University, 12 Oxford St., Cambridge, MA 02138, USA

Chapter 1

Self-assembly of Surfactant Molecules on Solid Surfaces

ABRAHAM ULMAN
Department of Chemistry and the Herman Mark Polymer Research Institute, Polytechnic University, Brooklyn, NY, USA

1. INTRODUCTION

In 1946 Zisman published the preparation of monomolecular layers by adsorption (self-assembly) of a surfactant onto a clean metal surface[1]. At that time, surface analytical tools were undeveloped, and the locus of chemistry was in the formation of covalent bonds. As the potential of self-assembly was not well recognized, this publication initiated only a limited level of interest. Later, studies of molecular interactions in two dimensions were carried out using the Langmuir–Blodgett technique[2], and one of the emerging centers of activity in the 1960s was Kuhn's laboratory in Göttingen. For years Kuhn's group has reacted glass surfaces with chlorosilane derivatives to form hydrophobic substrates; therefore, it is not surprising that the first self-assembled monolayer (SAM) of octadecyltrichlorosilane ($C_{18}H_{37}SiCl_3$, OTS) on a hydroxylated surface (SiO_2/Si) was prepared in Kuhn's laboratory[3]. However, the more recent revolution in the area came when Nuzzo and Allara showed that SAMs of alkanethiolates on gold can be prepared by adsorption of di-*n*-alkyl disulfides from dilute solutions[4]. Getting away from the moisture-sensitive alkyl trichlorosilanes as well as working with crystalline gold surfaces were two important reasons for the success of these SAMs. Monolayers of alkanethiolates on gold are the most studied SAMs to date.

Organised Molecular Assemblies in the Solid State Edited by James K. Whitesell
© 1999 John Wiley & Sons Ltd.

The formation of monolayers by self-assembly of surfactant molecules at surfaces is one example of the general phenomena of self-assembly, a dominating process in nature. There, self-assembly results in supermolecular hierarchical organizations of interlocking components, thus providing very complex systems[5]. In recent years, as chemistry became more involved in designing the molecular world[6], chemists have learned to take advantage of materials and processes in nature. In recent years, chemistry has been moving away from traditional disciplines and into the less defined interdisciplinary area. Research in nano, meso, micro, and macro scales has replaced conventional activities, and chemists have become more engaged in research at the interfaces of chemistry with physics, biology, and engineering. These important changes can find their origin partly in the areas of Langmuir–Blodgett and self-assembled monolayers.

The number of researchers actively involved in the study of SAMs has grown exponentially in recent years. This interest in the general area of self-assembly, and particularly in SAMs, stems partially from their perceived relevance to science and technology. In contrast to ultrathin films made by, for example, molecular beam epitaxy (MBE), and chemical vapor deposition (CVD), SAMs are highly ordered and oriented and can incorporate a wide range of groups both in the alkyl chain and at the chain termina[7]. Thus, a variety of surfaces with specific interactions can be produced with fine structural control[8,9] Due to their dense and stable structure, SAMs have potential applications in corrosion prevention, wear protection, and more. In addition, the biomimetic and biocompatible nature of SAMs makes their applications in chemical and biochemical sensing exciting. Their high molecular order parameter in SAMs makes them ideal as components in molecular electronics and electrooptic applications[10].

The ability to tailor both head and tail groups of the constituent molecules makes SAMs excellent systems for a more fundamental understanding of phenomena affected by competing intermolecular, molecular–substrate and molecule–solvent interactions like ordering and growth, wetting, adhesion, lubrication, corrosion and more. That SAMs are well-defined and accessible makes them good model systems for studies of physical chemistry and statistical physics in two dimensions, and the crossover to three dimensions. SAMs offer unique opportunities to increase fundamental understanding of self-organization, structure–property relationships, and interfacial phenomena. They provide the needed design flexibility, both at the individual molecular, and at the material levels.

The approach of building and understanding complicated systems by studying increasingly more complex, but well-controlled, model systems, is complementary to the cell biology approach of studying entire cells *in vitro*. Using increasingly more complex molecules leads to understanding of the relationship between the structure of the individual molecule and the structure

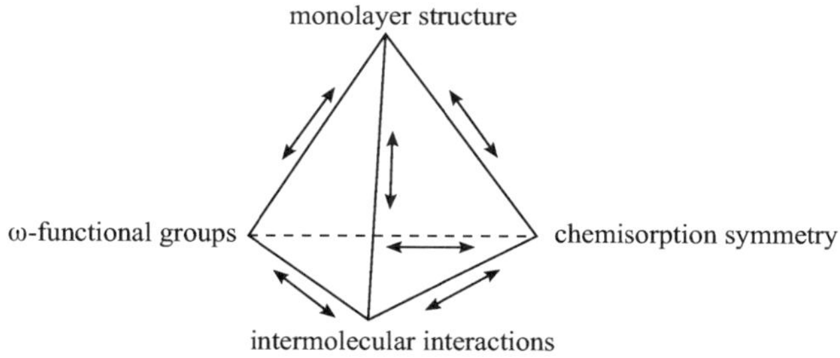

Figure 1.1 Mutual interactions among chemisorption symmetry, intermolecular interactions, ω-functional groups and monolayer structure in self-assembled monolayers.

of the assembly it forms, as well as of the relationship between a given microscopic molecular property and its manifestation in the appropriate macroscopic property. This, however, requires understanding of the mutual relationships among chemisorption symmetry, intermolecular interactions, the ω-functional groups and monolayer structure (Figure 1.1).

In this chapter, we discuss briefly the different groups of SAMs and their unique features. We then attempt to provide a general picture of self-assembly on surfaces, as it emerges from the consideration of the interplay of different forces that control this process. Finally, we suggest areas of research where SAM may have unique contributions.

2. SELF-ASSEMBLED MONOLAYERS

SAMs are ordered molecular assemblies formed by the adsorption of an active surfactant on a solid surface (Figure 1.2). This rather simple process makes SAMs technologically attractive for building superlattices and for surface engineering. The order in these two-dimensional systems is produced by a *spontaneous* chemical synthesis at the interface, as the system approaches equilibrium. Although the area is not limited to long-chain molecules[11], SAMs of functionalized long-chain hydrocarbons are most frequently used as building blocks of supermolecular structures.

2.1 Monolayers of Fatty Acids

Spontaneous adsorption of long-chain *n*-alkanoic acids ($C_nH_{2n+1}COOH$) has been studied in the past few years. This is an acid–base reaction, and the driving force is the formation of a surface salt between the carboxylate anion and a surface metal cation. A number of groups studied the adsorption of

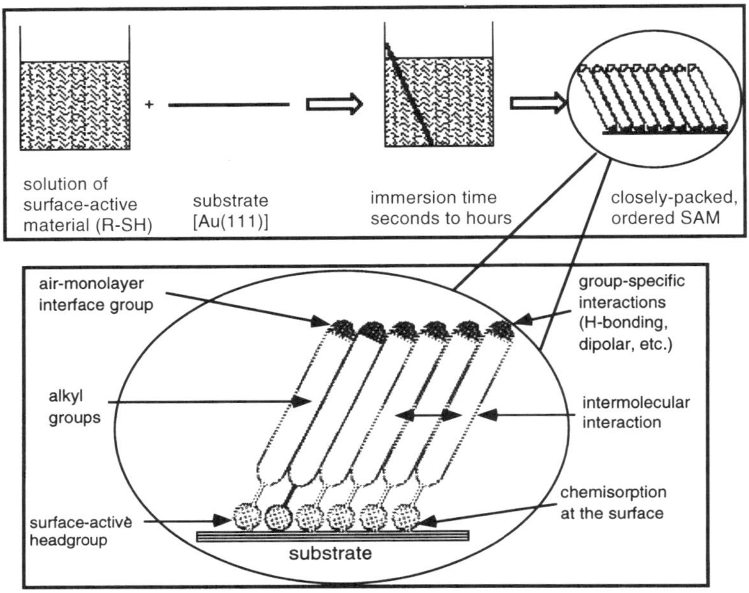

Figure 1.2 Self-assembled monolayers are formed by simply immersing a substrate into a solution of the surface-active material. The driving force for the spontaneous formation of the 2-D assembly includes chemical bond formation of molecules with the surface, and intermolecular interactions.

n-alkanoic acids on aluminum oxide[12–14], and on silver oxide[15]. When a diacetylenic group is introduced into the alkyl group[16], a rigid, rod-like, π-system breaks the cylindrical symmetry of the all-*trans* chain. Thus, the diacetylene units introduce a stratum of defects into the two-dimensional packing of the ordered all-*trans* chains[17]. As a result, monolayers in which the diacetylene group is connected to the carboxylic head group, i.e., the stratum of defects is as far as possible from the air–monolayer interface, exhibit the lowest surface free energy, since high ordering and close-packing can be restored.

Grazing incident X-ray diffraction (GIXD) of docosanoic acid ($CH_3(CH_2)_{20}COOH$) monolayers on AgO reveals that the carboxylate anion adsorbs in a $p(2 \times 2)$ overlayer structure, with a lattice spacing of 5.78 Å (Figure 1.3)[18]. The alkyl chains are in an all-*trans* extended configuration, and are tilted 26.7° from the surface normal. Furthermore, the plane the bisects the methylene groups is twisted with respect to the plane of the carboxylate group. Polarized infrared external reflection spectroscopy (PIERS) studies confirm this structure[15,19]. Studies show that the structure of monolayers formed from solution and from the gas phase on AgO are identical[20].

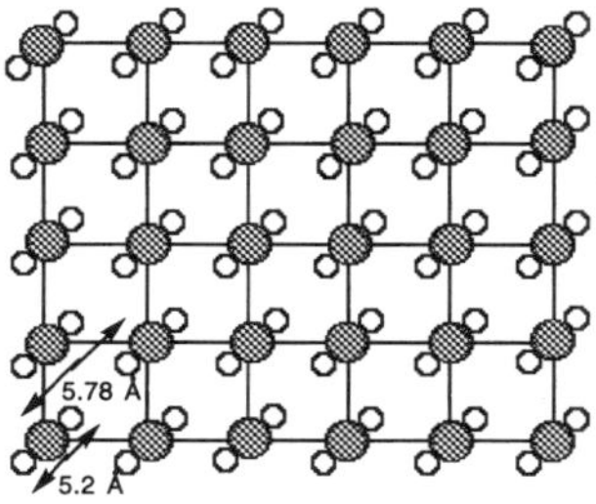

Figure 1.3 The $p(2 \times 2)$ adsorption scheme of fatty acids on AgO. The filled circles are the carboxylate carbon atoms, while the small, open circles are the carboxylate oxygen atoms.

2.2 Monolayers of Organosilicon Derivatives

SAMs of alkylchlorosilanes, alkylalkoxysilanes, and alkylaminosilanes require hydroxylated surfaces as substrates for their formation. The driving force for this self-assembly is the *in situ* formation of polysiloxane, which is connected to surface silanol groups (—Si—OH) via Si—O—Si bonds. Substrates on which these monolayers have been successfully prepared include silicon oxide[21–25], aluminum oxide[26,27], quartz[28–30] glass[25], mica[31–33], zinc selenide[25,26], germanium oxide[25] and gold[34–36]. Recently Allara *et al.* have compared OTS monolayers on silicon oxide and on gold activated by UV-ozone exposure. Using IR spectroscopy, ellipsometry, and wetting measurements show identical average film structures[37].

High quality SAMs of alkyltrichlorosilane derivatives are not simple to produce, mainly because of the need to carefully control the amount of water in solution[21,38,39]. Thus, while incomplete monolayers are formed in the absence of water[22], excess water results in facile polymerization in solution and polysiloxane deposition of the surface[28]. Recently, McGovern *et al.* have shown that extraction of surface moisture, followed by OTS hydrolysis and subsequent surface adsorption may be the mechanism of SAM formation[40].

Temperature has been found to play an important role in monolayer formation[21]. Here, the issue is the competition between a reaction of hydrolyzed (or partially hydrolyzed) trichlorosilyl groups with other such groups in solution to form a polymer, and the reaction of such groups with surface Si—OH moieties to form a SAM. As temperature decreases, the preference of surface reaction increases. Moreover, as temperature decreases, reaction kinetics decreases as well, resulting in the diminution of thermal disorder in the forming monolayer, and therefore, the formation of an ordered assembly with the gain of van der Waals (vdW) energy is facilitated. The threshold temperature below which an ordered monolayer is formed was found to be a function of the chain length (higher for octadecyl, 18°C, than for

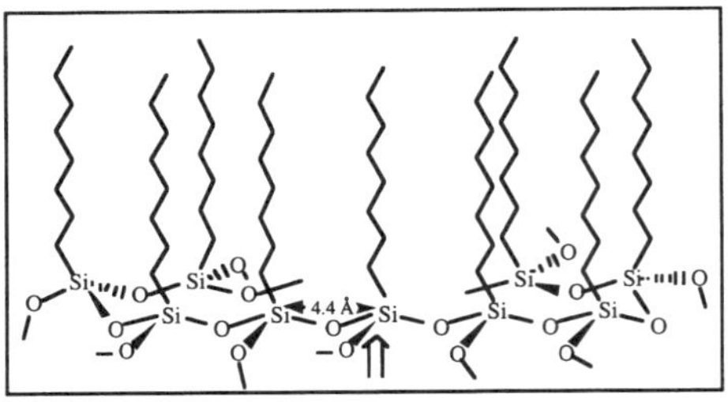

Figure 1.4 Schematic description of a polysiloxane at the monolayer–substrate surface. The arrow points to an equatorial Si—O bond that can be connected either to another polysiloxane chain or to the surface. (Reproduced by permission from Ulman[7] ©1991, Academic Press.)

tetradecyl chain, 10°C)[22]. Recent solid-state ^{13}C NMR studies of OTS monolayers deposited on fumed silica particles confirm these results[41].

Since substrates used in the formation of silane SAMs are amorphous, the packing and ordering of alkyl chains in SAMs of alkyl silanes is determined by the underlying structure of the surface polysiloxane chain. Figure 1.4 shows a polysiloxane model, where the alkyl chains can occupy either axial or equatorial positions[7]. If alkyl chains are connected to axial positions, the distance between them is ~4.4 Å. This leaves very little free volume and should require very little or no chain tilt. This connection between free volume and tilt is a general one, since the driving force for tilt is the reestablishment of vdW contact among chains. Allara *et al.* found that alkyl chains in OTS monolayers of SiO_2 and oxidized gold are tilted at $10 \pm 2°$ from the normal, with significant *gauche* defect content at the chain termina[37]. Based on both ellipsometry, and the concentration of *gauche* defects, it was concluded that the monolayer is $\sim 96 \pm 4\%$ of the theoretical maximum coverage, which explains the observed average tilt. An important conclusion of this study is that surface hydration is responsible for decoupling of film formation from surface chemistry, and the observed high film quality. Increased interaction of the forming siloxane chain with surface Si—OH groups introduces disorder and film defects.

Construction of multilayers requires that the monolayer surface be modified to a hydroxylated one. As with the original substrate, the quality of additional monolayers will be determined by the competition between chemisorption and chain–chain interactions. Hydroxylated surfaces can be prepared by a chemical reaction and the conversion of a nonpolar terminal group to a hydroxyl group. Examples of such reactions are the $LiAlH_4$ reduction of a surface ester group[42], the hydroboration–oxidation of a terminal vinyl group[23], and the conversion of a surface bromide using silver chemistry[43]. Different surface reactions may give different defects concentration; however, it has not been established yet

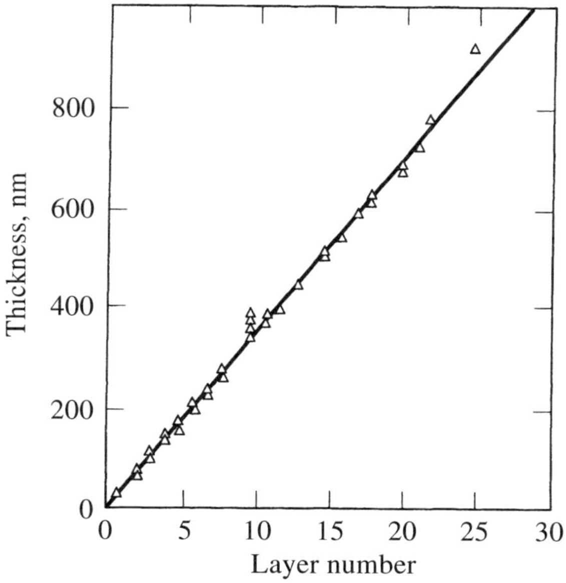

Figure 1.5 Film thickness vs layer number. (Reproduced by permission from Tillman *et al.*[42] © 1988, American Chemical Society.)

what is the minimum number of surface OH groups that is required for the formation of highly-ordered SAM, or of surface defects to impede the formation of a good alkyltrichlorosilane monolayer. Figure 1.5 shows film thickness vs layer number for ~0.1 mμ film prepared in our laboratory from methyl 23-trichlorosilyltricosanoate[39].

2.3 Organosulfur Adsorbates on Metal and Semiconductor Surfaces

Sulfur and selenium compounds have a strong affinity to transition metal surfaces[44–48], probably because of the possibility to form multiple bonds with surface metal clusters[49]. The number of reported surface active organosulfur compounds that form monolayers on gold has increased in recent years. These include, among others, di-*n*-alkyl sulfide[50,51], di-*n*-alkyl disulfides[4], thiophenols[52,53], mercaptopyridines[75], mercaptoanilines[54], thiophenes[55], cysteines[56,57], xanthates[58], thiocarbaminates[59], thiocarbamates[60], thioureas[61], mercaptoimidazoles[62–64], and alkaneselenoles (Figure 1.6)[65].

Organosulfur compounds coordinate very strongly also to silver[66–70], copper[64,66,67,71], platinum[72], mercury[73,74], iron[75,76], colloidal gold particles[77], gallium arsenide (GaAS) surfaces[78] and recently on indium phosphide (InP)[79].

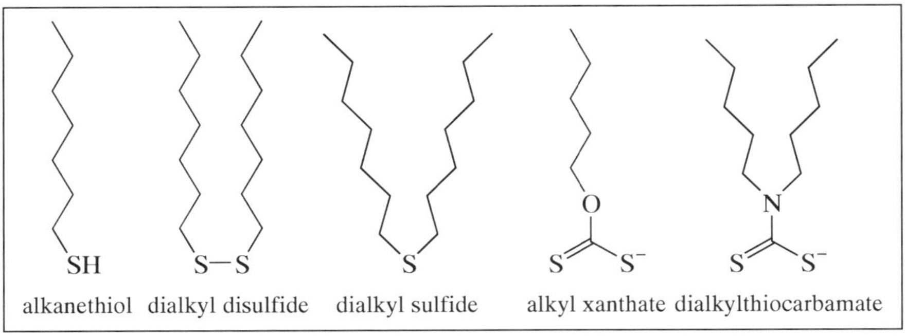

Figure 1.6 Sulfur compounds have a strong affinity to transition metal surfaces.

However, most investigations have been carried out on SAMs of thiolates on Au(111) surfaces.

Kinetic studies of alkanethiol adsorption onto Au(111) surfaces have shown that at relatively dilute solutions (10^{-3}M), two distinct adsorption kinetics can be observed: a very fast step, which takes a few minutes, by which the contact angles are close to their limiting values and the thickness ~80–90% of its maximum, and a slow step, which lasts several hours, at the end of which the thickness and contact angles reach their final values[80]. The initial step is chemisorption controlled. It is described well by diffusion controlled Langmuir adsorption, and was found to strongly depend on thiol concentration. The second step can be described as a surface crystallization process, where alkyl chains get out of the disordered state and into unit cells, thus forming a two-dimensional crystal. Hence, the kinetics of the first step is governed by the transition state of the surface-head-group reaction. On the other hand, the kinetics of the second step is related to chain disorder (e.g. *gauche* defects), the different compenents of chain–chain interaction (vdW, dipole–dipole, etc.), and the surface mobility of chains. It was found that the overall kinetics is faster for longer alkyl chains, probably due to the increased vdW interactions[74]. Second-harmonic generation, and XPS measurements[74,81], as well as near edge X-ray absorption fine structure (NEXAFS) studies[82] confirm this mechanism. Studies also showed pronounced differences between the short ($n < 9$) and long ($n > 9$) alkanethiolates[83]. This is probably due to the decreased rate of the second step resulting from the diminution of the interchain vdW attraction energy. In the case of simple alkyl chains, the masking of adsorption sites by disordered chains is not a serious problem. However, if the chain contains a bulky group, the two steps are coupled, and the chemisorption kinetics is greatly impeded by the chain disorder[84].

Early electron diffraction studies (both high-[85,86] and low-energy[87]) of monolayers of alkanethiolates on Au(111) surfaces show that the symmetry of

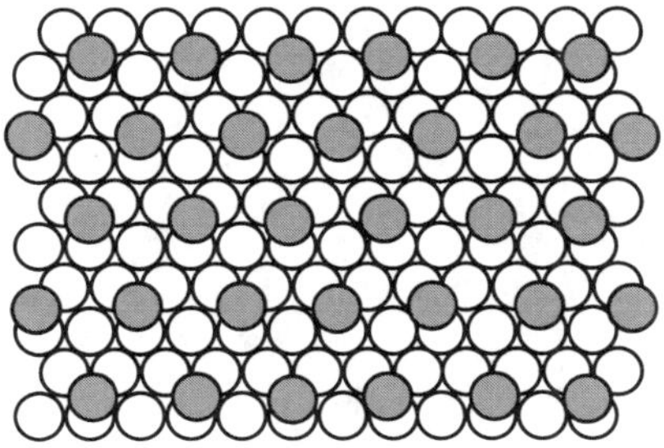

Figure 1.7 Hexagonal coverage scheme for alkanethiolates on Au(111). The open circles are gold atoms and the filled circles are sulfur atoms.

sulfur atoms is hexagonal with an S···S spacing of 4.97 Å, and calculated area *per* molecule of 21.4 $Å^2$. Helium diffraction[88] and atomic force microscopy (AFM)[89] studies confirmed that the structure formed by docosanethiol on Au(111) is commensurate with the underlying gold lattice and is a simple $\sqrt{3} \times \sqrt{3}R30^\circ$ overlayer (Figure 1.7). However, ultra high vacuum STM studies added very important information on the mechanism of SAM formation. It revealed the coexistence of a two-dimensional liquid phase at room temperature of butane-($CH_3(CH_2)_3S^-$) and hexanethiolate ($CH_3(CH_2)_5S^-$) monolayers on Au(111)[90]. No two-dimensional liquid was observed for octane-($CH_3(CH_2)_7S^-$) and decanethiolate ($CH_3(CH_2)_9S^-$) monolayers. The short chain homologues exhibited slow desorption of surface thiolate that led to the nucleation and growth of ordered domains having a unit cell of $p \times \sqrt{3}(8 \leq p \leq 10)$. On the other hand, both octane- and decanethiolate form densely-packed SAMs with a $c(4 \times 2)$ superlattice of a $\sqrt{3} \times \sqrt{3}R30^\circ$ lattice[91–93].

The above STM study also discovered a facile transport of surface gold atoms in the presence of the liquid phase. This suggests that the two-step mechanism does not provide a complete picture of the surface reactions, and that adsorption/desorption processes may have an important role in the formation of the final equilibrium structure of the monolayer. A support for the importance of a desorption process comes from atomic absorption studies showing the existence of gold in the alkanethiol solution. STM studies suggest that this gold comes from terraces, where single atomic deep pits are formed[89,94,95].

Alkanethiolate monolayers on Ag(111) have been less studied. Low energy electron diffraction (LEED) experiments suggest a commensurate ($\sqrt{7} \times \sqrt{7}R10.9^\circ$ overlayer of methanethiuolate (CH_3S^-) on Ag(111), with an S···S separation of 4.4 Å (Figure 1.8). For octadecanethiolate ($CH_3(CH_2)_{17}S^-$) monolayers, on the other hand, GIXD shows a lattice constant of 4.6–4.7 Å, with alkyl chains that are hardly tilted, and an overlayer

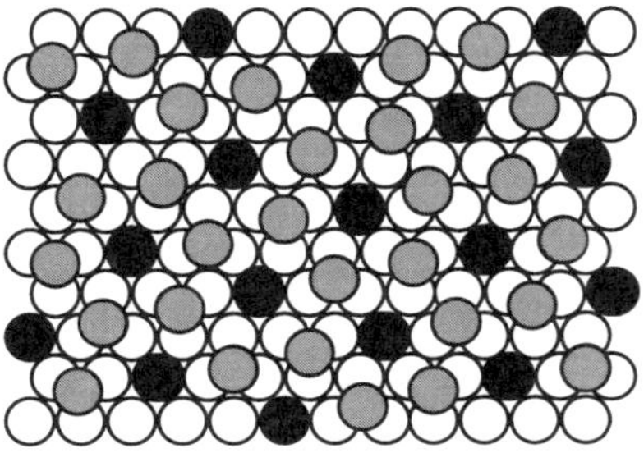

Figure 1.8 Diagram showing one of the two possible $\sqrt{7} \times \sqrt{7}$ structures of alkanethiolates on Ag(111). The open circles represent the silver atoms, while the dark- and light-shaded circles represent thiolates at the on-top and hollow sites, respectively.

very similar to $(\sqrt{7} \times \sqrt{7}R10.9^\circ$, but with 12° rotation, and an outmost Ag(111) layer slightly expanded[97]. Notice that the $(\sqrt{7} \times \sqrt{7}R10.9^\circ$ requires that the thiolates at the on-top site will be ~ 0.5 Å higher than those residing at the hollow site. Recent STM studies did not confirm this requirement, concluding that an ideal $(\sqrt{7} \times \sqrt{7}R10.9^\circ$ probably is not a correct description of the monolayer structure[98]. Thus, there is a significant difference between the structure of methylthiolate, and long alkanethiolates on Ag(111), with the latter controlled by the chain–chain vdW attraction. When a chain in an assembly of all-*trans* perpendicular chains moves vertically, there is a significant loss of vdW energy, unless the chain moves a $C_n - C_{n+2}$ distance[99]. It was suggested before, that vdW energy may be a driving force for reorganization of the substrate surface[48]. The case of alkanethiolates on Ag(111) is an excellent example of this phenomenon, where molecules remain at the same distance from the surface to maximize lateral interactions, thus bringing about reorganization of the silver outmost layer.

The alkyl chains in SAMs of thiolates on Au(111) usually are tilted ~ 26–28° from the surface normal, and display ~ 52–55° rotation about the molecular axis. This tilt is a result of the chains reestablishing vdW contact in an assembly with ~ 5 Å S···S distance, that is larger than the distance of ~ 4.6 Å, usually quoted for perpendicular alkyl chains in a close packed layer. On the other hand, thiolate monolayers on Ag(111) are more densely-packed due to the shorter S···S distance[90]. There were a number of different reports on chain tilt in SAMs on Ag(111), probably due to different amounts of oxide, formed on the clean metallic surface. Laibinis *et al.* suggested a value of 11–14°, while Ulman[100] suggested 7°, and Nemetz *et al.*[101] and Fenter and Eisenberzer[102] suggested that the chains are perpendicular to the surface. It is well accepted today, that in carefully prepared SAMs of alkanethiolates on a clean Ag(111) surface the alkyl chains are practically perpendicular to the surface.

2.4 Alkyl Monolayers on Silicon

Linford and Chidsey demonstrated for the first time that robust monolayers can be prepared where the alkyl chains are covalently-bound to the silicon substrate mainly by C—Si bonds[103,104]. In the first experiments they used hydrogen-terminated silicon (H—Si(111) and H—Si(100)), and diacetylperoxide[97]. The adsorption of alkyl chains was attributed to a series of free-radical reactions. It was found that these monolayers, although exhibiting thickness, wettability, and methylene stretching frequencies indicative of highly-packed chains, lost ~30% of the chains when exposed to boiling water. The apparent conclusion was that hydrolyzable acyloxy groups are removed, leaving the robust alkyl chains bound to the surface by the C—Si bonds. Reaction of alkynes also yielded strong, closely packed monolayers, and chlorine-terminated olefines gave monolayers with wettability indicative of Cl-terminated alkyl chains. The resulting monolayers are ~90% olefine-driven as shown by deuterium labeling experiments.

The monolayer density, as measured by X-ray reflectivity, is only ~90% of the value of a crystalline paraffin such as $n\text{-}C_{33}H_{68}$, suggesting a significant number of defects. These defects can be understood based on the fact that formation of the Si—C surface bonds are not reversible. Therefore, there is no equilibrium between molecules in solution and molecules at the surface. Such an equilibrium is necessary for annealing of surface defects. Again, the nature of the chemisorption reaction has a profound effect on the quality of the monolayer. Elipsometry and IR spectroscopy suggest that the chains are tilted ~45° from the surface normal, and that a twist angle of ~53° exists between the plane that bisects the methylene groups and the plane of the tilt. This tilt angle is not surprising since the interchain distance is 6.65 Å[97]. Hence, as in the case of alkanethiolates on gold, the chemisorption scheme defines the chain–chain distance, the free volume, and the tilt angle.

2.5 Multilayers of Diphosphates

One logical way to finding surface reactions that may lead to the formation of SAMs is to look for reactions that result in insoluble salts. This is the case of phosphate monolayers, based on their highly insoluble salts with tetravalent transition metal ions. In these salts, the phosphate forms layer structures, with one OH group sticking to either side. Thus, the idea was that replacing the OH for an alkyl chain—forming the alkyl phosphonic acid—should result in a bilayer structure with alkyl chains sticking at both sides of the metal phosphate sheet[105]. Interestingly, when zirconium is used (Zr^{4+}) the distance between next-neighbour alkyl chains is ~5.3 Å, which forces either chain disorder or chain tilt so that vdW attractive interactions can be reestablished.

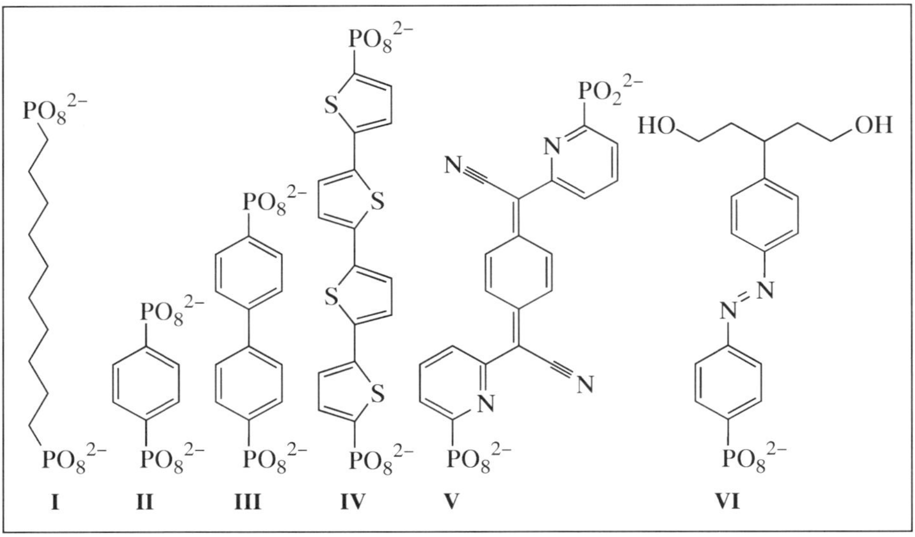

Figure 1.9 Some diphosphonic acids utilized in the formation of self-assembled multilayers.

Mallouk pioneered this area when he discovered that self-asembled multilayers can be prepared simply by alternating adsorption of Zr^{4+} ions and α,ω-alkylidenediphosphate (**I**) on a phosphorylated surface[106,107]. Other diphosphates have since been investigated, for examle, 1,4-benzene[108,109], 4,4′-biphenyl (**II**)[101], and quaterthenyl (**III**)[101,110]. More complex molecules also were investigated, for example, 7,8-dicyano-7.8-bis-*o*-pyridylquinonedimethide (**IV**)[111], bipyridine salts[105], and porphyrins[105]. Figure 1.9 displays some of these structures. These examples show the complexity of systems that can be introduced into SAMs, even when alkyl chains are completely absent.

3. STRUCTURAL ISSUES IN SAMS

3.1 Chemisorption: Bonding, Symmetry, and Free Volume

Chemisorption is the primary event in the formation of SAMs. The absorbate interacts with the surface through its head group, which is pinned to the surface as a result of the newly formed bond(s). There are two cases of monolayer formation. The first is on amorphous surfaces (e.g. glass), where intermolecular interactions define the packing and ordering, and the second is on crystalline surfaces, where the structure of the overlayer is determined by the surface.

SAMs of alkyltrichlorosilanes ($C_nH_{2n+1}SiCl_3$) belong to the first case. There, adsorbates are connected to each other—through covalent Si—O bonds—thus forming a polysiloxane chain. This chain is anchored to the surface via Si—O—Si bonds, which restricts its free motion, and hence its ability to accommodate an ordered, close-packed assembly of the alkyl chains. The Si···Si spacing—determined by the polysiloxane chain—is shorter than the $CH_2 \cdots CH_2$ spacing in closely-packed hydrocarbons[97]. This may result in a layer of *gauche* defects near the polysiloxane chains; however this has not been verified either experimentally or by molecular dynamics (MD) simulations. What has been found is that the alkyl chains in monolayers are locked perpendicular to the surface[112], and are hexagonally close packed[113]. Hence, when the surface is amorphous, interchain interactions determine the packing symmetry, which probably has an effect on the conformation of the underlying polysiloxane chain.

The chemisorption of alkanoic acids ($C_nH_{2n+1}COOH$) on amorphous metal oxide surfaces in not unique, and depends on the chemical properties of the metal ion. On AgO, both of the carboxylate oxygen atoms bind to the surface nearly symmetrically, and ordered monolayers are formed with a chain tilt angle from the surface normal of 15 to 25°[114]. On surfaces of CuO and Al_2O_3, the carboxylate binds asymmetrically to the surface displaying tilt angles close to zero (Figure 1.10)[19]. This is a nice demonstration of how the interplay between chemisorption geometry and the packing and ordering of the alkyl chains defines the equilibrium structure of the monolayer.

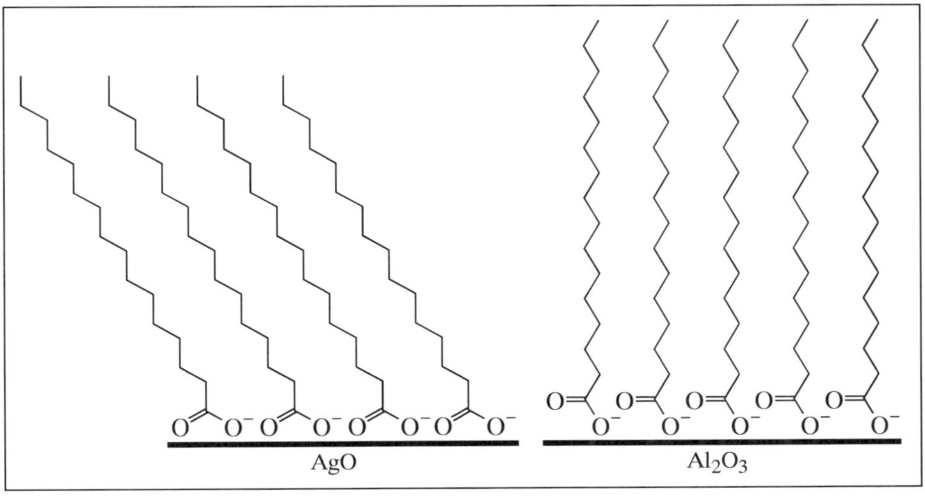

Figure 1.10 Schematic description of fatty acid monolayers on AgO and on Al_2O_3.

SAMs of biphenyl and naphthyl-containing alkanoic acids were also investigated[115]. It was found, that a biphenyl group or a naphthyl group, due to their larger cross-sectional area, causes higher tilt of molecular chains in carboxylic acid monolayers on silver and a correlation of wetting properties and chain tilt. Different behavior observed on silver and copper has been attributed to site specificity, binding geometry, and binding strength, among other factors.

The structure of SAMs of alkanethiolates is profoundly affected by the chemistry and symmetry of the underlying lattice. On Au(111), every sixth hollow site is occupied, resulting in a $\sqrt{3} \times \sqrt{3}R30°$ overlayer[81–83] that is commensurate with the underlying Au(111) lattice, and is the equilibrium structure of the monolayer (Figure 1.6)[116]. Selection of surface sites occurs because Au(111), like any surface, is energetically heterogeneous, due to different surface–adsorbate bond energies in different surface sites. In the case of Au(111), *ab initio* calculations suggest that a thiolate adsorbed at the hollow site is more stable by $\sim 6\,\text{kcal}\,\text{mol}^{-1}$ than when it is adsorbed at the on-top site[46]. Another important difference between the adsorption at the hollow- and the on-top sites is the nature of the sulfur surface bond. At the hollow site, the sulfur is connected to three gold atoms at the topmost gold layer, and to one gold atom, directly beneath the hollow site, and its charge[46,117] is -0.4 e. At the on-top site, the sulfur is connected to one gold atom, thus forming a gold thiolate salt, and its charge[46] is ~ -0.7 e. This difference has a profound effect not only on the final structure of the monolayer, but also on the process by which it forms. If the cleavage of the S—H bond, which occurs at the on-top site[118], is the rate determining step, the adsorption rate should be faster in polar solvents, due to the stabilization of the forming dipole. On the other hand, if the migration of a thiolate from the on-top to the hollow site is the slow step, the reaction should be faster in nonpolar solvents, due to the diminished charge separation. Second-harmonic generation studies showed that while the rate constant in ethanol is $1.3 \times 10^6\,\text{cm}^3\text{mol}^{-1}\text{s}^{-1}$, it is $4.7 \times 10^6\,\text{cm}^3\text{mol}^{-1}\text{s}^{-1}$ in hexane[119], suggesting the S—H bond cleavage is not the rate determining step.

Migration of thiolates between neighboring hollow sites is essential for healing of defects, and for two-dimensional crystallization. Such migration must occur either through the on-top or the bridge sites. In both cases, the transition state is more polar than the ground state, and hence should be sensitive to dielectric constant. Indeed, ethanol has been found to yield consistently highly ordered monolayers[76].

The formation of a neutral gold thiolate molecule (RS—Au) at the on-top site, may result in its desorption before the thiolate moves to the hollow site, thus leaving a defect. STM studies suggest that some of the pinholes observed in monolayers on Au(111) may be a result of such an etching process[91]. However, these pinholes dissapeared after annealing the monolayers at 77 or 100°C[120,121], suggesting that migration gold thiolate molecules (RS—Au) should be considered as a possible healing mechanism.

Alkanethiolates have two binding modes at the Au(111) hollow site, one with a bend angle around the sulfur of 180° (*sp*) and the other of 104° (sp^3), the latter being more more stable by 0.41 kcal mol^{-1}[46]. Therefore, packing requirements may dictate the final surface—S—C angle. Many studies have suggested that this angle in monolayers on Au(111) surfaces must be tetrahedral[46]. However, recent studies on triphenylthiolate (C_6H_5—C_6H_4—C_6H_4—S^-) monolayers on Au(111) suggest a tilt angle of ~6° from the surface normal[122], thus providing the first evidence that a second chemisorption mode is possible.

The energy barrier between the two chemisorption modes on Au(111) is very small (2.5 kcal mol^{-1})[46], suggesting that the thiolate may easily cross from one of these minima to the other, thus enabling a facile annealing mechanism (Figure 1.11). This predicts that the process of changing tilt direction may occur well below the melting point of the monolayer, and should be chain length dependent. Recent X-ray data show narrowing of the diffraction peak when monolayers of alkanethiolates on Au(111) were annealed[87]. The development

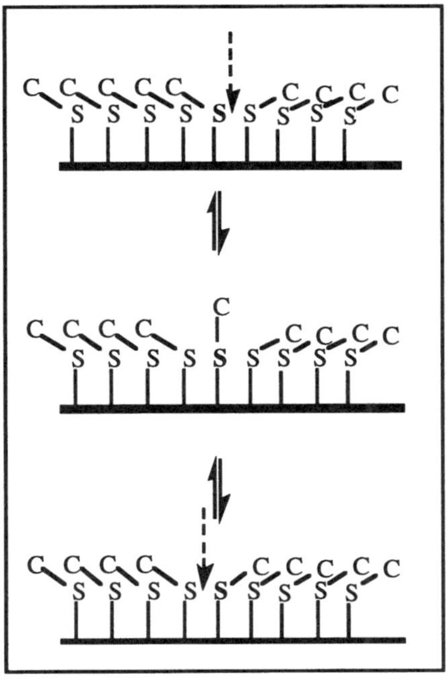

Figure 1.11 Possible scenario for annealing of alkanethiolate monolayers. The hybridization change results in the motion of only one chain. The arrow represents the grain boundary. Notice that after the molecule has moved, the grain boundary moved from the right of, to the left of, that molecule.

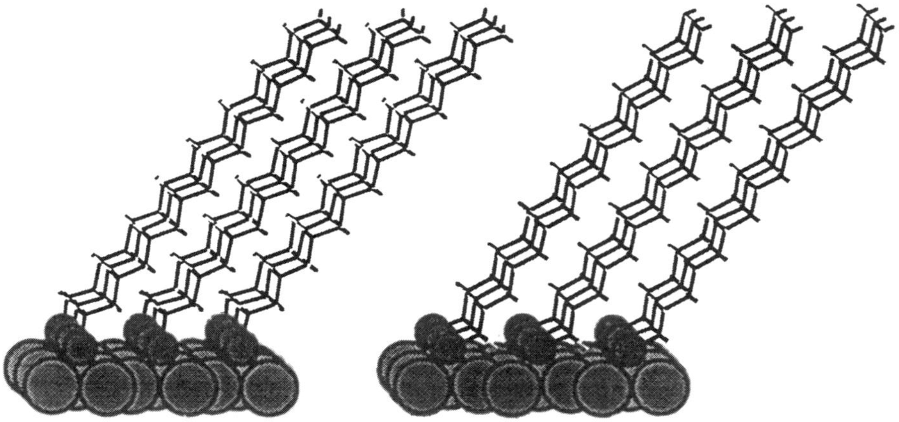

Figure 1.12 View of the tilt in a nine-molecule section of a fully covered $SC_{16}H_{33}$ monolayer on Au(111) minimized with a modified MM2 force field including the sp^3 chemisorption parameters (left, calculated stabilization energy -23.29 kcal mol^{-1}), and with the *sp* chemisorption parameters (right, calculated stabilization energy -22.25 kcal mol^{-1}). The gold atoms are shown in light gray having their full van der Waals radii. The sulfur atoms are shown in dark gray and are not drawn to scale.

of larger domain size was the apparent result of heating and cooling. Thus, close packing and high ordering of alkanethiolates on Au(111) apparently result from the relatively easy two-dimensional "recrystallization" process.

Molecular mechanics (MM) energy minimization indicates that the two modes lead to monolayers that exhibit different types of packing arrangements (Figure 1.12), but that are comparable in their ground state energies. (The monolayer resulting from the sp^3 mode are more stable by 0.6 kcal mol^{-1})[46]. Therefore, monolayers may consists of two different chemisorption modes ordered in different domains—simultaneously coexisting homogeneous clusters—each characterized by a different conformer in their unit cell. Annealing may result either in a unique chemisorption mode, or in a much more complex structure, as the different modes are redistributed. This may explain the observation of two-dimensional liquid in butane- and hexanethiolate monolayers on gold[86], where vdW interactions do not provide enough cohesive energy to allow for small domains to coexist as a two-dimensional solid.

The chemisorption of alkanethiolates on Ag(111) can be described as $(\sqrt{7} \times \sqrt{7})R10.9°$, with an S···S distance of 4.41 Å, almost optimal for the packing of chains that are perpendicular to the surface. This arrangement has been observed for S atoms[123], for SH[124], and SCH_3 groups[125] on Ag(111). Notice, that interchain repeat distance in crystalline paraffins is ~4.65 Å[126], while that in orthorhombic polyethylene is 4.45 Å[127]; hence, intermolecular separation in polymethylene assemblies may be a function of chain length. This suggests that alkyl chains may be "squeezable", and that a small separation between them may be partly forced by a strong chemisorption potential.

3.2 Intermolecular Interactions: the Interassembly Epitaxy Concept

In trying to understand the packing and ordering in SAMs, the fundamental question is how adsorbate–adsorbate interactions are affected by the chemisorption scheme at the surface and vice versa. However, we consider first formation of a two-dimensional assembly *as if chemisorption is irrelevant*, and discuss spontaneous formation of a two-dimensional assembly from complex molecules. Here, two-dimensional assemblies are considered as a structure of *commensurate intraassembly planes*[128], similar to three-dimensional molecular crystals that have been viewed and analyzed as an assembly of commensurate two-dimensional molecular layers[129].

Simple planar π-systems tend to spontaneously aggregate in layers, usually with herringbone structures. Similarly, paraffins spontaneously assemble in pseudohexagonally close-packed layer structures. In both cases there usually is more than one possible stable packing arrangement, and in some cases these arrangements have about the same stabilization energy[124]. Forming a *chemical* bond between the π-system and the alkyl chain results in a more complex molecule, and it is not clear *a priori* if such a material will spontaneously self-assemble in a layered structure. Some molecules do form stable two-dimensional layers, while others undergo bulk crystallization.

The factors that determine formation of layered structures of complex molecules are: (i) that cross-sectional area mismatch of different molecular parts is minimal; (ii) that intraassembly planes that are defined by different molecular parts are commensurate, i.e. an epitaxial matching of the sublattices of different intraassembly layers exists; and (iii) that bonds between different molecular parts do not perturb (i) and (ii). *A complex molecule will spontaneously form a layer structure if this results in no or very little perturbation in the layer structure of its components.* This principle is best manifested in smectic liquid crystals of biphenyl derivatives. Calculations show that cross-sectional area match, that two-dimensional intraassembly planes are commensurate[123], and that the chemical bonds between the π-system and alkyl chains do not cause any disruption of the assembly.

In reality, many molecules consist of parts that make lattices that are incommensurate. In such cases, the stability of the layered assembly can be augmented by chemisorption, or physisorption, to a solid surface, or by applying pressure, as in a Langmuir trough. Thus, although the unconstrained layered structure is not in a global minimum of free energy, there are mechanisms that may yield a layered assembly that is in a global minimum. In the following sections, we discuss such mechanisms.

3.3 The Brick-wall Model

If one goes to a higher level of complexity, an alkyl chain may be viewed as a segment of a one-dimensional crystal, simply because of the constant distance (2.52 Å) found between second-nearest-neighbor carbon atoms. Likewise, two-dimensional assemblies of packed alkyl chains, where coplanarity is enforced, might be viewed as stacks of microlayers or *strata*, where each stratum is comprised of closely packed atomic groups, i.e. methylene (CH_2) units (Figure 1.13). Consider an ideal monolayer system composed of straight alkyl chains, attached to an *atomically flat* substrate, with a repeat distance of ~4.4 Å (Figure 1.13a). In this assembly, intrastratum packing energy is minimized without putting any strain on individual molecules. The system can be perturbed by introducing a single stratum with *spacing* larger than 4.4 Å, with *symmetry* different than triagonal, or with blocks that have a smaller/larger *unit size*. For example, one could create a perturbed system by moving the attachment points from ~4.4 Å—as for alkanethiolates on Ag(111)—to ~5 Å—as for alkanethiolates on Au(111) (Figure 1.13b). As a result, close packing can no longer be present for undistorted vertical chains. In the perturbed system, packing requirements of the perturbing stratum (bottom in this case) and those of other strata conflict. The system accommodates the

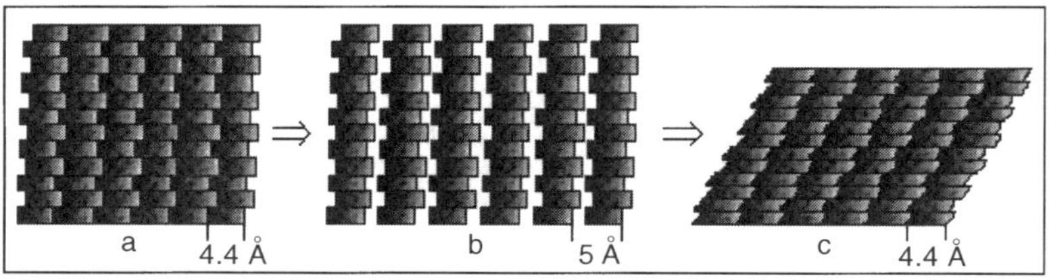

Figure 1.13 Schematic representation (not drawn to scale) of brick strata perturbed from close packing by the introduction of a single stratum with *spacing* larger than the block size. Intrastrata interactions are weaker than interstrata interaction, which are, essentially, chemical bonds. The free volume can be eliminated by tilting the brick columns in (b) thus reestablishing contact. Notice that the shape of the bricks changed, and while their volume is the same, their thickness is now the projection of the interstrata bond on the surface normal. The broken line represents monolayer thickness.

conflicting demands by tilting the chains over in such a way as to reestablish optimum vdW contacts within and between strata (Figure 1.13)[11].

3.4 What Makes Alkanethiolate Monolayers on Gold and Silver Surfaces Different?

There are a number of tradeoffs that result in the different self-assembled structures on *ideal* (i.e. atomically smooth) gold and silver (111) surfaces. The first one is the chain···chain distance. As suggested by Outka *et al.*[130], and discussed by Ulman *et al.*[131], only certain combinations of chain–chain separation (i.e. lattice spacing) and tilt angles permit *truly* effective packing of alkyl chains. The most effective packing is a trigonal lattice with spacing near ~4.4 Å with a molecular axis normal to the surface. The second most effective has a spacing near 5.0 Å and the molecular axis tilted ~30°—in the plane that bisects the methylene units—such that the distance between the chains is, again ~4.4 Å and the fit of bulges into depressions is again perfect. It can be thought of as a ratchet that has slipped by precisely one notch. In alkanethiolate monolayers on GaAs, the ratchet has slipped by two notches, resulting in a tilt angle of 57°[75]. This tilt solves the free volume problem in one line of equally-spaced chains. However, if there is additional free volume between the lines, and out-of-plane tilt, or lean, takes place. These two angles in the Cartesian space coordinates, can be transformed to molecular coordinates of tilt in the plain that bisects the methylene units and a rotation about the molecular axis. In a case where the chain–chain distance is between the above numbers, close packing may require that the chains would have different distance from the surface, so that the ratcheting effect is reestablished. If energy gain by this process is large enough, it may be a driving force for surface reconstruction. A second consideration is that crowding more molecules per unit surface area will result in both more chemisorption energy and vdW attraction for the entire system simply because there are more chains involved. This is a lot of energy, since if the S···S distance on Ag(111) is 4.4 Å, there are 26% more chains per unit area than in Figure 1.6. Why then does chemisorption of thiolates on Au(111) not follow the Ag(111) model? This is because the above effects are partially offset by the charge–charge repulsions both among sulfur and gold atoms. On Ag(111), not all thiolates occupy hollow-site positions and, as a result, the charges on the sulfur and gold atoms are larger. Furthermore, and most importantly, chemisorption potentials with significant lateral discrimination could counter the favorable effects of this crowding. Lateral discrimination arises because interaction energy varies for different positions on that surface. Gold differs from silver—due to relativistic effects—primarily in the on-top to hollow site roughness, 6.0 kcal mol^{-1} for Au(111), and only 3.3 kcal mol^{-1} for Ag(111)[46]. It is the combination of lateral discrimination and electrostatic effects that distinguishes the chemisorption of thiolates on Au(111) from that on Ag(111).

3.5 Alkanethiolates on Gold (001): a Different Symmetry

Docosanethiolate monolayers on Au(001) ($CH_3(CH_2)_{21}S$/Au(001)) were studied using electron[85], and helium diffraction[132]. Electron diffraction suggests that the thiolates adsorb in a simple square symmetry with S···S spacing of 4.54 Å. However, an early helium diffraction report indicated that the structure of the CH_3 surface is not simple, and cannot be explained by the simple square lattice symmetry. We decided to use molecular dynamics (MD) simulations to study the possible structure of $CH_3(CH_2)_{21}S$/Au(001). We observed that the chains in the assembly are tightly packed and exhibit very little thermal disorder[17]. Time evolution of selected energy components (Figure 1.14) shows that the onset of equilibration coincides with an abrupt decrease in both the total potential and the total vdW energies. This is the result of CH_2 units locking into an optimal close-packed arrangement within, and between, the strata. On the other hand, a gradual increase in total angle bend energy and torsional energy—resulting form increased chain distortions—is evident. The nature of these distortions can be seen in Figure 1.15, where Figure 1.15a exhibits both a concerted lean (~30°), and a mild bending distortion of the chains, while Figure 1.15b—a view parallel to the other face of the simulation cell—shows that alternating rows of chains acquire a staggered ~5° tilt. While in alkanethiolate monolayers on Au(111) there is a *tilt** of ~30° and a *lean* of

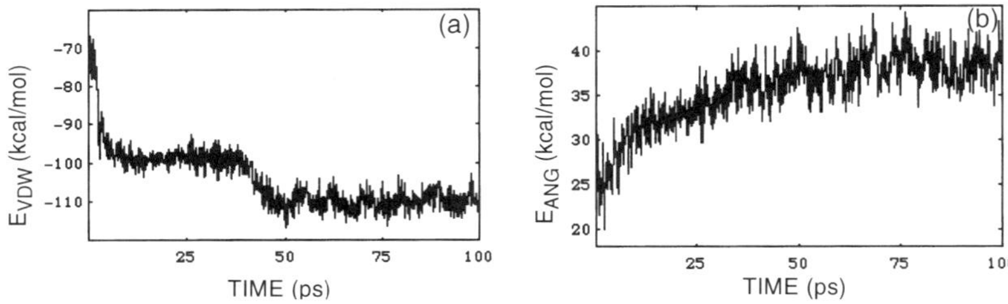

Figure 1.14 Time evolution of various contributions to total energy in the MD trajectory of docosanethiolate monolayer on Au(001). The total vdW energy and total bond stretch energy are represented by (a) and (b), respectively.

* We define the tilt as an angle from the normal within the plane that bisects the methylene groups, and the lean as the angle from the normal in a plane perpendicular to it. Notice, that this definition is valid for a translational layer (e.g. monoclinic unit cell). In the case of a glide layer (e.g. orthorhombic unit cell), one needs three angles to define ordering of chains in the layer. The definition of tilt and lean can be used in that case to describe ordering in one row of molecules.

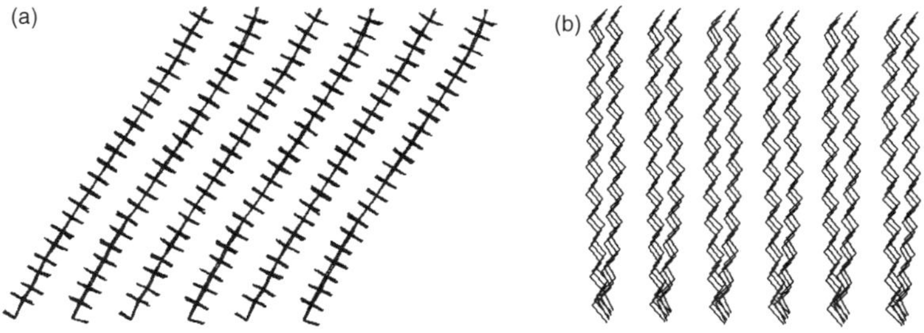

Figure 1.15 Two side views of a 6 × 6 assembly (three simulation cells by three simulation cells) of a typical low energy snapshot from the equilibrated portion of the MD trajectory ($t = 83.7$ ps).

$\sim 22°$ [17] in monolayers on Au(100) the *tilt* is $\sim 5°$ and the *lean* $\sim 22°$. The tilt and lean result from the relationship between molecular spacing and the dimensions of the methylene group, the latter being oval in shape with a ratio of ~ 1.13 between the two axes. It was argued before [120], that tilt of alkyl chains should obey the formula $\phi = n\text{R}/\text{D}$, where ϕ is the tilt angle, R = 2.52 Å, D is the minimum vdW separation between the chains (4.45 Å from crystallographic data [127]), and $n = 0, 1, 2$.

While carrying out only preliminary calculations, we have suggested that a corresponding formula should exist for the lean. *The choice of molecular orientation within the assembly is determined by the best combination of tilt and lean.* In the lean, the vdW interactions are mainly H···H, while in the tilt there is a significant contribution from C···C interactions. Therefore, it can be assumed that the larger contribution to the vdW energy comes from the tilt; hence, it should be the dominating factor in determining molecular orientation. In the present case, the S···S distance is 4.54 Å, leaving very little free volume in the tilt direction. In the other direction, where the dimension of the methylene group is smaller, the difference between the distance and the group dimension is the same as the distance between the rows of chains on Au(111). Both there, and in this case, using the same MD calculations, a lean of $\sim 22°$ was found, provided that the chains are undistorted and adopting an all-*trans* conformation [133]. A stratum-by-stratum examination reveals nearly hexagonal close packing in the middle strata—confirming that the minimum energy is a distorted *hcp* structure. It is conceivable that helium diffraction from such a surface, which does not have a unique symmetry, may result in quite complicated scattering data [132]. A stratum of methylenes packed in a square lattice can be deformed towards a trigonal *hcp* arrangement by sliding alternate rows of methylenes (Figure 1.16). The tilting of alternate rows seen in Figure 1.13b affects this kind of deformation in the individual strata. The chemisorption and intramolecular bonding constraints restrict the magnitude of sliding.

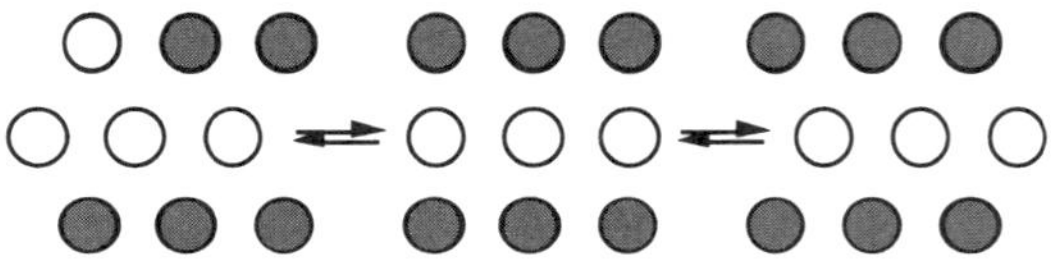

Figure 1.16 Deformation of a square lattice of methylene groups towards a trigonal *hcp* lattice by sliding alternate rows of methylene groups.

Helium diffraction on prepared monolayers revealed that the CH_3 surface has a lattice spacing of twice 2.885 Å in the [110] direction, and that the structure cannot be explained by a $p(2 \times 2)$ mesh[132]. Clearly, the introduction of a stratum of S atoms with a symmetry different than that preferred by alkanes resulted in molecular reorganizations that are reflected in the packing and ordering of the surface methyl groups. Thermal annealing at temperatures up to 135°C showed a significant increase in both diffraction and specular reflection intensities. These indicate that there was an increase in the ordered domain size on the topmost surface and a decrease in the number of defects. Based on the data, a $c(2 \times 8)$ structure (in surface notation) that can explain all the diffraction peaks observed can be suggested. This structure is commensurate with the ideal Au(001) truncation lattice, and no indication of hexagonal reconstruction on Au(001) could be detected. X-ray diffraction suggested a distorted hexagonal chain packing, thus verifying the MD prediction. Based on the X-ray data, a tilt of 33.08° in the plane that bisects the CH_2 groups, coupled with a rotation of 35.37° about the molecular axis could be suggested, again quite close to the MD predictions.

3.6 SAMS Containing Aromatic Groups: a Larger Unit Size

A model of a $CH_3(CH_2)_8—C_6H_4—O—(CH_2)_{11}—SiCl_3$ monolayer on SiO_2/Si suggests that for alkyl chains that are perpendicular to the surface, the phenyl ring has to exhibit a 35° tilt from the surface normal (Figure 1.17). However IR spectroscopy studies suggest a much smaller angle ($\sim 20°$) and an average tilt of $\sim 15°$ for the chains[134]. This difference is a result of the mismatch of cross-sectional area of the phenyl ring, and the alkyl chains. Because the packing in a monolayer cannot be closer than the spacing of the largest moiety, and because the cross-sectional area depends upon tilt and orientation, a clockwise rotation of 15° in the plane of the figure will result in a better cross-sectional area match between the different molecular parts. Thus, while the cross-sectional area of the alkyl chains increases from 19.6 Å^2 to 20.7 Å^2, that of the phenyl ring decreases from 25.1 Å^2 to 24.8 Å^2. No further information on the structure of the alkyl chain is available in this case, since no modeling was carried out.

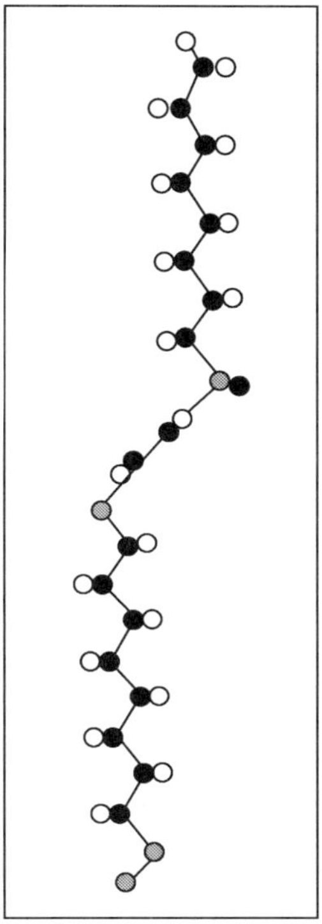

Figure 1.17 Side view of octyl 4-(8-mercaptooctyloxy)phenyl sulfone molecule minimized in isolation.

Self-assembled monolayers of $CH_3(CH_2)_7—SO_2—C_6H_4—O—(CH_2)_8—SH$ on gold were studied both experimentally[135], and using MD simulations[17]. Energy minimization shows that within the minimized assembly, the chains are nearly all-*trans*, and with an overall *lean* of $\sim 30°$ from the surface normal (Figure 1.18). This lean is in a plane that is about parallel to that of the phenyl ring. A very small *tilt*, in a plane that is about perpendicular to that of the phenyl ring, is observed, mainly in the upper chain. [Notice the similarity between this case and alkanethiolates on Au(001).] This is in agreement with the rectangular shape of the phenyl projection, with its different dimensions. Thus, since the phenyl–phenyl interactions determine the short chain-spacing, a small rotation of the alkyl chain with respect to the aromatic moiety will bring the spacing to that which is required for an effective tilt. On the other hand, the larger spacing in the perpendicular direction explains the lean angle of $\sim 30°$. Noticeable chain distortions can be seen near the sulfone groups in Figure 1.18a, again, in a striking similarity to alkanethiolates on Au(001). Hence, alkyl chains use a

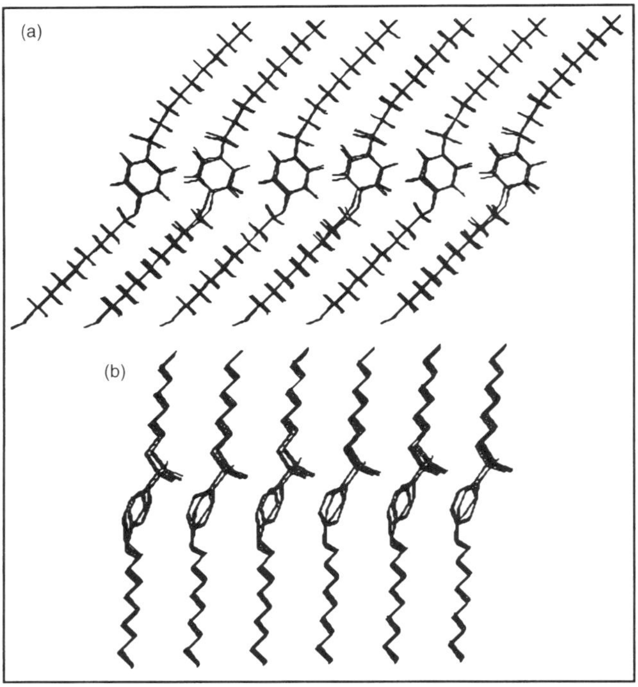

Figure 1.18 Two side views of a 6×6 assembly of octyl 4-(8-mercaptooctyloxy)phenyl sulfone molecules (three simulation cells by three simulation cells), after constant stress minimization under monolayer constraints.

Figure 1.19 Packing of the SO_2 group layer within the monolayer of Fig. 1.16. Note the two-dimensional ordering of the dipoles.

similar mechanism to heal intrastrata incommensurability whether it results from a stratum of different symmetry, or one with units of a different size. This general behavior probably exists in other systems and suggests that packing of atomic groups within the molecular assembly should be an important mechanism in determining the final equilibrium structure of the assembly.

From the beginning of the MD run, we noticed that the sulfone (—SO_2—) groups form a layer which practically does not change (Figure 1.19). The formation of this layer is driven by dipole–dipole interactions among the sulfone groups. Such interactions are an important driving force in the packing and ordering of the molecules, as has become apparent from a study of monolayers assembled from sulfone-containing alkannethiols (Figure 1.19).

3.7 SAMS of Sulfone-containing Alkanethiols

It was found that the introduction of sulfone groups with large in-plane dipoles has very little effect on molecular conformation within a monolayer, and the molecules are in the all-*trans* conformation (Figure 1.20). However, these in-plane dipoles have a profound effect on molecular tilt, i.e. the molecules tilt in one direction only. Such a molecular tilt is in contrast to that which exists in assemblies of simple alkanethiols on Au(111)[80]. Evidently, this molecular-tilt is the result of the strong $SO_2 \cdots SO_2$ electrostatic interactions that promote the formation of a plane of dipoles within the assembly. Hence, the introduction of a plane of dipoles within the assembly affects the packing and ordering of the monolayers and provides a means by which dipolar interactions within molecular assemblies may be studied.

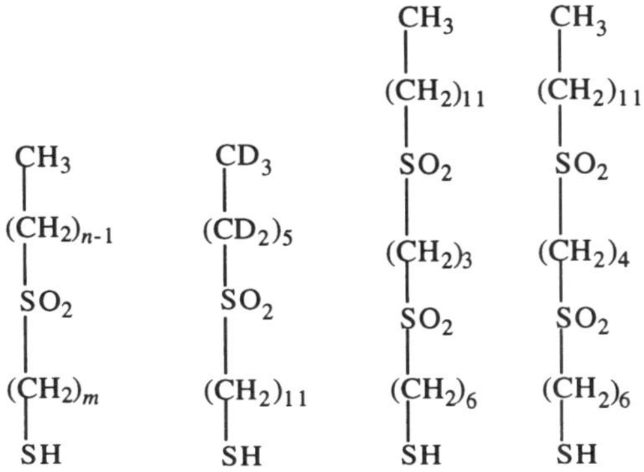

I $n = 6$, $m = 11$ IV   V   VI
II $n = 8$, $m = 9$
III $n = 11$ $m = 6$

Figure 1.20 Sulfone-containing alkanethiols.

Table 1.1 Contact angles of monolayers of **I–III**.

		θ_{H_2O}	
Molecule	θ_{HD}	*a*	*r*
I	40	106	97
II	40	108	97
III	41	109	101
ODT	45	110	105

From the wetting behavior of these films (Table 1.1), one finds the general trend that the contact angles decrease as the length of the chain above the sulfone group is reduced. This can be interpreted in two ways: (i) the disorder near the surface could be increasing; and/or (ii) the dipolar interactions due to the polar sulfone become more prominent. IR data indicate that the disorder resulting from the incorporation of the sulfone groups is the primary cause. In considering the differences in the wetting behavior between these molecules and the regular alkanethiol monolayers, a third factor must also be taken into consideration. The polarizability of a hydrocarbon chain is anisotropic, and if such molecules display a *preferred* direction of tilt, they will present a surface with a different degree of polarizability than that obtained from a surface with no preferred direction of tilt. Evidence for such a preferred molecular conformation, within these monolayers, has been established.

It seems quite remarkable that the presence of —SO_2—groups perturbs the monolayer so little relative to the monolayers without them. We feel, however, that ordering effects such as those reported here should not be restricted to sulfone groups, provided that the size and shape of the interacting groups are comparable to that of a methylene. Clearly, sulfone groups do perturb in-layer close packing of the alkyl chains due to the introduction of free volume, but this perturbation is compensated for by the ordering induced through the dipole–dipole interactions. On the other hand, if a group is too bulky, and there are no ordering mechanisms, it introduces a planar perturbation from close packing that results in a progressive distortion of the alkyl chains propagating from the plane of the perturbation. For example, this kind of perturbation exists in self-assembled alkanethiol monolayers containing polar aromatic groups, where an even bulkier group was introduced into the chain (Figure 1.18).

3.7.1 Surface Engineering Using Alkanethiolate SAMs

The use of functionalized alkanethiolate SAMs is important both for engineering of surface properties and for further chemical reactions. It is beyond the scope of this chapter to review all reported functionalities. Suffice is to mention that simple, e.g. CH_3, CF_3, $CH{=}CH_2$,$C{\equiv}CH$, Cl, Br, CN, OH,

OCH_3, NH_2, $N(CH_3)_2$, $N(CH_3)_3^+$, SO_3H, and $Si(OCH_3)_3$, COOH, $COOCH_3$, $CONH_2$[41–45,55,76,136–138] as well as more complex functionalities, e.g. ferrocenyl[139–147], biotinyl[148–151], 2,2-bipyridyl[152], tetrathiafulvalene-carboxylate[153], and tetraphenyl porphyrin[154,155], were attached to the chain termina of alkanethiolate monolayers. These monolayers are thus becoming the system of choice for studies of surface phenomena, electron transfer, molecular recognition, etc.

Surface OH and COOH groups are very useful for chemical transformations. The carboxylate group can be transformed to the corresponding acid chloride using $SOCl_2$[156]. Further reactions with amines and alcohols yield bilayer structures with amide and ester linkages, respectively. Reacting the acid chloride with a carboxylic acid-terminated thiol provides the corresponding thioester. This reaction has been recently used by Kim *et al.* to form polymeric self-assembled monolayers and multilayers from the diacetylene $HS(CH_2)_{10}C{\equiv}C{-}C{\equiv}C(CH_2)_{10}COOH$[157].

SAMs of OH-terminated alkanethiols have been used in many surface modifications reactions. For example, they have been reacted with OTS to yield a well ordered bilayer[158], with $POCl_3$ to form the phosphate surface[102,103,159], with trifluoroacetic anhydride to form the corresponding surface ester[160], with epichlorohydrin to form an ethylene oxide surface, and with chlorosulfonic acid to form a sulfate surface[161].

In order to appreciate the importance of surface engineering at the molecular level, the number of applications for which controlling the properties of interfaces is of primary importance should be considered. The lubrication of moving parts requires complete wetting by the lubricant. In the manufacturing of photographic films, adhesion of the gelatin emulsions to the polymeric film-base is vital. In xerography, the interaction of colloidal particles—polymer beads, toner particles—with surfaces determines the quality of the final print. In such applications, the interface is typically irregular, and the control of structure, chemical functionality, and roughness at the molecular level is practically impossible (Figure 1.21). As a result, the ability to modify surfaces in a predictable way and study their interfacial properties has been limited.

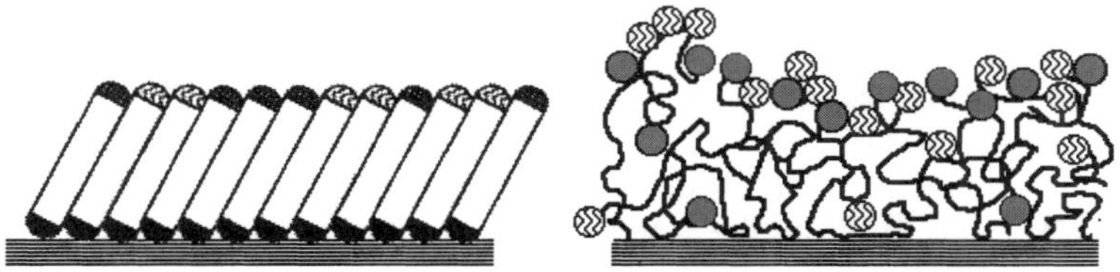

Figure 1.21 Schematic view of a polymer surface (right) and a monolayer surface (left). The different shaded circles represent different functional groups.

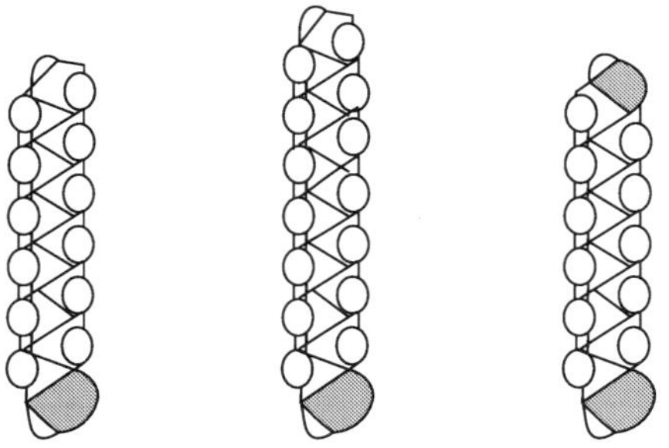

Figure 1.22 Alkanethiols used in surface engineering studies.

SAMs, particularly of alkanethiolates on gold surfaces, are especially suited for the study of interfacial phenomena. This is due to the fine control of the concentration of surface functional groups, achieved by simply changing the concentration in the solution from which the mixed monolayers are adsorbed. The difference between such monolayers and polymer surfaces is depicted in Figure 1.21. One of the more interesting examples of surface engineering is the mixed monolayer of dodecanethiol (DDT, $CH_3(CH_2)_{11}SH$), and 11-hydroxyundecanethiol (HUT, $HO(CH_2)_{11}SH$) (Figure 1.22)[162]. Wetting by water of freshly prepared HUT : DDT mixed SAMs showed linear behavior. However, wetting by *n*-hexadecane (HD, $C_{16}H_{34}$) of the same series of SAMs, showed a highly nonlinear behavior about a concentration threshold. In Figure 1.23, the cosines of the HD contact angles (θ—at ambient atmosphere—are plotted as a function of surface OH concentration. The sharp variation of the contact angle at $\sim$30–45% surface OH groups from partial to complete wetting reflects an abrupt change in the solid–vapor (γ_{SV}), and solid–liquid (γ_{SL}) interfacial free energies.

The jump in $\cos\theta$ (surface free energy) is due to the formation of a *condensed* water layer by a process of H-bonding (Figure 1.24). If one starts from a pure DDT monolayer (100% CH_3 surface), and randomly replaces CH_3 groups by OH groups, one should find that at low concentrations the OH groups are spaced far from each other. At such concentrations, there may be water clusters around the isolated OH centers, formed by correlated H-bonded water molecules. As the surface OH concentration increases, both the number and relative size of the water clusters increases. At a sufficiently large enough surface OH concentration, the water clusters unite to form a dense prewetting water layer, thus forming a "dressed" surface, and changing its macroscopic free energy. Hence, the wettabilities probe the transition from "bare" to "dressed" surfaces.

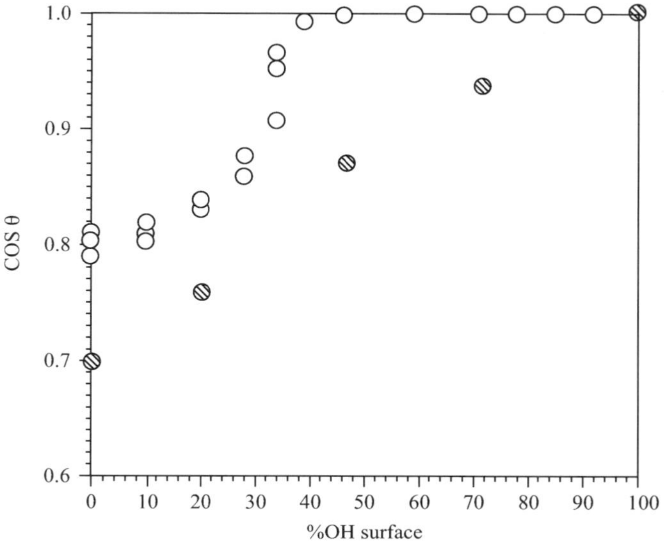

Figure 1.23 Cos θ of HD for HUT:DDT (o), and for HUT:TDT (ϕ) mixed alkanethiolate monolayers on gold, as a function of the surface OH concentration. Notice that for complete wetting $\theta = 0°$, and $\cos\theta = 1$.

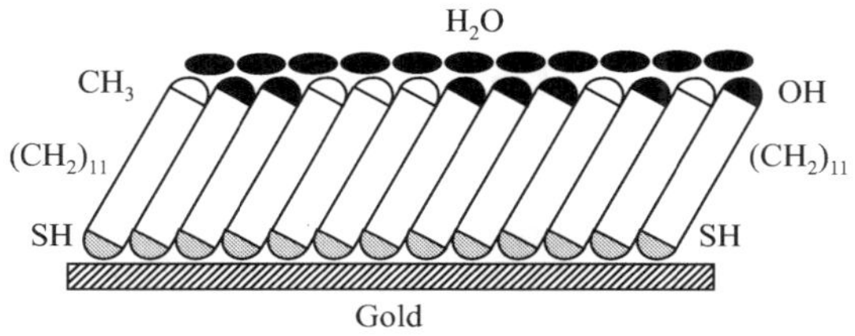

Figure 1.24 *Condensed* water layer, formed by a process of H-bonding, on a molecularly smooth HUT:DDT SAM surface, when surface OH concentration is >35%.

The surface was molecularly roughened to physically "break" the condensed water layer. Thus, instead of two molecules having the same length (HUT and DDT), the methyl-terminated component was made longer by two methylene (CH_2) units ($C_{14}H_{29}SH$, TDT, Figure 1.25). With these C_{14} molecules, the TDT : HUT mixed monolayer surfaces have ethyl groups (C_2H_5—) emerging from the surface into the second adsorbed water layer, thus disrupting its formation.

Hexadecane contact angles on the TDT : HUT mixed monolayers are presented in Figure 1.23. Indeed, *no* abrupt transition was found in the wetting properties of these monolayers. In fact, the cosines of the HD contact angles

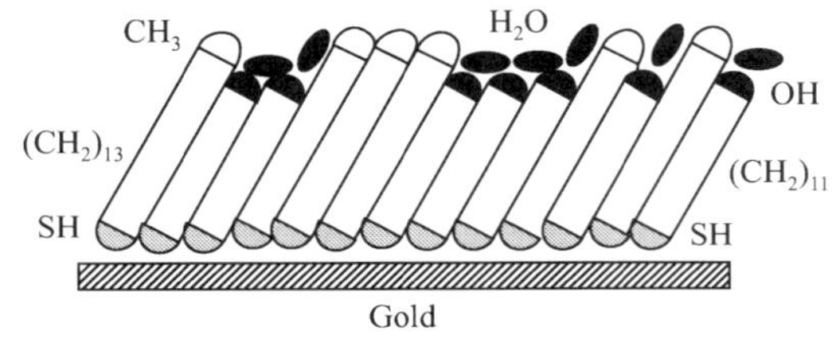

Figure 1.25 Roughening the surface at the molecular level destroys the observed transition.

exhibit a reasonably smooth relationship with the surface OH concentration. This is a remarkable change, considering that the maximum length of the emerging ethyl groups is only ~2.5 Å. Thus, although contact angles measure the *macroscopic* surface free energy, there are underlying molecular mechanisms, yet to be fully understood, that have dramatic effects on the macroscopic behavior. Figure 1.23 is therefore an excellent example of the advantage that molecularly engineered surfaces offer.

In the present case, the observed transition is a result of the interplay between water layering and wetting by HD. The underlying interactions between surface functional groups and HD molecules occur in the prewetting layer (Figure 1.26). This layer precedes the advancing liquid droplet and its dimensions extend far from the droplet edge (three-phase line)[163].

The prewetting layer has thickness of a few molecular layers (in this case hexadecane layers), and therefore, this is where the interactions of the liquid molecules with the surface are the strongest. This is why minute changes in surface functionality or roughness have a profound effect on the wetting behavior that is measured by a liquid droplet with dimensions many orders of magnitude larger that the thickness of the prewetting layer.

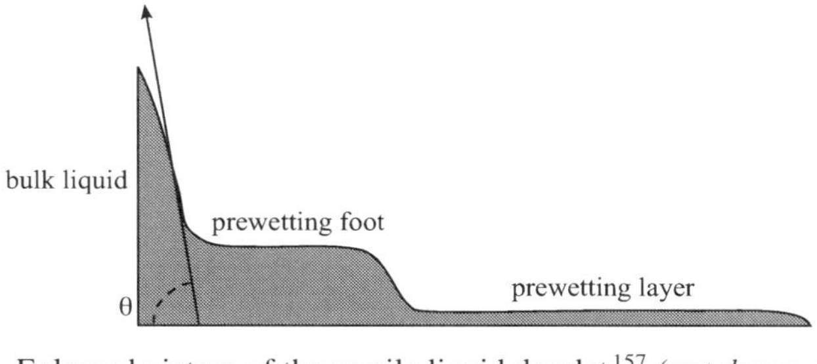

Figure 1.26 Enlarged picture of the sessile liquid droplet[157] (*not drawn to scale*). The prewetting foot is many molecular layers thick, while the prewetting layer is a few molecular layers thick. The arrow represents γ_{LV}, as in Figure 1.4, and θ is the contact angle.

4. ISSUES CONCERNING SURFACE STABILITY

The surface of the monolayer is responsible for properties such as wetting, adhesion, friction, selective binding, etc.[164,165]. Structural changes at the surface apparently penetrate into the monolayer bulk, and although it is not clear *a priori* to what depth, one could expect this depth to be a function of monolayer viscosity, i.e. molecular chain length. Degradation of surface structure may have a profound effect on the surface chemical reactivity and the ability to carry out quantitative surface reactions, for example, in the construction of multilayers from trichlorosilane derivatives[42]. This was our primary reason for studying surface stability, and for the choice of mixed monolayers of OH and CH_3 groups[162]. A second reason is that addressing the molecular mechanism behind interfacial phenomena can be achieved only if surfaces are prepared and characterized at the molecular level. OH and CH_3 groups represent two extremes—hydrophilic and hydrophobic—and therefore, one could expect that surface stability and interfacial phenomena will be very sensitive to the OH/CH_3 ratio.

Mixed monolayers $HO—(CH_2)_{11}—SH$, HUT, and $CH_3—(CH_2)_{11}—SH$, DDT on Au(111) were studied in great detail. It was found that HUT/Au(111) monolayers are unstable and a considerable increase in both advancing and receding water contact angles occurs with time[160]. The same surface instability was observed also for mixed monolayers. Experiments ruled out surface contamination, and showed that the decrease of surface free energy with time increases with the increasing number of surface OH groups. These observations support a mechanism in which surface free energy decreases due to the decrease of surface OH groups, resulting from conformational changes at chain termina. Thus, a 100% OH surface, after 60 min, yields the same contact angles as found for freshly made samples with surface OH/CH_3 ratio of $\sim 1:1$. This is equivalent to a decrease in surface free energy of $\sim 30\,mJ\,m^{-2}$. XPS studies of fresh, and 3-day aged OH surfaces that were exposed to an ambient of trifluoroacetic anhydride, showed that in the latter surface OH groups are less available to a chemical reaction, thus confirming the mechanism of surface reorganization. Studies of monolayers made of a longer chain derivative ($HO—(CH_2)_{21}—SH$, HHT) as a function of temperature showed that surface reorganization is a function of monolayer melting point. Hence, one way to pursue stable, molecularly-engineered surfaces is to utilize long-chain ω-substituted alkanethiols.

5. FUTURE DIRECTIONS

SAMs were originally investigated as building blocks for superlattices with engineered physical properties. However, the more immediate contribution to science and technology will come from their utilization in surface engineering. One example of a potential technology is semiconductor surface patterning. Potochnik *et al.* have used silane derivatives[166], while Whitesides *et al.* have demonstrated the utility of alkanethiol monolayers.

For many applications, controlling the properties of interfaces is of primary importance. The lubrication of moving parts requires complete wetting by the lubricant. In the manufacturing of photographic films, adhesion of the gelatin emulsions to the polymeric film-base is vital. In xerography, the interaction of colloidal particles—polymer beads, toner particles—with surfaces determines the quality of the final print. In such applications, the interface is typically irregular, and the control on structure, chemical functionality, and roughness at the molecular level is practically impossible. As a result, the ability to modify surfaces in a predictable way and study their interfacial properties has been limited. SAMs provide molecularly engineered surfaces, that with properly chosen molecules may offer the ability to decouple surface chemical properties from conformational disorder. In the following paragraph we discuss the potential contribution of SAMs to understanding the change in polymer properties near surfaces.

The wetting of surfaces by polymers is controlled by the polymer–surface interactions; however, the relative importance of the surface chemistry and its structure in this interfacial phenomena is still unknown. The friction coefficient f for the monomers in contact with a strongly attracting surface can be much higher than monomeric friction coefficient f_0 in the bulk, and the ratio between the two determines chain trapping near the surface[167]. Since f controls the reptation dynamics, it strongly influences both the polymer rheological properties near the surface, and the mode by which it wets the surface. For example, the appearance of slippage at the wall depends on the polymer-wall friction coefficient f, which is a function of interaction strength[168]. Silberzan and Leger[169] and Novotny[170] measured the spreading rate of a polymer droplet on a surface and found that the measured diffusion constant was at least an order of magnitude smaller than that of the bulk. Recently, Zheng *et al.* measured the monomer-surface friction coefficient for polystyrene on a number of surfaces and obtained excellent agreement with reptation theory modified to account for increased friction due to surface–monomer contact[171]. Using molecularly engineered surfaces, a better understanding of the surface–monomer interaction mechanism can be achieved. Such understanding is key to understanding polymer behavior near surfaces, which has relevance to practical problems of wetting, coating and adhesion. By using smooth, chemically homogeneous surfaces, and changing surface free energy in a systematic way, the contribution of chemistry to the polymer wetting mechanism can be addressed. Polymer properties (molecular weight and molecular weight distribution) can also be altered systematically, and the coupling between these two parameters can be investigated in detail.

Interfacial phenomena play a crucial role in adhesion[172]. Control of adhesion is central to a large number of industrial processes and problems. While there have been enormous efforts to solve practical problems, an understanding of adhesion at the molecular level is lacking. Prevailing macroscopic theoretical approaches model adhesion as arising from interplay between surface forces described simplistically either as contact forces[173] or by the equation:

$$P = -\mathrm{A}\ D/12z^2$$

where P is the force (negative for attraction), D is the diameter, z is the gap between the surfaces, and A is the Hamaker constant[174] with elastic stresses in the adhering bodies treated by the Hertz theory. Intricacies of specific molecular ordering and interactions at the surface are typically neglected in these theories. This may be why such theories fail to account for the experimentally observed hysteresis phenomena[175,176] that depend on the molecular details.

Surfaces used in adhesion at the engineering level are far from being in a molecularly ordered well-characterized state. This makes it difficult to disentangle the relative importance of the many different effects that are ultimately responsible for adhesion. The complexity of adhesion phenomena may be addressed by minimizing the number of uncontrolled parameters, thus reducing the ambiguity. This can be accomplished by using SAMs as model surfaces.

6. CONCLUSIONS

Organization of complex, semiflexible organic molecules within quasi two-dimensional assemblies is the result of a delicate interplay between substrate–adsorbate interactions, nonbond interactions between adsorbates—electrostatic and vdW forces— and intramolecular interactions such as bond stretches, angle bends, and torsions. Surface reorganization contributes to the final equilibrium structure of the assembly.

As molecular materials become a center of activity in chemistry, understanding packing and ordering at the molecular level, and the relationship between a molecular structure and property, and the structure and property of the bulk it forms becomes ever more urgent. Nanodevices are essential not only for information processing, memories, and communicative systems, but also for environmental sensors and biosensors that can detect changes in very small cell clusters. Such devices are based on the ability to manipulate the structure and properties of matter at the molecular level. Self-assembly offers one possible route for the construction of nanodevices. Therefore, molecular design and self-assembly may become key to molecular-based nanodevices. However, the fundamental understanding of the process of self-organization, and the relationship between the kinetics of assembly formation and its final equilibrium structure are prerequisites to any technological advances in this area.

7. REFERENCES

1. W. C. Bigelow, D. L. Pickett and W. A. Zisman, *J. Colloid Interface Sci.* **1**, 513 (1946).
2. G. L. Gaines, *Insoluble Monolayers Liquid–Gas Interfaces*. Interscience, New York (1966).
3. J. Sagiv, *J. Am. Chem. Soc.* **102**, 92 (1980).
4. R. G. Nuzzo and D. L. Allara, *J. Am. Chem. Soc.* **105**, 4481 (1983).
5. H. Kuhn and A. Ulman, In *Thin Films*, ed. by A. Ulman, Vol. 20. Academic Press (1995).
6. P. Ball, *Designing the Molecular World*. Princeton University Press, Princeton (1994).
7. A. Ulman, *An Introduction to Ultrathin Organic Films: From Langmuir–Blodgett to Self-Assembly*. Academic Press, Boston (1991).
8. J. A. Zasadzinski, *et al., Science* **263** , 1726 (1994).
9. Sukenik and references therein.
10. G. G. Roberts, *Adv. Phys.* **34**, 475 (1985).
11. D. A. Tirrell, *et al.*, *MRS Bull.* **July**, 23–28 (1991).
12. D. L. Allara and R. G. Nuzzo, *Langmuir* **1**, 45 (1985).
13. D. L. Allara and R. G. Nuzzo, *Langmuir* **1**, 52 (1985).
14. H. Ogawa, T. Chihera and K. Taya, *J. Am. Chem. Soc.* **107**, 1365 (1985).
15. N. E. Schlotter, M. D. Porter, T. B. Bright and D. L. Allara, *Chem. Phys. Lett.* **132**, 93 (1986).
16. D. Y. Huang and Y. T. Tao, *Bull. Inst. Chem., Acad. Sin.* **33**, 73 (1986).
17. Y. Shnidman, A. Ulman and J. E. Eilers, *Langmuir* **9**, 1071 (1993).
18. M. G. Samart, C. A. Brown and J. G. Gordon, *Langmuir* **9**, 1082 (1993).
19. Y.-T. Tao, *J. Am. Chem. Soc.* **115**, 4350 (1993).
20. E. L. Smith and M. D. Porter, *J. Phys. Chem.* **97**, 4421 (1993).
21. P. Silberzan, L. Léger, D. Ausserré and J. J. Benattar, *Langmuir* **7**, 1647 (1991).
22. S. R. Wasserman, Y.-T. Tao and J. M. Whitesides, *Langmuir* **5**, 1074 (1989).
23. J. D. Le Grange, J. L. Markham and C. R. Kurjian, *Langmuir* **9**, 1749 (1993).
24. R. Maoz and J. Sagiv, *J. Colloid Inteface Sci.* **100**, 465 (1984).
25. J. Gum and J. Sagiv, *J. Colloid Interface Sci.* **112**, 457 (1986).
26. J. Gum, R. Iscovici and J. Sagiv, *J. Colloid Interface Sci.* **101**, 201 (1984).
27. N. Tillman, A. Ulman, J. S. Schildkraut, T. L. Penner, *J. Am. Chem. Soc.* **110**, 6136 (1988).
28. S. Brandriss and S. Margel, *Langmuir* **9**, 1232 (1993).
29. K. Mathauser and C. W. Frank, *Langmuir* **9**, 3002 (1993).
30. K. Mathauser and C. W. Frank, *Langmuir* **9**, 3446 (1993).
31. G. Carson and S. Granick, *J. Appl. Polym. Sci.* **37**, 2767 (1989).
32. C. R. Kessel and S. Granick, *Langmuir* **7**, 532 (1991).
33. D. K., Schwartz, S., Steinberg, J., Israelachvili and J. A. N., Zasadzinski, *Science* **69**, 3554 (1992).
34. H. O. Finklea, L. R. Robinson, A. Blackburn, B. Richter, D. L. Allara and T. Bright, *Langmuir* **2**, 239 (1986).

35. I. Rubinstein, E. Sabatani, R. Maoz and J. Sagiv, *Proc. Electrochem. Soc.* **86**, 175 (1986).
36. I. Rubinstein, E. Sabatani, R. Maoz and J. Sagiv, *Electroanal. Chem.* **219**, 365 (1987).
37. D. L. Allara, A. N. Parikh and F. Rondelezm, *Langmuir* **11**, 2357 (1995).
38. C. P. Tripp and M. L. Hair, *Langmuir* **8**, 1120 (1992).
39. D. L. Angst and G. W. Simmons, *Langmuir* **7**, 2236 (1991).
40. M. E. McGovern, K. M. R. Kallury and M. Thompson, *Langmuir* **10**, 3607 (1994).
41. W. Gao and L. Reven, *Langmuir* **11**, 1860 (1995).
42. N. Tillman, A. Ulman and T. L. Penner, *Langmuir* **5**, 101 (1989).
43. C. N. Sukenik, personal communication.
44. L. H. Dubois and R. G. Nuzzo, *Ann. Phys. Chem.* **43**, 437 (1992).
45. C. D. Bain and G. M. Whitesides, *Adv. Mater.* **1**, 506 (1989).
46. J. P. Folkers, J. A. Zerkowski, P. E. Laibinis, C. T. Seto and G. M. Whitesides, In *Supramolecular Architecture*, ACS Symposium Series 499, ed. by T. Bein, pp. 10–23. American Chemical Society, Washington, DC (1992).
47. T. R. Lee, P. E. Laibinis, J. P. Folkers and G. M. Whitesides, *Pure Appl. Chem.* **63**, 821 (1991).
48. G. M. Whitesides and G. S. Ferguson, *Chemtracts-Organic Chemistry* **1**, 171 (1988).
49. H. Sellers, A. Ulman, Y. Shnidman and J. E. Eilers, *J. Am. Chem. Soc.* **115**, 9389 (1993).
50. E. B. Troughton, C. D. Bain, G. M. Whitesides, D. L. Allara and M. D. Porter, *Langmuir* **4**, 365 (1988).
51. E. Katz, N. Itzhak and I. Willner, *J. Electroanal. Chem.* **336**, 357 (1992).
52. E. Sabatani, J. Cohen-Boulakia, M. Bruening and I. Rubinstein, *Langmuir* **9**, 2947 (1993).
53. M. A. Bryant, S. L. Joa and J. E. Pemberton, *Langmuir* **9**, 753 (1992).
54. W. Hill and B. Wehling, *J. Phys. Chem.* **97**, 9451 (1993).
55. T. T.-T. Li, H. Y. Liu and M. J. Weaver, *J. Am. Chem. Soc.* **106**, 1233 (1984).
56. J. M. Cooper, K. R. Greenough and C. J. McNeil, *J. Electroanal. Chem.* **347**, 267 (1993).
57. K. Uvdal, P. Bodö and B. Liedberg, *J. Colloid Interface Sci.* **149**, 162 (1992).
58. A. Ihs, K. Uvdal and B. Liedberg, *Langmuir* **9**, 733 (1993).
59. Th. Arndt, H. Schupp and W. Schepp, *Thin Solid Films* **178**, 319 (1989).
60. J. A. Mielczarski and R. H. Yoon, *Langmuir* **7**, 101 (1991).
61. T. R. G. Edwards, V. J. Cunnane, R. Parsons and D. Gani, *J. Chem. Soc. Chem. Commun.* 1041 (1989).
62. A. J. Arduengo, J. R. Moran, J. Rodriguez-Paradu and M. D. Ward, *J. Am. Chem. Soc.* **112**, 6153 (1990).
63. G. Xue, X.-Y. Huang, J. Dong and J. Zhang, *J. Electroanal. Chem.* **310**, 139 (1991).
64. S. Bharathi, V. Yegnaraman and G. P. Rao, *Langmuir* **9**, 1614 (1993).
65. M. G. Samanat, C. A. Broen and J. G. Gordon, *Langmuir* **8**, 1615 (1992).
66. A. Ulman, *J. Mater. Educ.* **11**, 205 (1989).
67. P. E. Laibinis, G. M. Whitesides, D. L. Allara, Y.-T. Tao, A. N. Parikh and R. G. Nuzzo, *J. Am. Chem. Soc.* **113**, 7152 (1991).
68. M. W. Walczak, C. Chung, S. M. Stole, C. A. Widrig and M. D. Porter, *J. Am. Chem. Soc.* **113**, 2370 (1991).
69. P. E. Laibinis and G. M. Whitesides, *J. Am. Chem. Soc.* **112**, 1990 (1992).
70. A. Ihs and B. Liedberg, *Langmuir* **10**, 734 (1994).
71. P. E. Laibinis and G. M. Whitesides, *J. Am. Chem. Soc.* **114**, 9022 (1992).

72. K. Shimazu, Y. Sato, I. Yagi and K. Uosaki, *Bull. Chem. Soc. Jpn.* **67**, 863 (1994).
73. A. Demoz and D. J. Harrison, *Langmuir* **9**, 1046 (1993).
74. D. Mandler, preprint.
75. M. Stratmann, Adv. Mater. **2**, 191 (1990).
76. M. Volmer, M. Stratmann and H. Viefhaus, *Surf. Interf. Analysis* **16**, 278 (1990).
77. M. Brust, M. Walker, D. Bethell, D. J. Schiffrin and R. Whyman, *J. Chem. Soc. Chem. Commun.* 801 (1994).
78. C. W. Sheen, J. X. Shi, J. Martensson, A. N. Parikh and D. L. Allara, *J. Am. Chem. Soc.* **114**, 1514 (1992).
79. Y. Gu, Z. Lin, R. A. Butera, V. S. Smentkowski and D. H. Waldeck, *Langmuir* **11**, 1849 (1995).
80. C. D. Bain, E. B. Troughton, Y.-T. Tao, J. Evall, G. M. Whitesides and R. G. Nuzzo, *J. Am. Chem. Soc.* **111**, 321 (1989).
81. M. Buck, F. Eisert, J. Fischer, M. Grunze and F. Träger, *Appl. Phys.* **A53**, 552 (1991).
82. M. Buck,F. Eisert and M. Grunze, *Ber. Bunsenges. Phys. Chem.* **97**, 399 (1993).
83. G. Hähner, Ch. Wöll, M. Buck and M. Grunze, *Langmuir* **9**, 1955 (1993).
84. Y. Shnidman, S. D. Evans, A. Ulman, J. E. Eilers *Langmuir*, **9**, 1071 (1993).
85. L. Strong and G. M. Whitesides, *Langmuir* **4**, 546 (1988).
86. C. E. D. Chidsey and D. N. Loiacono, *Langmuir* **6**, 709 (1990).
87. L. H. Dubois, B. R. Zegarski and R. G. Nuzzo, *J. Chem. Phys.* **98**, 678 (1993).
88. C. E. D. Chidsey, G. Liu, Y. P. Rowntree and G. Scoles, *J. Chem. Phys.* **91**, 4421 (1989).
89. C. A. Alves, E. L. Smith and M. D. Porter, *J. Am. Chem. Soc.* **114**, 1222 (1992).
90. G. E. Poirier, M. J. Tarlov and H. E. Rushneier, *Langmuir* **10**, 3383 (1994).
91. P. Fenter, P. Eisenberger and K. S. Liang, *Phys. Rev. Lett.* **70**, 2447 (1993).
92. N. Camillone, C. E. D. Chidsey, G.-Y. Liu and G. Scoles, *J. Phys. Chem.* **98**, 3503 (1993).
93. G. E. Poirier and M. J. Tarlov, *Langmuir* **10**, 2859 (1994).
94. G. Edinger, A. Gölzhäuser, K. Demota, Ch. Wöll and M. Grunze, *Langmuir* **9**, 4 (1993).
95. C. Schönenberger, J. A. M. Sondag-Huethorst, J. Jorritsma and L. G. J. Fokkink, *Langmuir* **10**, 611 (1994).
96. A. L. Harris, L. Rothberg, L. Dhar, N. J. Levinos and L. H. Dubois, *J. Phys. Chem.* **94**, 2438 (1991).
97. P. Fenter, P. Eisenberger, J. Li, N. Chamillone, S. Bernasek, G. Scoles, T. A. Ramanarayanan and K. S. Liang, *Langmuir* **7**, 2013 (1991).
98. A. Dhirani, M. A. Hines, A. J. Fisher, O. Ismail and P. Guyot-Sionnest, *Langmuir* **11**, 2609 (1995).
99. A. Ulman, J. E. Eilers and N. Tillman, *Langmuir* **5**, 1147 (1989).
100. A. Ulman, *J. Mat. Ed.* **11**, 207 (1989).
101. A. Nemetz, T. Fischer, A. Ulman and W. Knoll, *J. Chem. Phys.* **98**, 5912 (1993).
102. P. Fenter and P. Eisenberger, personal communication.
103. M. R. Linford and C. E. D. Chidsey, *J. Am. Chem. Soc.* **115**, 12631 (1993).
104. M. R. Linford, C. E. D. Chidsey, P. Fenter and P. M. Eisenberger, *J. Am. Chem. Soc.* **116**, 0000 (1995).
105. G. Cao, H.-G. Hong and T. E. Mallouk, *Acc. Chem. Res.* **25**, 420 (1992).
106. H. Lee, L. J. Kepley, H.-G. Hong and T. E. Mallouk, *J. Am. Chem. Soc.* **110**, 618 (1988).
107. H. Lee, L. J. Kepley, H.-G. Hong, S. Akhter and T. E. Mallouk, *J. Phys. Chem.* **92**, 2597 (1988).

108. M. L. Schilling, H. E. Katz, S. M. Stein, S. F. Shane, W. L. Wilson, S. Buratto, S. B. Ungahse, G. N. Taylor, T. M. Putvinski and C. E. D. Chidsey, *Langmuir* **9**, 2156 (1993).
109. S. F. Bent, M. L. Schilling, W. L. Wilson, H. E. Katz and A. L. Harris, *Chem. Mater.* **6**, 122 (1994).
110. H. E. Katz, M. L. Schilling, S. B. Ungahse, T. M. Putvinski and C. E. D. Chidsey, In *Supramolecular Architecture*, ACS Symposium Series 499, ed. by T. Bein pp. 24–32. American Chemical Society, Washington, DC (1992).
111. H. E. Katz, M. L. Schilling, C. E. D. Chidsey, T. M. Putvinski and R. S. Hutton, *Chem. Mater.* **3**, 699 (1991).
112. R. Dahint, K. Bierbaum and M. Grunze, *Springer Series Surf. Sci.* (in press).
113. K. Robinson, A. Ulman, J. B. Lando and A. J. Mann, unpublished results.
114. Y.-T. Tao, *J. Am. Chem. Soc.* **115**, 4350 (1993).
115. Y.-T. Tao, M.-T. Lee and S.-C. Chang, *J. Am. Chem. Soc.* **115**, 9547 (1993).
116. N. Camillone III, *et al., J. Chem. Phys.* **94**, 8493 (1991).
117. M. Majda, personal communication.
118. H. Sellers, *Surf. Sci.* **294**, 99 (1993).
119. M. Grunze, *Phys. Scripta* (in press).
120. R. L. McCarley, D. J. Dunaway and R. J. Willicut, *Langmuir* **9**, 2775 (1993).
121. J.-P. Bucher, L. Santesson and K. Kern, *Langmuir* **10**, 979 (1994).
122. E. Sabatani, J. Cohen-Boulakia, M. Bruening and I. Rubinstein, *Langmuir* **9**, 2974 (1993).
123. K. Schwaha, N. D. Spencer and R. M. Lambert, *Surf. Sci.* **81**, 273 (1979).
124. G. Rovida, F. Pratesi, *Surf. Sci.* **104**, 609 (1981).
125. A. L. Harris, C. E. D. Chidsey, N. J. Levinos and D. N. Loiacono, *Chem. Phys. Lett.* **141**, 350 (1987).
126. V. Ryckaert *et al.*, *Phys. Rev. Lett.* **58**, 698 (1987).
127. T. Seto, T. Hara and K. Tanaka, *Jpn J. Appl. Phys.* **7**, 31 (1968).
128. A. Ulman and R. P. Scaringe, *Langmuir* **8**, 894 (1992).
129. R. P. Scaringe, In *Electron Crystallography of Organic Molecules*, ed. by J. R. Fryer and D. L. Dorset, pp. 85–113. Kluwer, Dordrecht (1990).
130. D. A. Outka, J. Stöhr, J. P. Rabe, J. D. Swalen and H. H. Rotermund, *Phys. Rev. Lett.* **59**, 1321 (1987).
131. A. Ulman, J. E. Eilers and N. Tillman, *Langmuir* **5**, 1147 (1989).
132. N. Camillone III, C. E. D. Chidsey, G.-Y. Liu and G. Scoles, *J. Chem. Phys.* **98**, 4234 (1993).
133. J. E. Eilers, unpublished results.
134. N. Tillman, A. Ulman, J. S. Schildkraut and T. L. Penner, *J. Am. Chem. Soc.* **110**, 6136 (1988).
135. S. D. Evans, E. Urankar, A. Ulman and N. Ferris, *J. Am. Chem. Soc.* **113**, 4121 (1991).
136. R. G. Nuzzo, B. R. Zegarski and L. H. Dubois, *J. Am. Chem. Soc.* **109**, 733 (1987).
137. C. E. D. Chidsey and D. N. Loiacono, *Langmuir* **6**, 682 (1990).
138. K. Dohlhofer, J. Figura and J.-H. Fuhrhop, *Langmuir* **8**, 1811 (1992).
139. C. E. D. Chidsey, C. R. Bertozzi, T. M. Putvinski and A. M. Mujsce, *J. Am. Chem. Soc.* **112**, 4301 (1990).
140. K. Uosaki, Y. Sato and H. Kita, *Langmuir* **7**, 1510 (1991).
141. C. E. D. Chidsey, *Science* **251**, 919 (1991).
142. D. D. Popenoe, R. S. Deinhammer and M. D. Porter, *Langmuir* **8**, 2521 (1992).
143. Y. Sato, B. L. Frey, R. M. Corn and K. Uosaki, *Bull. Chem. Soc. Jpn* **67**, 21 (1994).
144. D. M. Collard and M. A. Fox, *Langmuir* **7**, 1192 (1991).

145. S. E. Creager and G. K. Rowe, *Anal. Chim.* **246**, 233 (1991).
146. G. K. Rowe and S. E. Creager, *Langmuir* **7**, 2307 (1991).
147. G. K. Rowe and S. E. Creager, *Langmuir* **10**, 1186 (1994).
148. L. Häussling, H. Ringsdorf, F.-J. Schmitt and W. Knoll, *Langmuir* **7**, 1837 (1991).
149. L. Häussling, B. Michel, H. Ringsdorf and H. Rohrer, *Angew. Chem. Int. Ed. Engl.* **30**, 679 (1991).
150. F.-J. Schmitt, L. Häussling, H. Ringsdorf and W. Knoll, *Thin Solid Films* **210/211**, 815 (1992).
151. J. Spinke, J. Liley, H.-J. Guder, L. Angermaier and W. Knoll, *Langmuir* **9**, 1821 (1993).
152. Y. S. Obeng and A. J. Bard, *Langmuir* **7**, 195 (1991).
153. C. M. Yip and M. D. Ward, *Langmuir* **10**, 549 (1994).
154. J. Zak, H. Yuan, K. Woo and M. D. Porter, *Langmuir* **9**, 2772 (1993).
155. J. E. Hutchinson, T. A. Postlethwaite and R. W. Murray, *Langmuir* **9**, 3277 (1993).
156. R. V. Duevel and R. M. Corn, *Anal. Chem.* **64**, 337 (1992).
157. T. Kim, R. M. Crooks, M. Tsen and L. Sun, preprint.
158. A. Ulman and N. Tillman, *Langmuir* **5**, 1418 (1989).
159. L. Bertilsson and B. Liedberg, *Langmuir* **9**, 141 (1993).
160. S. D. Evans, R. Sharma and A. Ulman, *Langmuir* **7**, 156 (1991).
161. S. Löfås and B. Johnsson, *J. Chem. Soc. Chem. Commun.* 1526 (1990).
162. A. Ulman, S. D. Evans, Y. Shnidman, R. Sharma, J. E. Eilers and J. C. Chang, *J. Am. Chem. Soc.* **113**, 1499 (1991).
163. P. G. deGennes, *Rev. Mod. Phys.* **57**, 827 (1985).
164. J. F. Padday, *J. Adhesion Sci. Tech.* **6**, 1347 (1992).
165. R. J. Good, *J. Adhesion Sci. Tech.* **6**, 1269 (1992).
166. S. J. Potochnik, P. E. Pehrsson, D. S. Y. Hsu and J. M. Calvert, *Langmuir* **11**, 1842 (1995) and references therein.
167. P. G. deGennes, *Conference Proceedings Liquids at Interfaces*, ed. by J. Charvolin, J. F. Joanny and J. Zinn-Justin. North Holland, Amsterdam (1990).
168. E. Manias, G. Hadziioannou, I. Bitsanis and G. ten Brinke, *Europhys. Lett.* (in press).
169. P. Silberzan and L. Leger, *Macromolecules* **25**, 1267 (1992).
170. V. J. Novotny, *J. Chem. Phys.* **92**, 3189 (1990).
171. X. Zheng, B. B. Sauer, J. G. Van Alsten, S. A. Schwarz, M. H. Rafailovich, J. Sokolov and M. Rubinstein, *Phys. Rev. Lett.* **74**, 407 (1995).
172. K. Kendall, *Science* **263**, 1720 (1994).
173. K. L. Johnson, K. Kendall and D. A. Roberts, *Proc. R. Soc. London Ser. A* **324**, 301 (1971).
174. B. V. Derjaguin, V. M. Muller and Y. P. Toporov, *J. Colloid Interface Sci.* **53**, 314 (1975).
175. M. K. Chandhury and G. M. Whitesides, *Science* **255**, 1230 (1992).
176. M. K. Chandhury and G. M. Whitesides, *Langmuir* **7**, 1013 (1991).

Chapter 2

Design and Synthesis of Complex Conjugated Polymers

TIMOTHY M. SWAGER
Department of Chemistry, Massachusetts Institute of Technology, Cambridge, MA, USA

1. INTRODUCTION

Chemists have long been intrigued by organic molecules and materials with extensive electronic delocalization. Research in the field of conjugated polymers was ignited 20 years ago by the historic discovery of metallic levels of conductivity in polyacetylene[1]. Since this genesis, detailed physical understanding and synthetic methods have been developed that allow for the design of ever more complex and useful materials[2,3]. This chapter reviews progress, with an emphasis on results from the author's laboratory, which illustrates methods for the synthesis and design of conducting polymers containing novel structural elements that exhibit special electronic or chemical properties.

2. METHODS OF POLYMER SYNTHESIS

The vast majority of procedures for the synthesis of polymers can be classified in two broad categories, step-growth and chain-growth polymerizations[4,5]. Step-growth, or condensation polymerizations, involve the pair-wise coupling of monomers, A and B, to form a polymer, $(\text{—A—B—})_n$. A disadvantage of this method is that near quantitative reactions are required for the production of high molecular weight (M_n) materials, since the degree of polymerization (DP)

Organised Molecular Assemblies in the Solid State Edited by James K. Whitesell
© 1999 John Wiley & Sons Ltd.

is determined by the well-known formula:

$$DP = 1/(1 - \text{extent of reaction})$$

As the extent of reaction approaches unity, DP approaches infinity. In contrast, chain-growth reactions wherein monomers are sequentially added at a reactive site, can provide high M_n materials with even low monomer conversion. These latter methods have often produced materials with broad molecular weight distributions, however there have been many advances that now afford the controlled polymerization of a vast array of monomers[6]. If a controlled chain-growth polymerization displays a rapid initiation, has a slower chain propagation which displays uniform kinetics, and is not prone to termination or chain transfer reactions, then the polymerization is said to be living. In a living polymerization the polydispersity index ($PDI = M_w/M_n$) is close to unity at high molecular weight. A slightly less stringent but accepted definition is: a living polymerization displays a linear relationship between the M_n and the percent conversion of the monomer.

The majority of procedures for the synthesis of conjugated polymers involve step-growth methods. This fact is largely due to the development of palladium[7] and nickel[8] catalyzed coupling procedures which are extremely effective for the synthesis of unsaturated compounds. A feature that has limited the utility of most chain-growth polymerizations is that these methods often sacrifice a degree of unsaturation and as a consequence produce polymers which do not have extended conjugation. Nevertheless, chain-growth methods have proved to be effective for the synthesis of conducting polymers by way of precursor polymers, which can isomerize or undergo elimination reactions to produce extended conjugation[9,10].

The work detailed in this chapter will focus on step-growth polymerizations rather than chain-growth methods. Transition metal mediated coupling reactions have a number of attractive features that have resulted in their extensive use for the synthesis of conjugated polymers. First, they are tolerant of many functional groups and thereby allow the incorporation of interesting structural features into the polymers. Second, they can directly produce structures with extended conjugation and avoid the need for additional transformations on the polymer after it has been synthesized. Third, these reactions offer the potential for the creation of highly diverse families of materials (libraries) through the use of different combinations of monomers. Coupling reactions can be conducted in a homo-coupling or a cross-coupling fashion. As shown in Reactions 1 and 2, transition metal mediated homo-coupling can be performed by way of oxidative or reductive transformations, respectively. $Ni^{\circ}L_4$ complexes have been extensively used to reductively couple aromatic halides[11]. This reaction can be made pseudo-catalytic in $Ni^{\circ}L_4$ if conducted in the presence of Zn which functions to reduce the Ni^{2+} back to a Ni° complex. The oxidative coupling of organo-lithium compounds by CuX_2

and Fe(acac)$_3$ is also well established, however these procedures have only recently been applied to conjugated polymers (Reaction 1)[12].

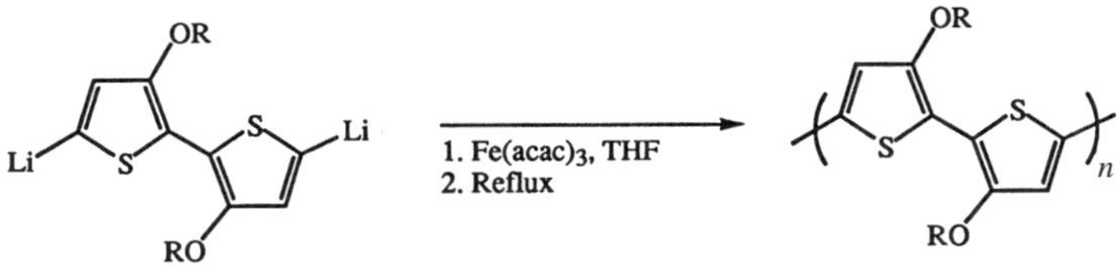

(1)

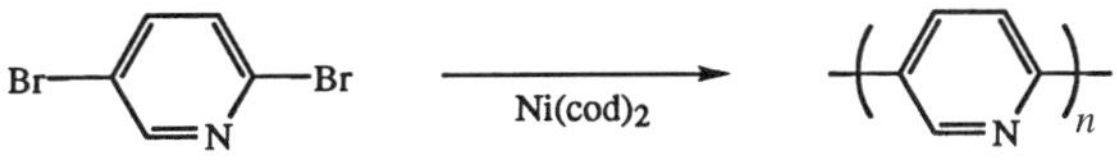

(2)

Cross-coupling reactions can be used to combine pairs of unsaturated monomers to produce alternating copolymers (Reaction 3)[12]. Homopolymers may also be produced by cross-coupling procedures if the monomer has the proper functionality (Reaction 4)[13]. These procedures require that a monomer contain a halogen or halogen equivalent (R—X) which is capable of undergoing an oxidative addition reaction with the transition metal, typically nickel or palladium. The RX group is then matched with a coupling partner that contains an electro-positive substituent such as Sn, B, Li, Cu, Hg, Zn, or Mg (R′—M). The R′—M moiety reacts to affect the replacement of the halide on the transition metal with a carbon fragment (R′). The R—X and R′—M functional groups may be contained in different monomers (Reaction 3) or the same monomer (Reaction 4). Once the R and R′ groups are both bound to the transition metal center, reductive elimination occurs to form a new C—C bond (i.e. R′—R). The ability to match different combinations of monomers introduces diversity. For example, single component or pair-wise combinations of four dihalide monomers (X—R—X) and four M—R—M monomers can potentially afford eight homopolymers and sixteen different alternating copolymers. Formulations beyond simple binary mixtures present considerably more diversity.

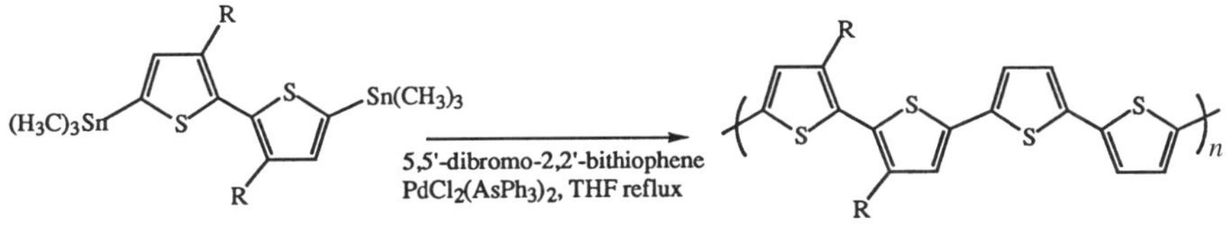

(3)

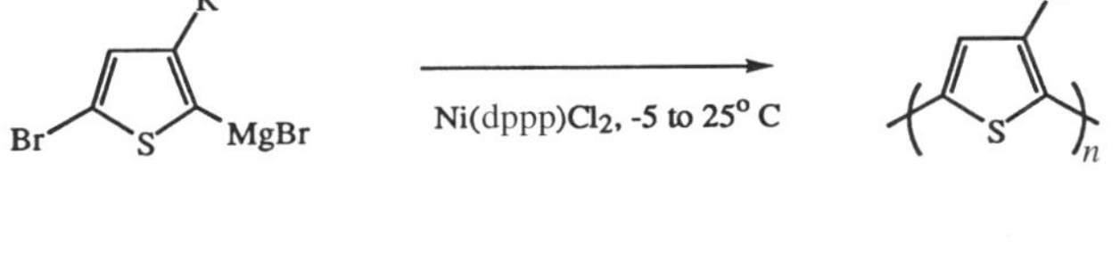

(4)

A shortcoming of cross-coupling chemistry, and indeed any step-growth method, is that depending upon the substrates and conditions the yields may not always be quantitative. This feature often leads to low molecular weight materials. Optimization of transition metal based coupling procedures is often complicated by the fact that a number of different processes in the catalytic cycle have the potential to limit the rate and reaction yield. For example, in palladium catalyzed cross-coupling reactions, any of the oxidative addition, transmetallation, or reductive elimination steps can be rate limiting. The M_n of polymers synthesized by these methods are often reported to be less than 20 000 Daltons. Depending on the intended use of the polymer, this may or may not be problematic. A polymer's mechanical properties are very sensitive to molecular weight. However their optical, electrochemical, and conducting properties are not as sensitive. We have found that improvements in molecular weights can be made by the careful control of ligand stoichiometry, the amount of catalyst, and the addition of rate-enhancing ligands. The following examples illustrate these features.

In step-growth polymerizations the highest molecular weights are generally achieved with careful control of the stoichiometry. A typical step-growth polymerization of monomers A and B, to produce (—A—B—), requires a one-to-one stoichiometry. In contrast for the synthesis of poly(phenylene ethynylene)s, we found that the addition of a slight excess of the acetylene containing component can be extremely effective at producing high molecular weight materials[14]. As shown in Scheme 2.1, extremely high absolute molecular weights can be produced (>1 000 000 Daltons) if the optimal stoichiometry is determined[15]. The need for excess acetylene is due to a small, but ubiquitous, amount of oxidant which leads to the production of diacetylene units in the presence of Cu^+ and amine bases. The excess acetylene monomer compensates for the small amounts of diacetylene linkages formed. Although these linkages introduce some irregularity into the polymers, they do not

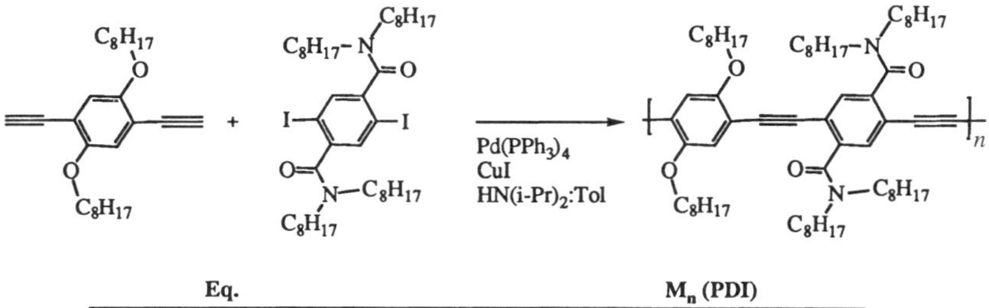

Eq.		M_n (PDI)
1.00	1.00	2.68×10^5 (2.9)
1.03	1.00	1.5×10^6 (3.1)
1.05	1.00	1.1×10^6 (3.1)

Scheme 2.1

present disruptions in conjugation or detectable changes in the material's physical properties[14,16].

In the case of Suzuki-type polymerizations as shown in Reaction 5, we found that optimal procedures require mixed solvents and very low catalyst loading[17]. Our results suggest that at higher catalyst loading, the reaction is interrupted by the formation of colloidal palladium particles. By keeping the catalyst dilute, the rate at which palladium centers condense to nucleate colloidal particles is greatly reduced. However this procedure has the drawback that the lower catalyst loading necessitates longer reaction times. By use of these carefully optimized conditions, we were able to increase the number average molecular weight of the polymers in Reaction 5 from the typical range of 6000–20000 Daltons to the range of 50 000 Daltons[17].

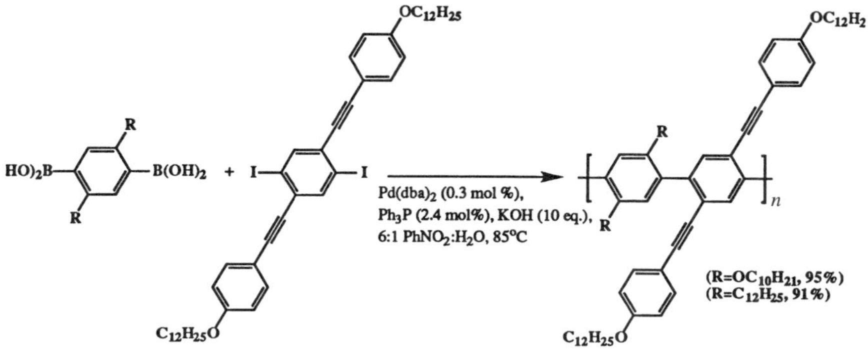

(5)

Ligand accelerated polymerizations have also proved to be useful. In Pd catalyzed Heck-type reactions, the use of $(o\text{-Tol})_3P$ can result in dramatic rate increases relative to $(Ph)_3P$. In the polymerization shown in Reaction 6, we found substitution of $(o\text{-Tol})_3P$ for $(Ph)_3P$ increased the molecular weight from 20 000 to 40 000 Daltons[18]. Additionally in the case of Stille procedures we have utilized $(Ph)_3As$[12] which has been demonstrated to accelerate transmetallation reactions between vinyl-tin reagents and palladium. The transmetallation step is generally rate limiting in Stille coupling reactions.

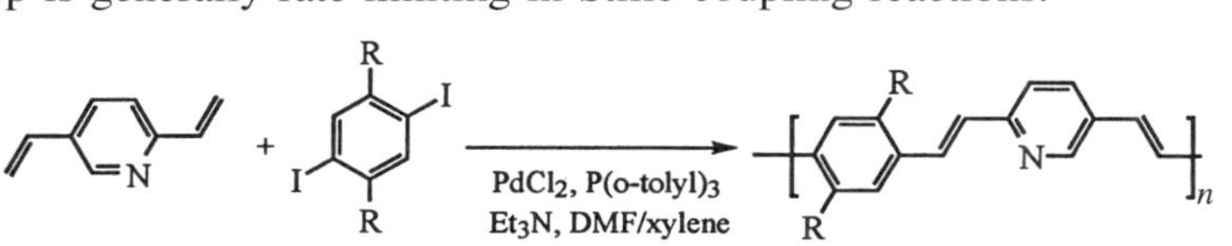

(6)

A common method for the synthesis of electron-rich polyaromatic conducting polymers, is oxidative electrochemical polymerization (Scheme 2.2)[19]. In this process the monomers are initially oxidized to form radical cations which upon condensation and loss of H^+, form new C—C bonds

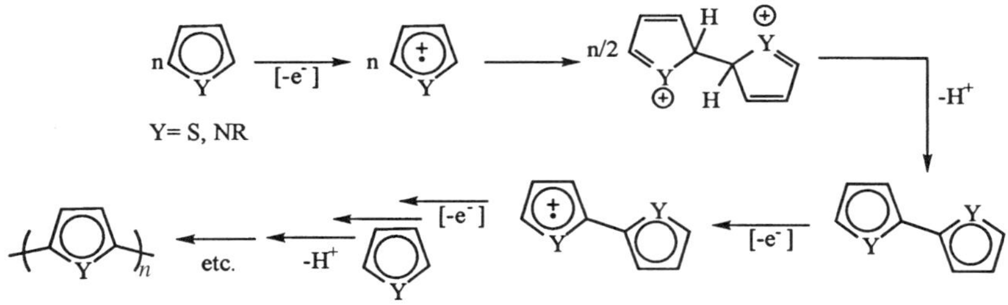

Scheme 2.2

between sp^2 carbons. Once the polymer chains are initiated, the oxidized polymer, which oxidizes at a lower potential than the monomer, precipitates on the electrode and reacts with additional monomers from the solution. Hence, this method involves a combination of step-growth and chain-growth processes. For polymers to be effectively synthesized by an electrochemical polymerization the growing polymers must be insoluble in the electrolyte solution and the polymers must be stable in their oxidized (doped) state.

Electrochemical procedures have both advantages and limitations. The limitations are that typically only small amounts of material may be synthesized and the polymer films are often insoluble and are not readily characterized by typical solution methods such as NMR. An advantage of electrochemical deposition is that only small amounts of a monomer are required to carry out a polymerization. Additionally, the polymers need not be soluble since they can be synthesized directly on to a functional assembly/device for analytical studies or applications. Typical solution methods are greatly restricted when the products are insoluble polymers. As will be discussed, electrochemical polymerizations are particularly useful for the synthesis of highly crosslinked network materials which have no prospect for solubility.

3. POLYMERS AND PROPERTIES

Conjugated polymers in their pristine (neutral) state are generally insulators or wide band-gap semiconductors. These materials have completely occupied and completely empty valence and conduction bands, respectively, and as a result cannot transport charge. This fact can be understood from the classic analogy of a filled versus partially filled parking lot of cars. If a parking lot is filled completely such that there is no open space whatsoever, none of the cars can move. Removing a few cars creates "holes" which allow the cars to be shuttled such that a hole can eventually be moved from one end of the parking lot to the other. This is analogous to what is observed in conducting polymers. The oxidation of a conducting polymer removes electrons from the valence band thereby creating "holes". Conduction by holes gives characteristics similar to what is observed in p-type doping of a classical semiconductor such as silicon. As a result, oxidative injection of charges into conducting polymers is referred to as p-doping. Likewise, the addition of electrons to the conduction band of these materials creates n-doped materials. In the vast majority of systems the most stable form is achieved when the polymers are oxidized, and most studies have been performed on p-doped conducting polymers. An alternative mechanism to enable conduction in a conjugated polymer is optical excitation, which promotes an electron from the valence band to the conduction band. Effective screening of the electrostatic attraction between the electron-hole pair allows charge separation and these materials can function as photoconductors[20].

Although the initial interest in conducting polymers was principally spurred by their novelty as electrical conductors, the current technology developing around these materials is multi-faceted. The focus has shifted from simply maximizing electrical conductivity, to creating advanced materials with tailored optical, structural, and electrical properties. The following sections discuss representative examples from a few different classes of polymers. A single chapter cannot possibly provide a comprehensive discussion, therefore the goal is to illustrate some current aspects of conducting polymers.

3.1 Poly(phenylenes ethynylene)s

Early investigations of conjugated polymers and the theoretical models developed demonstrated that polymers comprised of alternating aromatic rings and C—C triple bonds (ethynylene linkages) did not show the level of delocalization displayed by other more heavily investigated conducting polymers. The concept of "the degree of bond alternation", provides a useful qualitative framework to understand the degree of delocalization (band-width) and electronic band-gaps in organic polymers[21]. In this description, predictions of the electronic properties can be made by consideration of the bond length heterogeneity between the unsaturated carbon centers in the polymer backbone. In the case of poly(phenylene ethynylene), the triple bond is considerably shorter than the double bonds in the aromatic ring and there is a large bond length alternation. Interconnecting the aromatic rings with double bonds, as in the case of poly(phenylene vinylene)s, gives a lower degree of bond length alternation, greater delocalization, and a lower band-gap. In other words, the double bonds are closer in length to the bonds in the phenyl rings. The differences can be rationalized by considering the relative energetics of the aromatic and quinoid resonance forms[21]. The lower degree of bond length alternation in poly(phenylene vinylene) produces a nonaromatic resonance form that is lower in energy than that of poly(phenylene ethynylene). Indeed, UV-vis spectroscopy reveals that poly(phenylene vinylene) has a more delocalized electronic structure with broader absorption bands and a absorbtion edge which is red shifted from poly(phenylene ethynylene). Poly(phenylene ethynylene) also requires a higher oxidation potential for p-doping[22]. This larger oxidation potential arises from less extensive delocalization, rendering these materials less effective at stabilizing charge. As will be discussed, this decrease in stability was likely a major reason for the low conductivities observed in early investigations of these materials[23–25].

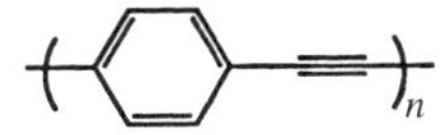

Poly(phenylene ethynylene)

Poly(phenylenes ethynylene)s have some particular properties which make them interesting as conductors. Specifically, their smaller band-widths produce a higher density of states[26]. In other words, the lower energy dispersion of poly(phenylene ethynylene)'s conduction and valence bands produces a higher density of orbitals over a given energy range than is observed in many other conducting polymers. This higher density of states should enhance redox or hopping mediated conduction processes. Another interesting feature of poly(phenylene ethynylene)s is their rigid-rod structure, which makes them particularly interesting structures for organized assemblies and/or the formation of electrical innerconnects in molecular electronics applications[27–29]. The rigid-rod structure does however have the drawback of low solubility. This decreased solubility is due to a reduced conformational flexibility which limits the entropic stabilization achieved when polymers are dissolved in a solvent. Hence to afford solubility in these materials, either large enthalpic interactions with the solvent or long side-chains that provide an entropic driving force for solubilization, must be incorporated into the polymer structure.

An additional consideration for achieving higher solubility in poly(phenylene ethynylene)s, is disorder in the alkyl-oxy side-chains[22]. For example, polymer **1** with uniform hexadecyl-oxy ($C_{16}H_{33}O$) side-chains is prone to gelation and the formation of insoluble aggregated materials. In **1** the side-chains are powerful promoters for the self-organization of the polymer into layered structures. Indeed as shown in Figure 2.1, these systems have very high degrees of order in the solid state, and the resulting lamellar structures display X-ray diffraction peaks out to seventh order (700). By introducing alternating lengths of alkyl side-chains in **2** and **3**, higher solubility is achieved[22]. This is a result of

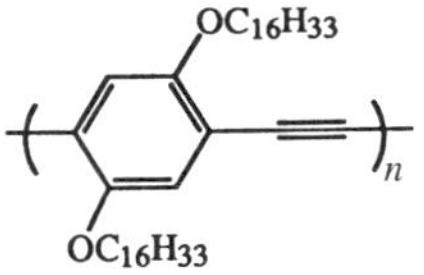

1

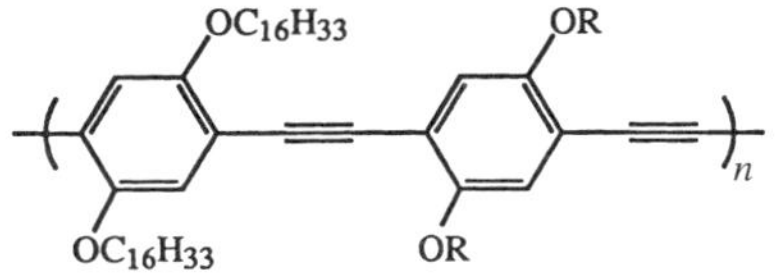

2 R= C_8H_{17}
3 R= random mixture of 0.5 CH_3 and 0.5 C_8H_{17}

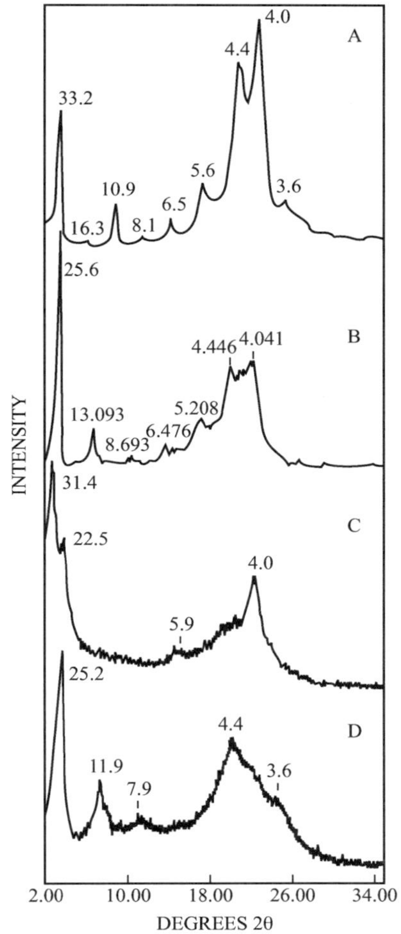

Figure 2.1 X-ray diffraction (Cu Kα) for annealed samples of **1** (A), **2** (B), **3** (C) and **4** (D). Peaks are labeled with *d*-spacings in Å. (Reproduced by permission from The American Chemical Society.)

increased side-chain disorder with mixed lengths. Clearly if the side-chains in both systems are fully disordered the smaller side-chains must provide a smaller entropic contribution than the longer side-chains. Consistent with this higher disorder, polymers **2** and **3** with different length side-chains are less ordered in the solid state and display weaker and less defined X-ray diffraction patterns (Figure 2.1). The incorporation of other aromatic rings without side-chains, such as in **4**, also produces a less ordered state (Figure 2.1). However in this last case the solubility is also greatly reduced.

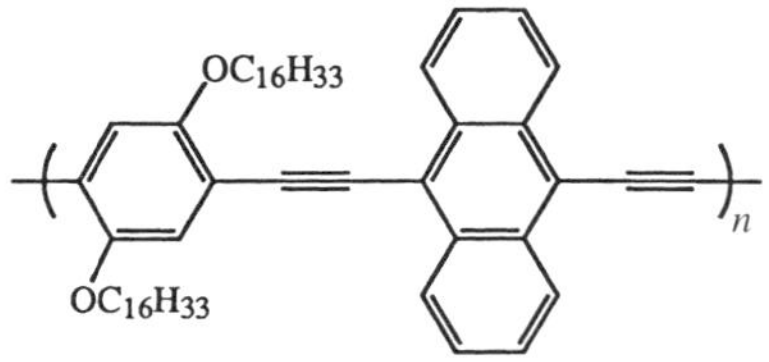

4

Earlier studies concluded that poly(phenylene ethynylene) could only display low conductivity[23–25]. Recently however, more recently poly(arylene ethynylene)s have been shown to exhibit high conductivity, if examined under the proper conditions. Their large bandgaps have generally made them difficult to oxidize and attempts to electrochemically oxidize polymers **1–4** in acetonitrile solutions results in irreversible decomposition[30]. This result is a consequence of the fact that acetonitrile is a nucleophilic solvent and reacts with the cations generated in the polymer backbone. Using special electrochemical conditions, with sulfur dioxide as the electrochemical solvent and low temperature, a stable conducting state could be produced[22]. Sulfur dioxide is critical because it is a very non-nucleophilic electrochemical solvent with a large window of oxidative stability. Electrochemical doping, followed by *in situ* conductivity measurements, indicated that polymers **1–4** displayed conductivities which range from ~0.2 to ~5.0 S/cm. These values are most impressive when it is considered that polymer composition is mainly saturated hydrocarbon. For example in **1**, 80% of the carbon centers are sp^3 hybridized.

One of the more striking features of poly(phenylene ethynylene) is their large fluorescence quantum yields. These materials typically display fluorescence quantum yields in excess of 0.5 and are visibly emissive under normal incandescent light. Simple inspection of the absorption and emission bands in these materials suggest that emission selectivity occurs from a few sites in the polymer's backbone[31]. This is shown in Figure 2.2 wherein the emission band of **1** is noticeably narrow when compared to its absorption manifold, and hence does not obey a typical mirror image relationship. It is important to note that this is not an intrinsic property of the local chromophore, since model compound **5** displays a broader emission band (Figure 2.2). These observations led to the design of polymer **6** which has anthracene containing moieties at its terminus[31]. The anthracene moieties create low energy trapping sites and, as shown in Figure 2.3, excitations migrate to these traps. The result of this migration is that greater than 95% of the emission is at longer wavelengths than that of **1** and indicating selective emission from states associated with the anthracene moieties[31].

Energy migration in poly (phenylene ethynylene)s also presents opportunities for the design of more sensitive fluorescence chemosensors[14,16]. In this scheme, trapping sites are formed by the binding of guest molecules (analytes) to receptors which are periodically attached to the polymer and optical excitations

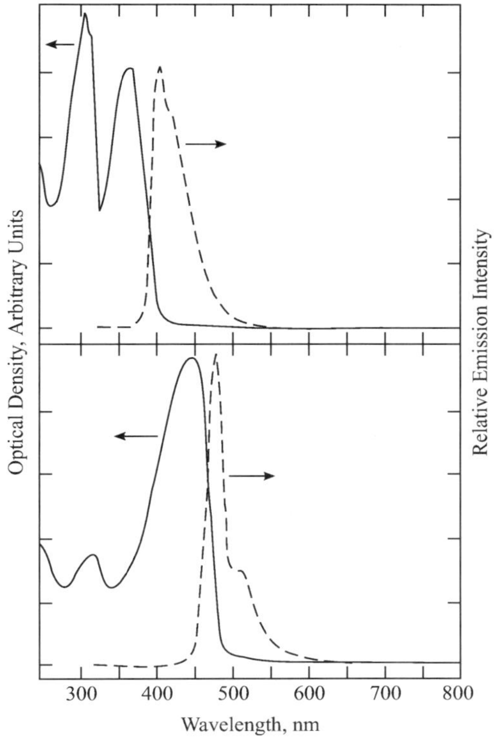

Figure 2.2 Absorption and emission spectra of **5** (top) and **1** (bottom). (Reproduced by permission from The American Chemical Society.)

are created randomly along the polymer backbone. These excitations then migrate throughout the polymer's backbone, and are trapped upon encountering an occupied receptor site. The trapping sites have different characteristics depending on the interaction with the analyte. If the guest molecule has filled or unoccupied states at energies which allow electron-transfer reactions to occur upon excitation, quenching will be observed. Alternatively, the occupied receptor site can produce a new emissive state at lower energy and selective emission from these types of traps will shift the emission band to the red. This latter situation is analogous to what was observed for polymer **6**.

Energy migration in poly(phenylene ethynylene)s provides a mechanism for the amplification of chemosensory signals[14,16]. This was demonstrated by making comparisons between monomeric fluorescent receptors and their polymeric counterparts. If energy migration is present in the polymeric system, then only fractional binding is necessary to totally extinguish the emission and thus provide an amplified response. This effect can be readily seen from

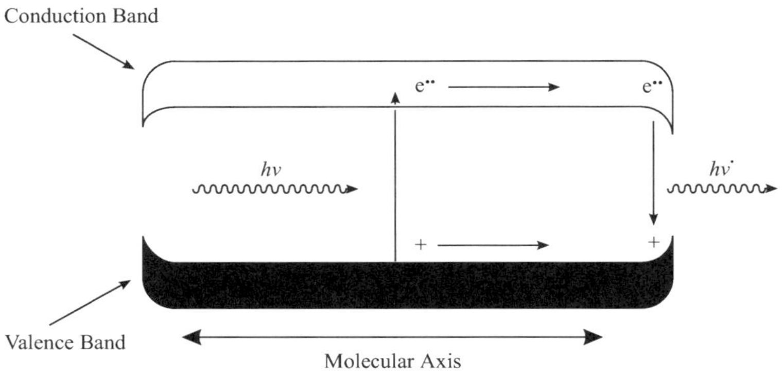

Figure 2.3 Schematic representation of energy migration in a semiconductive molecular wire with a decrease in the band gap at the terminus. (Reproduced by permission from The American Chemical Society.)

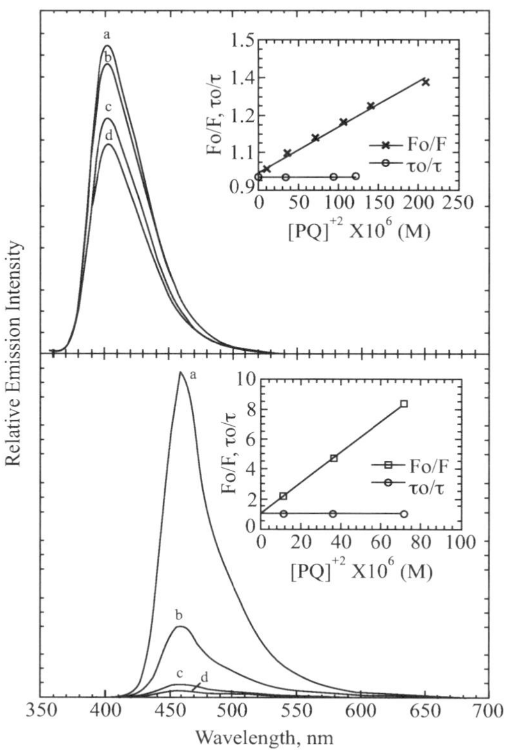

Figure 2.4 Emission spectra of **7** (top) and **8** (bottom) as a function of added $[PQ^{2+}]$. The curves are labeled to indicate the amount of PQ^{2+} added: (a) $[PQ^{2+}] = 0$; (b) $[PQ^{2+}] = 3.56 \times 10^{-5}$M; (c)$[PQ^{2+}] = 2.10 \times 10^{-4}$ M; (d) $[PQ^{2+}] = 3.45 \times 10^{-4}$M. The Stern–Volmer plots are shown as insets and the plots show τ_0/τ to be invariant with added PQ^{2+}. (Reproduced by permission from The American Chemical Society.)

inspection of the data shown in Figure 2.4. In these experiments the emission from both the monomer, **7**, and the polymer, **8**, were measured as a function of added paraquat. Care was taken to maintain a constant concentration of the receptor to ensure that the equilibrium concentration of occupied receptors is similar for both **7** and **8**. As shown from the insets in Figure 2.4, the fluorescence lifetimes are not decreased with added paraquat, and the systems display strictly static quenching. The dramatic reduction in the polymer's fluorescence intensity relative to that of the monomer indicates that the fluorescence chemosensory event has been amplified.

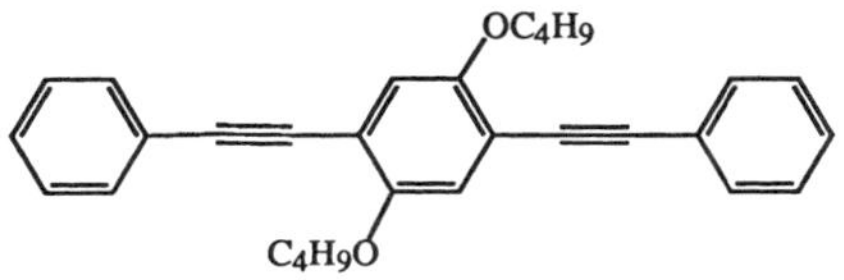

5

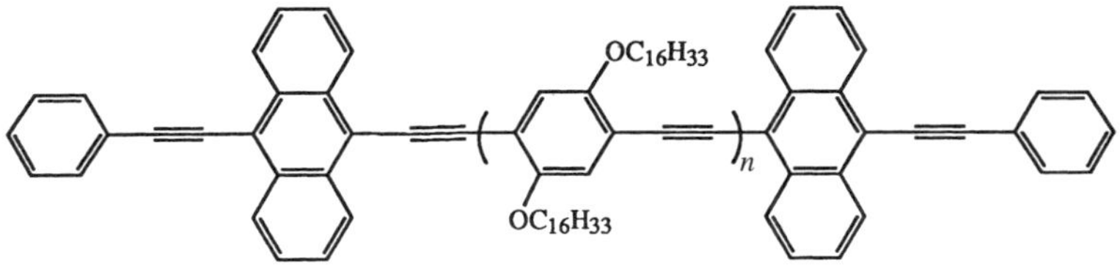

6

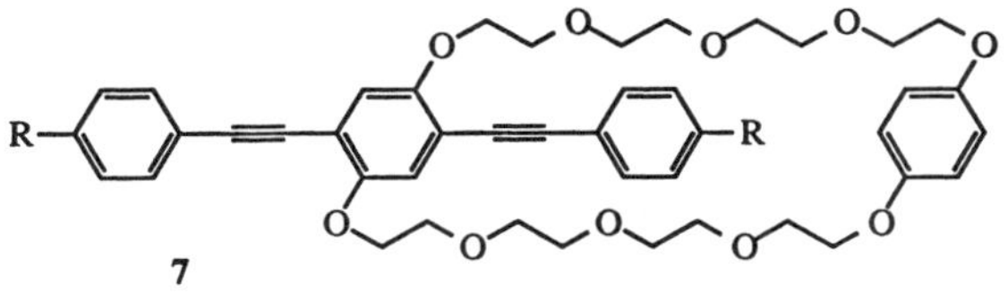

7
$R=OC_{12}H_{25}$

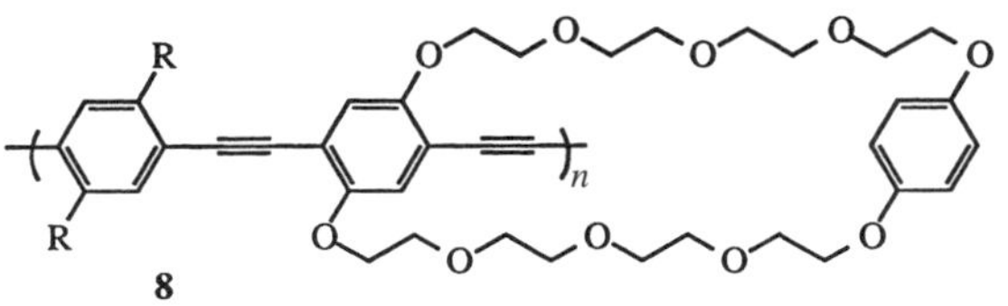

8
$R=CON(C_8H_{17})_2$

3.2 Poly(arylene vinylene)s

Poly(phenylene vinylene)s have been studied throughout the development of the field of conducting polymers. While they can display extremely high conductivity[32,33], their doped forms lack the stability required for many applications. The discovery that poly(phenylene vinylene) displays electroluminescence (EL)[34] initiated a broad-based effort to develop conducting polymer EL devices. In these devices the polymer is placed between two electrodes. Application of a voltage inserts electrons and holes which can produce an emission upon recombination. However, poly(phenylene vinylene) is not necessarily an ideal material for EL applications. First, this material is insoluble, and can only be processed by employing a soluble precursor polymer. Second, the quantum yield for luminescence from the parent polymer is generally low due to strong interactions between polymer chains. The attachment of bulky side-chains and the formation of cyclophanes which prevent stacking of the polymers, both produce more emissive materials[18,35].

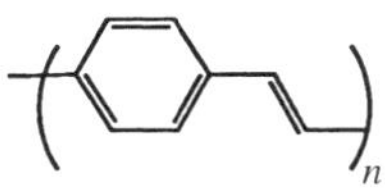

Poly(phenylene vinylene)

Another problem with EL has been that very low work function metals must be used as the electrodes. It was reasoned that modification of the poly(phenylene vinylene) structure to give a material with a higher electron-affinity (larger work-function) would enable less reactive metals to be used as the electron-injecting electrodes in EL devices[36]. Indeed, the nitrile containing polymer **9** was synthesized and shown to be compatible with more stable metal electrodes[36]. Subsequent to these findings, the substitution of an electron efficient heterocyclic moiety such as pyridine for the phenyl moiety in poly(phenylene vinylene) was investigated in an effort to increase the electron affinity[37,38]. This substitution produced poly(pyridylene vinylene), **10**, which is a blue emitter like poly(phenylene vinylene) but has the added bonus that it can be made soluble via protonation or alkylation. The origin of **10**'s solubility in its ionic forms is due to both enthalpic interactions with the solvent and electrostatic repulsions between polymer chains. The pyridine substitution also provides a non-symmetric repeat unit which results in different regioisomeric forms. The regioisomeric content was predicated to have large consequences, particularly when the polymers were alkylated with methyl triflate[37]. The head-to-tail structure of the methyl triflate form **11**, was particularly interesting since when reduced it produces a non-Kekulé electronic configuration with one unpaired electron per repeating unit (Scheme 2.3). Such an open shell structure is interesting since it may constitute a rare example of an uncharged conducting polymer. Indeed, experiments on reduced materials showed the head-to-tail structure to have the highest conductivity of the different regioisomers when reduced[37].

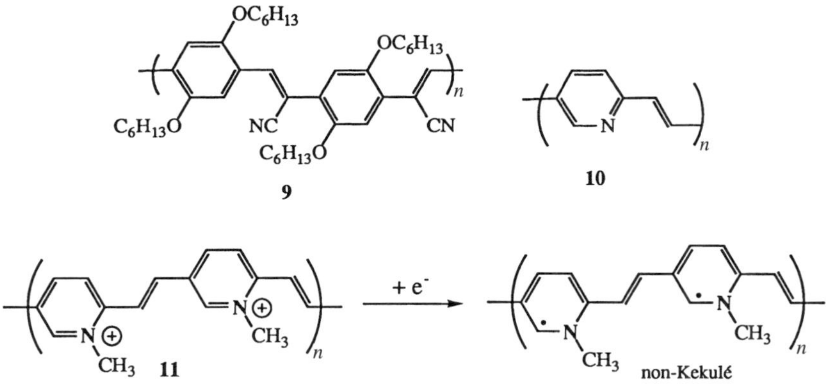

Scheme 2.3

The EL and photophysical properties of poly(pyridylene vinylene) have been studied in great detail[39–43], and indeed this polymer was shown to be an alternative to poly(phenylene vinylene). These two systems can be combined forming alternating copolymers **12**, **13**, and **14**, which are synthesized by Heck-type polymerizations (Reaction 6)[18]. These polymers are soluble in standard organic solvents and the polymers containing cyclophane moieties (**14**) were designed to deter or prevent aggregation[44,45]. Decreased aggregation produces more highly emissive bulk films. Copolymers **12**, **13**, and **14** are also of interest due to the donor–acceptor nature of the phenylene and pyridylene groups, respectively. Indeed the solid-state organization of these materials was found to be highly sensitive to protonation or alkylation of the pyridine moieties which further enhance its acceptor character. The solid-state structure organizes into a staggered geometry creating favorable charge-transfer interactions between chains illustrated in Figure 2.5. The optical transitions also display charge-transfer characteristics from the phenyl ring to the pyridine moiety. Consistent with this character, protonation of the pyridine produces a red shift.

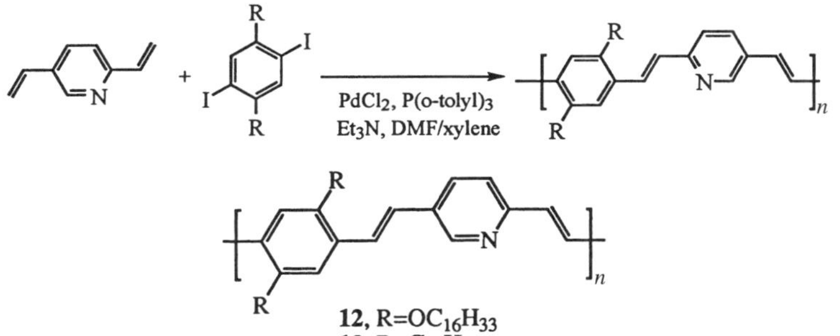

12, R=$OC_{16}H_{33}$
13, R=$C_{12}H_{25}$

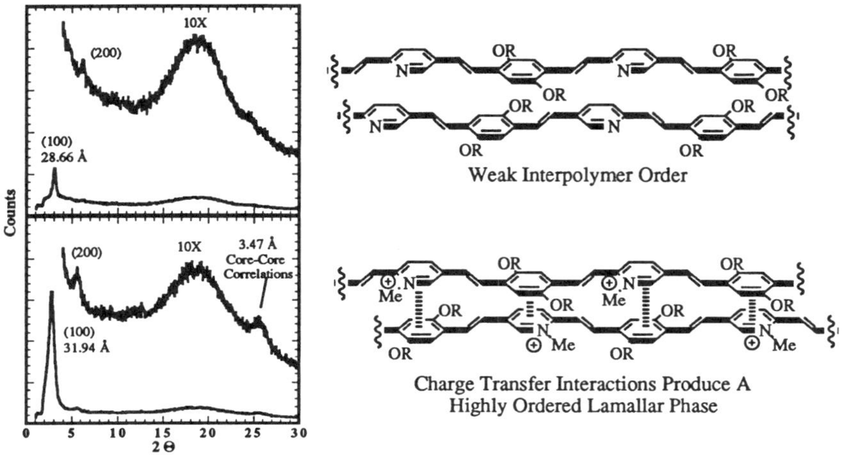

Figure 2.5 (Left) X-ray diffraction (200°C) of polymer **12** (top) and its methylated form (TfO^- salt) (bottom). (Right) Schematic representation of the polymer ordering showing a less ordered layered structure for **12** (top) and a more ordered structure for the methylated analog (bottom).

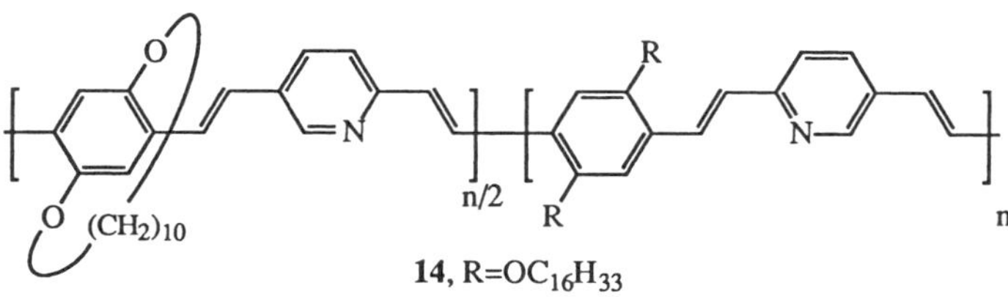

14, $R=OC_{16}H_{33}$

4. POLYARYLENES

Polyphenylene has been of interest as both a conducting polymer as well as a structural material[46]. The aromatic stability of the phenyl unit is unparalleled and provides this organic polymer with extraordinary stability to oxidative and high temperature environments. Polyphenylene has been synthesized by precursor routes and also directly by coupling methods. The coupling methods are of limited utility due to the extremely low solubility of this polymer that restricts the reaction conversion to only produce oligomeric materials. The steric requirements in polyphenylenes are more severe than the other polymers discussed. Indeed even the simplest member of this family, biphenyl, exhibits a non-planar twisted conformation. Hence any attempts to solublize polyphenylene by adding substituents will result in extremely non-planar conformations which have little or no electronic delocalization between repeating units. The related polymer polypyridylene (Reaction 2, *vide supra*) is soluble in strong acid[11], however the synthetic coupling procedures cannot be conducted under these conditions, and similarly the synthesis is also greatly restricted by the low solubility of these materials.

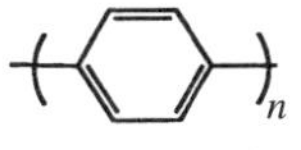

Poly-p-phenylene

An alternative route to soluble delocalized polyphenylenes involves the incorporation of fused cyclic structures which force coplanarity between the phenyl rings[47–49]. For example **15** which has saturated centers bridging the *ipso*-carbons of biphenyl units has been shown to provide solubility as well as high degrees of conjugation. The synthesis of these structures is more involved and relies on the initial formation of a precursor polymer that is subsequently cyclized to form **15**. Other polymers having conjugated bridging groups such as **16**[50] and **17**[20] have also been synthesized. These structures are of interest due to their fused polycyclic aromatic character, can be considered to have a one-dimensional graphitic (graphite ribbon) structure[51].

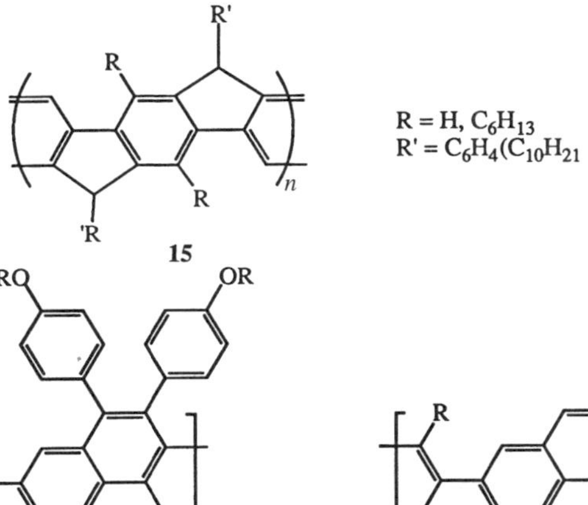

15

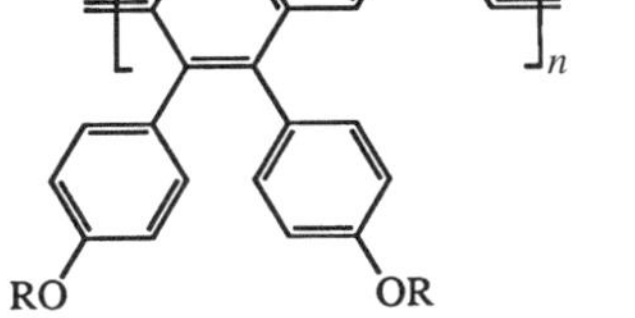

16 (R = -*n*-$C_{10}H_{21}$)

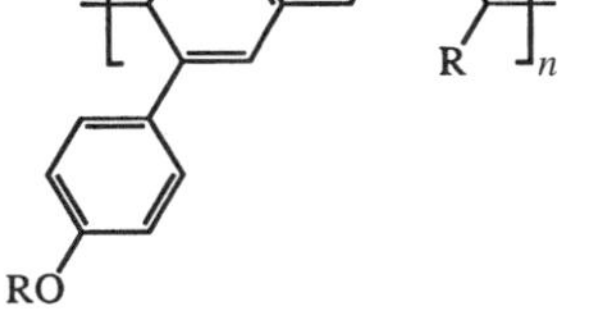

17 (R = -*n*-$C_{12}H_{25}$)

Polypyrrole and polythiophene are also polyarylenes. An advantage of these polymers is that the five-membered ring provides greater steric tolerance than polyphenylene for solubilizing substituents. Soluble materials belonging to these polymer families have been studied in great detail[2,3]. Polypyrroles functionalized by the attachment of substituents to the nitrogen are readily prepared. Other substitution patterns have been studied; however as a consequence of the more complicated synthetic procedures required, these modifications are not frequently used. Thiophenes, on the other hand, exhibit an extremely diverse synthetic chemistry which allows for facile functionalization[52]. Thiophene chemistry has been used for the synthesis of polymers which contain receptor units for metal ions[12,53] and/or electron poor organic species[54,55]. This interest in receptor containing polymers has been motivated for the design of sensory materials. The key to the sensory function is that the binding of a guest in the receptor must cause a local perturbation to the polymer's electronic structure or redox potential.

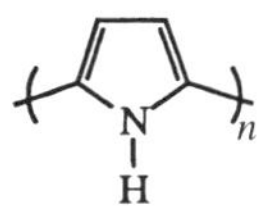

Polypyrrole

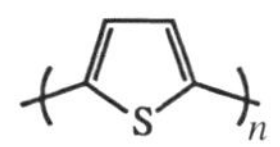

Polythiophene

The crown ether containing polymers, **18**, display a response to the binding of alkali metal ions[12]. In particular these systems display selective iono-chromic behavior based on the fact that interaction of the crown ether with a specific ion induces a non-planar conformation of the polymer backbone as shown in Reaction 7. These effects are sensitive to both the size of the macrocycle and of the ion. For example, **18** ($y = 0$), with five oxygen atoms, can be considered a 15-crown-5 analog and consequently displays the largest response with Na^+. Polymer **18** ($y = 1$), is an 18-crown-6 analog and displays the largest response with K^+. More elaborate ionophores utilizing calixarene scaffoldings have also been used. Polymer **19** is particularly selective[53] and shows large blue shifts when exposed to a Na^+, but no optical changes in response to Li^+ or K^+. The blue shift with Na^+ is suggestive of a non-planer conformation in the complexed form. In this case the calixarene receptor is complex and its behavior is dependent upon the polymer. For example **20** displays a slight red shift with Na^+ complexation which is a totally different optical behavior than that of **19**. The differences between these materials is the result of electronic contributions from the groups which interconnect the bithiophene containing receptor units. Although polymer **20** displays only slight optical changes with Na^+, it does display extremely large conductivity responses[53]. The Na^+ binding produces an electrostatic perturbation, and as shown in Figure 2.6 the occupation of the receptor with the Na^+ increases the local redox potential of the polymer. This binding creates a barrier to transport of the cationic carriers, presumed to be bipolarons[56]. This mechanism creates an extremely sensitive sensory response due to the fact that Na^+ occupation of a single thiophene site has the potential to influence the transport of many carriers. Hence the response is magnified or amplified relative to that of a simple monomeric sensory

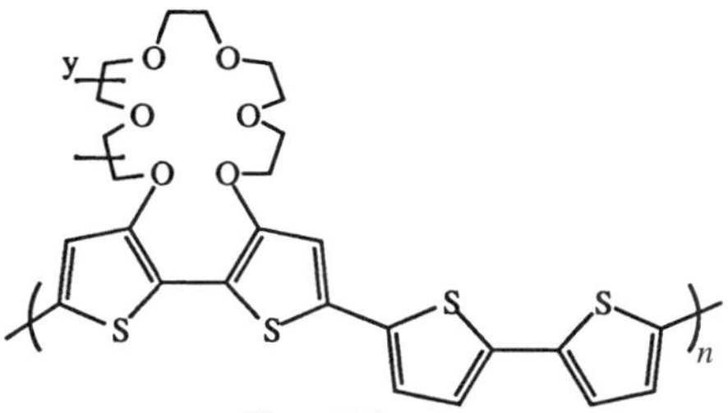

18 y = 0, 1

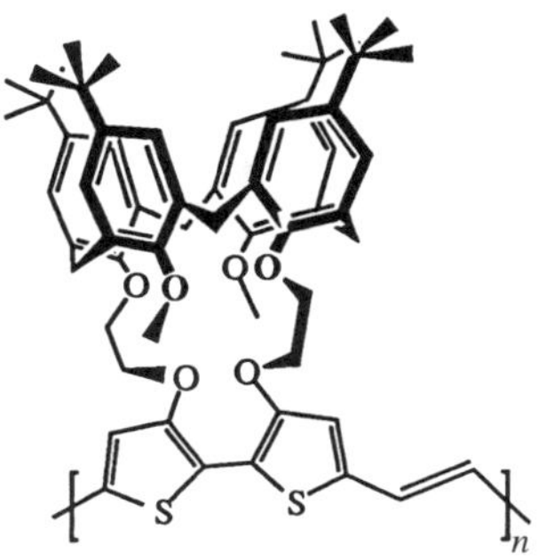

19

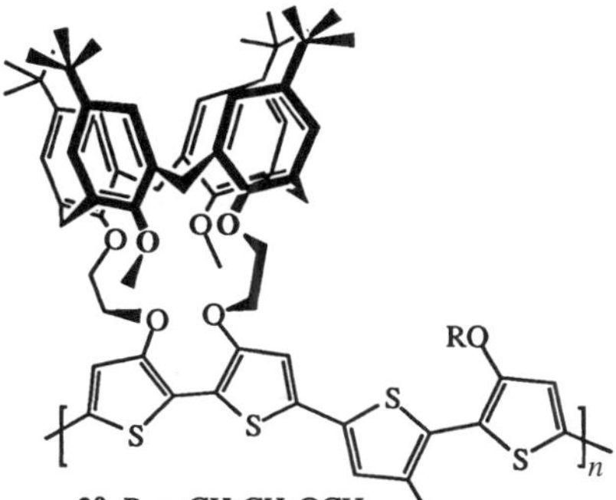

20 R = $-CH_2CH_2OCH_3$

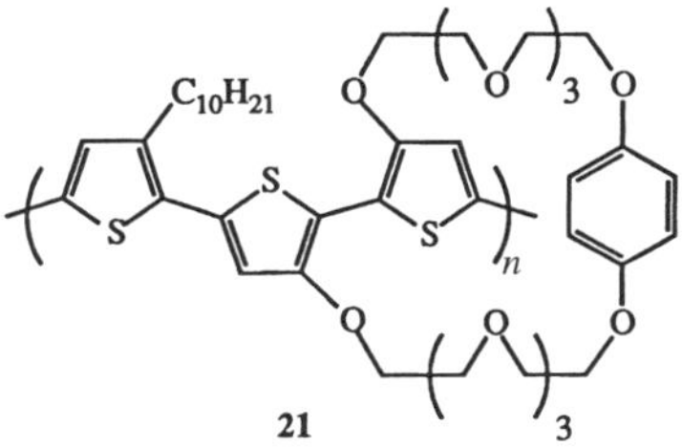

21

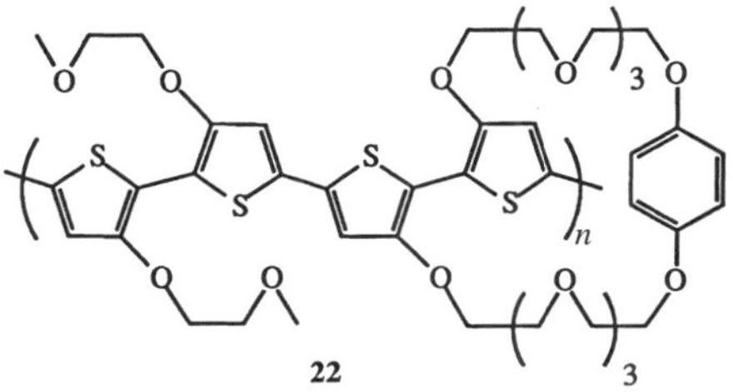

22

molecule. Similar conductivity changes have been observed in polymers **21** and **22** which bind the electron deficient herbicide paraquat. These polymers possess a large macrocyclic moiety which facilitates charge-transfer interactions between the donor(Polythiophene) and the acceptor(Paraquart). These charge-transfer interactions in turn raise the local redox potential of the polymer and an attenuation in the conductivity is observed.

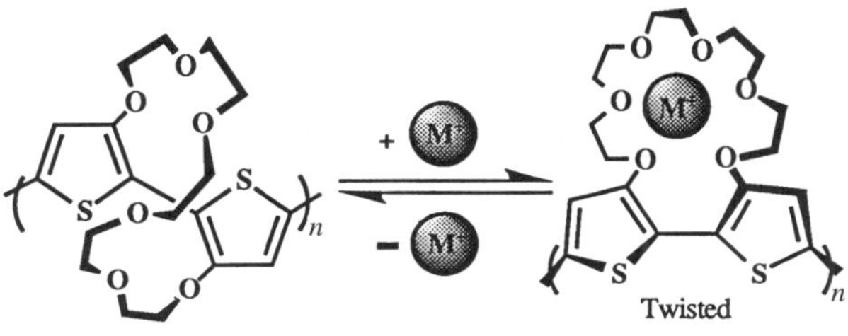

(7)

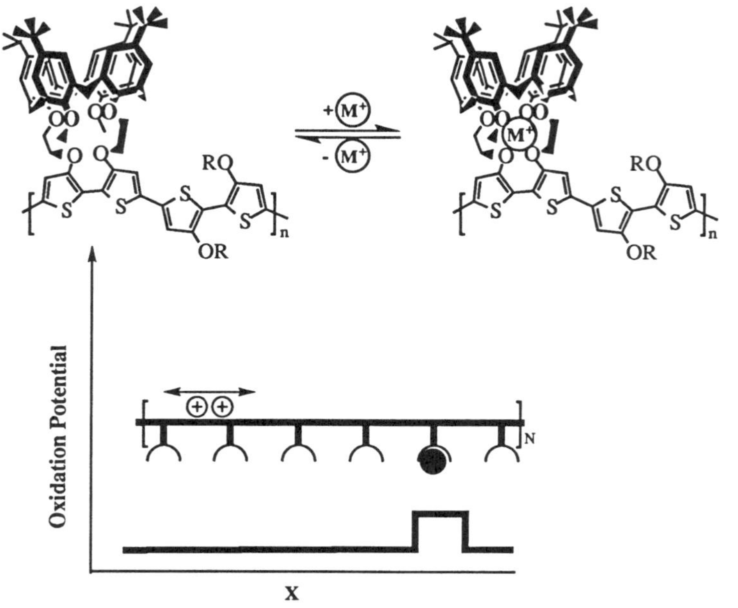

Figure 2.6 Scheme showing reversible binding of metal ions by **19** and the conceptual mechanism by which conductivity is attenuated. Dicationic carriers migrate along the polymer's backbone, however when they encounter a receptor which is occupied by an ion their path is blocked by a barrier. This barrier can be thought of as an increase in the local redox potential of the polymer.

5. HIGHLY CROSSLINKED NETWORK POLYMERS

As illustrated in the preceding section the versatility of thiophenes and their ability to be electrochemically polymerized has made them attractive candidates for the synthesis of other unconventional polymeric systems. In particular, thiophene residues have been utilized for the formation of polymers which contain transition metal ions. Of particular interest to us has been the question; can conjugated polymers facilitate electron transfer between localized redox centers? If designed correctly, conducting polymers provide a conduit which functions to create highly conducting redox systems[57,58]. An example of a redox polymer is **23** and was investigated to determine if a conjugated structure would facilitate electron transfer between $Ru(bpy)_3$ centered redox processes[57]. Polymer **23**'s highly crosslinked structure precludes synthesis by standard solution methods. Hence polymerization of the octahedral monomer with six different coupling sites can only be accomplished electrochemically with the insoluble polymer precipitating *in situ* on the electrode surface. The structure of **23** is most likely highly irregular, however the immediate coordination sphere of all of the Ru centers are the same and the electrochemistry exhibits well-defined redox waves. Conductivity studies of these systems as a function of electrochemical potential have shown **23** to be an excellent redox conductor[57]. Hence the conjugated oligo-thiophene interconnects facilitate electron transfer between the $Ru(bpy)_3$ sites.

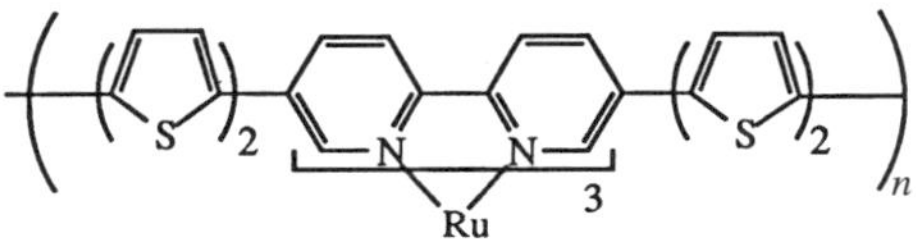

23

Other exotic structures have also been synthesized using thiophene elaborated bipyridyl ligands. In particular polymetallorotaxane ("wheel and axle") structures, **24** and **25**, assembled around copper and zinc ions have synthesized[58]. These polymers are formed by electrochemical polymerization of rotaxane monomers, and once synthesized the templating copper or zinc ions can be removed. In the metal-free polyrotaxane **26** the macrocyclic ligands are physically entrapped around the polymer backbone, most likely by physical or chemical crosslinks. This non-covalent assembly creates receptor sites for tetrahedral metal ions and indeed the polymer readily binds either copper or zinc metal ions to reconstitute the polymetallorotaxane[58]. the complexation with zinc ion in **25** creates a red shift in the maximum of the UV-vis spectrum relative to **26**. Similar to what was seen in the alternating copolymers **12**, **13**, and **14**, we believe this red shift indicates charge-transfer character.

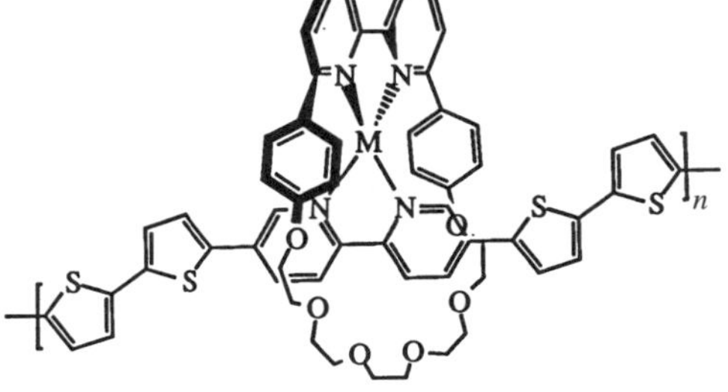

24 M= Cu^+
25 M= Zn^{2+}

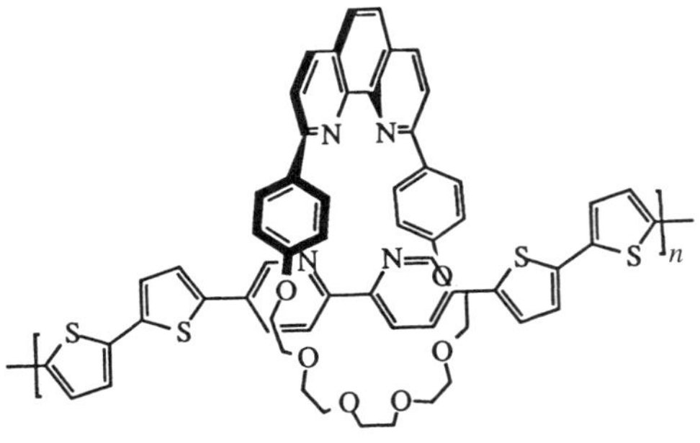

26

6. SUMMARY

At present the field of conducting polymers is poised for major developments resulting from the integration of current concepts from the fields of molecular recognition, catalysis, sensors, and electro-optics. The synthetic chemist will play a major role in this innovative cross-fertilization which promises many future discoveries and technologies. In particular greater diversity in structures and synthetic methods are necessary to create these opportunities. The author hopes that the reader will join in this exciting and challenging area of research.

7. REFERENCES

1. C. K. Chiang, Y. W. Park, A. J. Heeger, H. Shirakawa, E. J. Louis and A. G. MacDiarmid, *Phys. Rev. Lett.* **39**, 1098 (1977).
2. T. A. Skotheim, (ed.), *Handbook of Conducting Polymers*, Vol. 1 and 2. Marcel Dekker, Inc., New York (1986).
3. A. J. Epstein and Y. Yang, (eds), *MRS Bull.* **22**(6), 13–15 (1997).
4. F. W. Billmeyer Jr, *Textbook of Polymer Science*, 3rd Edn. John Wiley & Sons, Inc., New York (1984).
5. G. Odian, *Principles of Polymerization*, 3rd Edn. John Wiley & Sons, Inc., New York (1993).
6. O. W. Webster, *Science*, **251**, 887 (1991).
7. R. F. Heck, *Palladium Reagents in Organic Syntheses, Best synthetic Methods*. Academic Press, California (1987).
8. M. Kumada, *Pure Appl. Chem.* **52**, 669–679 (1980).
9. J. H. Edwards and W. J. Feast, *Polymer* **21**, 595 (1980).
10. T. M. Swager and R. H. Grubbs, *J. Am. Chem. Soc.* **111** , 4413–4422 (1989).
11. T. Yamamoto, T. Maruyama, T. Ikeda and M. Sisido, *J. Chem. Soc. Commun.* 1306–1307 (1990).
12. M. J. Marsella and T. M. Swager, *J. Am. Chem. Soc.* **115**, 12214–12215 (1993).
13. R. D. McCullough, R. D. Lowe, M. Jayaraman and D. L. Anderson, *J. Org. Chem.* **58**, 904–912 (1993).

14. Q. Zhou and T. M. Swager, *J. Am. Chem. Soc.* **117**, 12593–12602 (1995).
15. P. M. Cotts, T. M. Swager and Q. Zhou, *Macromolecules* **29**, 7323 (1996).
16. Q. Zhou and T. M. Swager, *J. Am. Chem. Soc.* **117**, 7017–7018 (1995).
17. M. B. Goldfinger and T. M. Swager, *J. Am. Chem. Soc.* **116**, 7895–7896 (1994).
18. D. -K. Fu, B. Xu and T. M. Swager, *Tetrahedron* **53**, 15487–15494 (1997).
19. J. Bargon, S. Mohmand and R. J. Waltman, *IBM, J. Res. Dev.* **27**(4), 330–341 (1983).
20. K. Pakbas, C. H. Lee, A. J. Heeger, T. W. Hagler and D. McBranch, *Synth. Met.* **64**, 295 (1994).
21. J. L. Brédas, *J. Chem. Phys.* **82**(8), 3808–3811 (1985).
22. D. Ofer, T. M. Swager and M. S. Wrighton, *Chem. Mater.* **7**, 418–425 (1995).
23. M.V. Lakshmikantham, J. Vartikar, K. -Y. Jen, M. P. Cava, W. S. Huang and A. G. MacDiarmid, *Polym. Prepr.* **24**, 75 (1983).
24. K. Sanechika, T. Yamamoto and A. Yamamoto, *Bull. Chem. Soc. Jpn* **57**, 752 (1984).
25. M. Tateishi, M. Nishihara and K. Aramaki, *Chem. Lett.* 1727 (1987).
26. R. Hoffmann, *Solids and Surfaces: a Chemists's View of Bonding in Extended Structures*. VCH Publishers, Inc., New York (1988).
27. F. L. Carter (ed.), *Molecular Electronic Devices*. Marcel Dekker, Inc., New York (1982).
28. F. L. Carter (ed.), *Molecular Electronic Devices II*. Marcel Dekker, Inc., New York (1987).
29. L. Jones II, J. S. Schumm and J. M. Tour, *J. Org. Chem.* **62**(5), 1388–1410 (1997).
30. T. M. Swager, unpublished results.
31. T. M. Swager, C. J. Gil and M. S. Wrighton *J. Phys. Chem.* **99**, 4886–4893 (1995).
32. M. A. Masse, R. J. Composto, R. A. L. Jones and F. E. Karasz, *Macromolecules* **23**, 3675–3682 (1990).
33. D. Chen and M. J. Winokur, *Phys. Rev. B* **41**(10) 6759–6767 (1990).
34. J. H. Burroughes, D. D. C. Bradley, A. R. Brown, R. N. Marks, K. MacKay, R. H. Friend, R. L. Burn and A. B. Holmes, *Nature* **347**, 539 (1990).
35. C. Zhang, S. Hoger, K. Pakbaz, F. Wudl and A. J. Heeger, *J. Electron Mater.* **22**(4), 413–417 (1993).
36. N. C. Greenham, S. C. Moratti, D. D. C. Bradley, R. H. Friend and A. B. Holmes, *Nature* **365**, 628 (1993).
37. M. J. Marsella, D. -K. Fu and T. M. Swager, *Adv. Mater.* **7**, 145–147 (1995).
38. J. Tian, C. -C. Wu, M. E. Thompson, J. C. Sturm, R. A. Register, M. J. Marsella and T. M. Swager, *Adv. Mater.* **7** 395–398 (1995).
39. J. W. Blatchford, S. W. Jessen, L. -B. Lin, J. J. Lih, T. L. Gustafson, A. J. Epstein, M. J. Marsella, D. -K. Fu, T. M. Swager, A. G. MacDiarmid, S. Yamaguchi and H. Hamaguchi, *Phys. Rev. Lett.* **76**, 1513–1516 (1996).
40. D. D. Gebler, Y. Z. Wang, J. W. Blatchford, S. W, Jessen, L. -B. Lin, T. L. Gustafson, H. -L. Wang, T. M. Swager, A. G. MacDiarmid and A. J. Epstein, *J. Appl. Phys.* **78**, 4264–4266 (1995).
41. A. J. Epstein, J. W. Blatchford, Y. Z. Wang, S. W. Jessen, D. D. Gebler, L. B. Lin, T. L. Gustafson, H.-L. Wang, Y. W. Park, T. M. Swager, A. G. MacDiarmid, *Synth. Met.* **78**, 253–261 (1996).
42. J. W. Blatchford, T. L. Gustafson, A. J. Epstein, D. A. Vanden Bout, J. Kerimo, D. A. Higgins, P. F. Barbara, D.-K. Fu, T. M. Swager and A. G. MacDiarmid, *Phys. Rev. B* **514**, 541 (1996).
43. J. W. Blatchford, S. W. Jessen, L.-B. Lin, T. L. Gustafson, D.-K. Fu, H.-L. Wang, T. M. Swager, A. G. MacDiarmid and A. J. Epstein, *Phys. Rev. B* **54**, 9180–9189 (1996).

44. U. Fahnenstich, K. Koch, M. Pollmann, U. Scherf, M. Wagner, S. Wegener and K. Muellen, *Makromol. Chem., Macromol. Symp.* **54/55** (Int. Symp. New Polym. React. React. Mech., 1991) 465–476 (1992).
45. J. Huber and U. Scherf, *Macromol. Rapid Commun.* **15**, 897–902 (1994).
46. D. L. Gin and V. P. Conticello, *Trend. Polym. Sci.* **4**(7), 217 (1996).
47. U. Scherf and K. Muellen, *Makromol. Chem. Rapid Commun.* **12**, 489 (1991).
48. U. Scherf and K. Muellen, *Synthesis* 23–38 (1992).
49. U. Scherf and K. Muellen, *Adv. Polym. Sci.* **123**, 1–40 (1995).
50. K. Chmil and U. Scherf, *Makromol. Chem. Rapid Commun.* **14**, 217 (1993).
51. A. Schlüter, *Adv. Mater.* **3**, 282 (1991).
52. J. Roncali, *Chem. Rev.* **92**, 711–738 (1992).
53. M. J. Marsella, R. J. Newland, P. J. Carroll and T. M. Swager, *J. Am. Chem. Soc.* **117**, 9842–9848 (1995).
54. M. J. Marsella, P. J. Carroll and T. M. Swager, *J. Am. Chem. Soc.* **116**, 9347–9348 (1994).
55. M. J. Marsella, P. J. Carroll and T. M. Swager, *J. Am. Chem. Soc.* **117**, 9832–9841 (1995).
56. J. L. Brédas and G. B. Street, *Accts Chem. Res.* **18**, 309–315 (1985).
57. S. S. Zhu and T. M. Swager, *Adv. Mater.* **8**, 497–500 (1996).
58. S. S. Zhu, P. J. Carroll and T. M. Swager, *J. Am. Chem. Soc.* **118**, 8713–8714 (1996).

Chapter 3

Zeolite as a Medium for Photochemical Reactions

V. RAMAMURTHY, REBECCA J. ROBBINS, K. J. THOMAS AND P. H. LAKSHMINARASIMHAN
Department of Chemistry, Tulane University, New Orleans, LA, USA

1. INTRODUCTION

Most of our understanding of the reactivity of organic molecules is based on experiments conducted either in a gas phase or in an isotropic liquid phase. Studies conducted in these media have provided a number of empirical rules that can be used to predict the behavior of molecules in a new environment. While the knowledge gained from these studies forms the basis of modern biology, there have been shortfalls in predicting the behavior of a molecule that is enclosed in a biological matrix. Extrapolation of knowledge based on isotropic liquid phase studies to complex biological systems is limited by the fact that the former ignores specific interactions between the reactant and the medium. Being inspired by and having realized the complexity of natural systems, chemists have utilized a number of organized media to study the behavior of molecules[1–11]. Examples of organized media which have been investigated include molecular crystals, inclusion complexes (both in the solid and solution states), liquid crystals, micelles and related assemblies such as vesicles, microemulsions and membranes, monolayers, Langmuir–Blodgett (LB) films, surfaces (silica, clay and zeolites) and more recently natural systems such as proteins and DNA. In this chapter we are concerned with one such

Organised Molecular Assemblies in the Solid State Edited by James K. Whitesell
© 1999 John Wiley & Sons Ltd.

organized/confined medium—zeolite. Particular attention will be paid to alkali and alkaline earth cation exchanged X and Y zeolites. The focus of this chapter is on photochemical and photophysical behavior of organic guest molecules included within X and Y zeolites.

2. STRUCTURAL FEATURES OF ZEOLITES

Zeolites are inorganic microporous and microcrystalline materials capable of complexing or adsorbing small and medium-sized organic molecules[12–16]. Typical particle sizes are in the 0.1 to 10 μm range and possess regular periodic structures that may be thought of as sponge-like material with channels and cages that extend periodically and regularly across their entire structure. Adsorption of organic compounds on zeolites can occur both on the external and internal surfaces of a zeolite crystal. Internal complexation (intracrystalline adsorption) occurs by diffusion of the guest into the channels and cavities within the zeolite crystal and is size and shape selective. While internal surface areas largely exceed external surfaces, intracrystalline adsorption may occur only if the kinetic diameter of the guest is smaller than the diameter of the intracrystalline cavities[17,18]. Similar to other supramolecular systems such as inclusion compounds, complexation of organic molecules with zeolites is strongly dependent on non-bonded interactions and electrostatic forces optimized by size and shape considerations. Molecular alignment and conformational restrictions may be expected when adsorption occurs. These structural limitations imposed on the substrate have important consequences on its intrazeolite chemical reactivity.

There are approximately 40 naturally occurring and over 100 synthetic forms of zeolites. The primary building blocks of the zeolites are the $[SiO_4]^{4-}$ and $[AlO_4]^{5-}$ tetrahedra[14]. These tetrahedra are linked by all their corners to form channels and cages or cavities of discrete size with no two aluminum atoms sharing the same oxygen. As a result of the difference in charge between the $[SiO_4]^{4-}$ and $[AlO_4]^{5-}$ tetrahedra, the total framework charge of an aluminum-containing zeolite is negative and hence must be balanced by a cation, typically an alkali or alkaline earth metal cation. As such, zeolites can be represented by the empirical formula $M_{2/n} \cdot Al_2O_3 \cdot xSiO_2 \cdot y\,H_2O$, where M is the cation of valence n (typically Na, Ca, Mg, etc.), $x = 2 - \infty$, and y varies from 0 to approximately 10. These cations can generally be exchanged by conventional methods[13,14]. The cations and water molecules present are located in the cages, cavities, and channels of the zeolites. The position, size, and numbers of cations as well as the position and number of water molecules can significantly alter the properties of the zeolite.

There are two types of internal structures: one provides an internal pore system comprised of interconnected cage structures; the other provides a system

Table 3.1 Size of pore openings and dimensionality of the pore system for selected medium-pore and large-pore molecular sieves[14].

Molecular sieve name	Pore (window) size (Å)	Channel/cage size
Faujasite (X and Y type)	7.4 with a cage ($d \sim 12$ Å)	Three-dimensional channel
Omega	7.5 (3.4 × 5.6)	Two non-interconnected channels
Linde Type L	7.1 ($d \sim 7.5$ Å)	Single channel with a lobe channels
Mordenite	7.0 × 6.7 and (2.6 × 5.7)	Two interconnected channel
Beta	7.5 × 5.7; 6.5 × 5.6	Three-dimensional channel
Offretite	6.7 and (3.6 × 4.9)	Three-dimensional channel
ZSM-34	6.7 and (3.6 × 4.9)	Three-dimensional channel
ZSM-11	5.3 × 5.4	Two interconnected channels
ZSM-5	5.3 × 5.6 and 5.1 × 5.5	Two interconnected channels
Theta-1	4.4 × 5.5	Single channel
4A	4.2 with a cage ($d \sim 12$ Å)	Three-dimensional channel
$AlPO_4$-5	7.3	Single channel
$AlPO_4$-8	7.9 × 8.7	Single channel
VPI-5	12.1	Single channel
Cloverite	30	Three-dimensional channel

of uniform channels. For example, zeolites where the channels are parallel to one another and where there are no connecting channels large enough for guest molecules to cross from one channel to the next are considered one-dimensional channel systems. The preferred zeolites for inclusion of reactive organic molecules have two- or three-dimensional channels to provide rapid intracrystalline diffusion of guest molecules. Access to these channels, cages or cavities is through a pore or a window which can be of the same size or smaller than the size of the channels, cages or cavities. It is this pore dimension which determines the size of molecules that can be adsorbed into these structures. The variety in internal structure types also results in a range of pore dimensions (Table 3.1) from the small-pore materials, with the largest pore opening consisting of eight-membered rings of the tetrahedra, to the medium-pore zeolites with 10-membered ring openings, and the large-pore materials with 12-membered ring openings. Large-pore zeolites of interest for photochemical studies include the zeolites X, Y, L[19], mordenite[20], offretite[21], omega[22], and beta[23]. Figure 3.1 provides relative dimensions of guest molecules and apertures of different zeolites. These data should be useful in choosing appropriate zeolites for chemical studies.

In addition to the above described zeolites, a name reserved for molecular sieves made up of $[SiO_4]^{4-}$ and $[AlO_4]^{5-}$ tetrahedra, there are a number of microporous materials in which the tetrahedral sites are occupied by cations

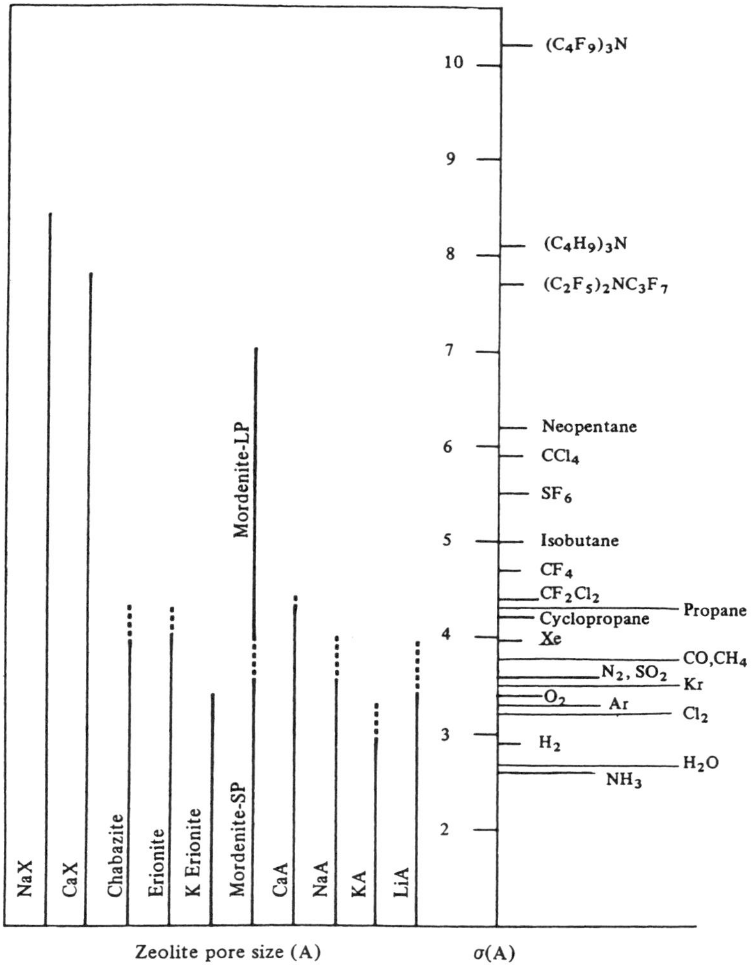

Figure 3.1 An approximate guideline for the entrance diameter of a zeolite with the size of an organic molecule. (Reproduced by permission from Breck[42].)

other than Al^{3+} and Si^{4+} (e.g. Be, Mg, Zn, Co, Fe, Mn, B, Ga, Cr, Ge, Ti, P and As). One class of materials which has attracted considerable attention and may serve as hosts for photochemical studies is $AlPO_4$. Such materials have no framework charge and thus have no ion exchange capacity. Aluminophosphate frameworks $AlPO_4$-5 (large pore)[24], $AlPO_4$-8 (extra-large pore), and VPI-5 (very-large pore)[25,26] possess one-dimensional channel systems with pore entrances of interest to organic chemists (Table 3.1). Cloverite, a gallophosphate system, possesses intersecting three-dimensional channel systems[27] of diameter ~30 Å. However, entrance to these channels is limited to ~13 Å. Even larger channel (tunable between 15 Å and 100 Å) zeolite-type materials (made up of only silicon, aluminum and oxygen in the framework) have been reported recently[28]. These mesoporous solids with one-dimensional channel systems (similar to VPI-5) and with intersecting three-dimensional channel systems (similar to cloverite) have been termed MCM-41.

Since this chapter deals mainly with X and Y zeolites, a description of their internal structure is provided in detail. The two synthetic forms of faujasite are referred to as zeolites X and Y and have the following typical unit cell composition[13,14]:

$$\text{X type} \quad M_{86}(AlO_2)_{86}(SiO_2)_{106} \cdot 264H_2O$$
$$\text{Y type} \quad M_{56}(AlO_2)_{56}(SiO_2)_{136} \cdot 253H_2O$$

where M is a monovalent cation. The faujasite framework has two main structural features (Figure 3.2). The main supercage is a result of assembly of the basic unit "sodalite cage". While sodalite cages are too small to accommodate organic molecules, the spherical supercages are ~13 Å in diameter. Access to the supercages is afforded by four 12-ring windows ~9 Å in diameter tetrahedrally distributed about the center of the supercage. The supercages form a three-dimensional network with each supercage connected tetrahedrally to four other supercages through the 12-membered ring window. Hence X and Y zeolites should not be viewed as crystalline materials consisting of independent supercages. All supercages are open and are interconnected to four others. A unit cell of X and Y zeolites consists of eight supercages.

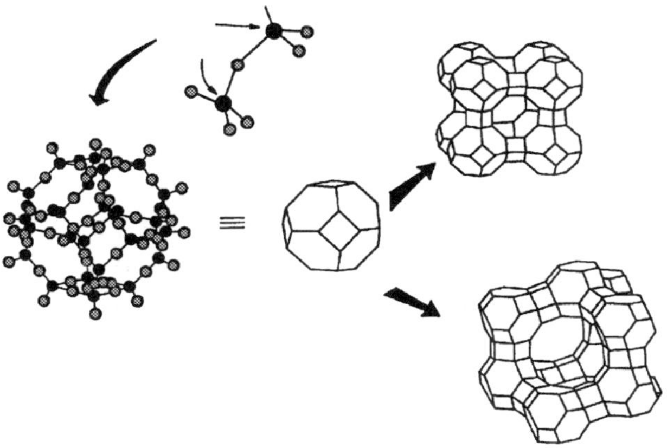

Figure 3.2 Construction of supercage from sodalite cages. Basic structural unit of X and Y zeolites. (The authors thank D. Corbin for the figure.)

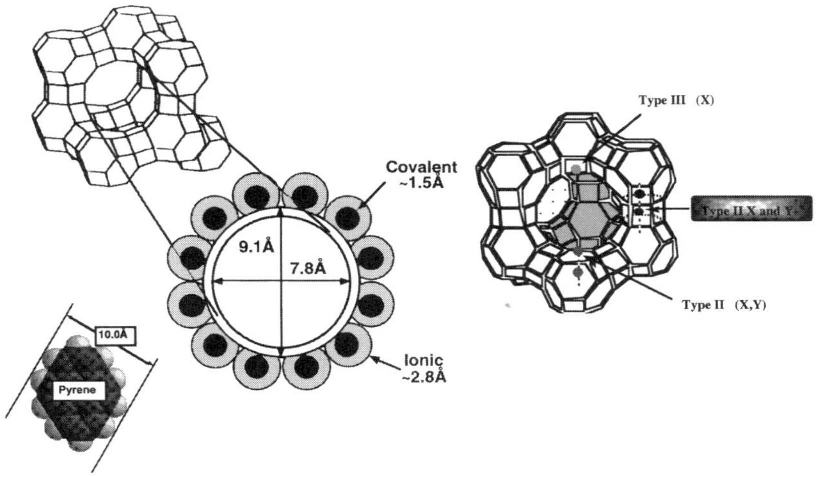

Figure 3.3 Dimensions of and location of cations within a supercage. (The authors thank D. Corbin for the figure.)

Charge-compensating cations present in the internal structure of zeolites are known to occupy three different positions in zeolites X and Y. As illustrated in Figure 3.3, the first type (site I), with 16 cations per unit cell (both X and Y), is located on the hexagonal prism faces between the sodalite units. The second type (site II), with 32 per unit cell (both X and Y), is located in the open hexagonal faces. The third type (site III), with 38 per unit cell in the case of X type and only eight per unit cell in the case of Y type, is located on the walls of the larger cavity. Only cations at sites II and III are expected to be readily accessible to the organic molecule adsorbed within a supercage. Cation locations for dry zeolites have been identified through X-ray crystallography[29–46]. Recently it has been found that the location of cations can depend on the adsorbed guest molecule[47]. Therefore, locations which are reported for guest free structures should be taken only as a first approximation for guest filled zeolites.

3. ZEOLITE HAS A "HARD" AND "ACTIVE" REACTION CAVITY

One is accustomed to carrying out reactions in large reaction vessels that are disproportionately larger than the size of a molecule. However, when the size of the reaction vessel is nearly the same as that of the reactant molecule, one will have to consider factors that might normally be ignored[48]. While a photochemical macromolecular reactor, such as a quartz cuvette, should play no role in the photochemical events occurring on the substrate of interest, a molecular-sized enclosure can influence the reactivity of the substrate. Zeolites similar to glasses and quartz vessels are made up of silica and alumina and therefore, it should be possible to excite an organic molecule without perturbing the electronic structure of the zeolite.

Reactions taking place within a zeolite can be envisioned to occur within an enclosed space[49,50] termed the "reaction cavity". A reaction cavity should be considered as the space in which excited state reactant molecules and their pre-product intermediates are confined (mobilized) during their lifetimes. Unlike isotropic solution media where molecules have limited but equal mobility and conformational flexibility in all dimensions, in a confined space their mobility and flexibility will be restricted or constrained in at least one if not in all three dimensions.

A reaction cavity is defined in terms of the factors such as "hard" and "soft" and "active" and "passive" and "free volume". The concept of free volume is introduced into the reaction cavity model to accommodate the shape changes that occur as the reactants transform themselves to products. For example, the shape and the free volume of the supercage within X and Y zeolites will determine to some extent the nature of the product that is obtained from a guest molecule. The volume available for an organic molecule within a supercage depends on the number and nature of the cation. As the calculated supercage volumes given in Table 3.2 show, the available volume for a guest decreases as the cation size increases[51] from Li to Cs. Since surfaces of zeolites, possess time independent structures like crystalline materials, the free volume needed to accommodate shape changes which occur during the course of a reaction must be present intrinsically within the fixed structure. Reaction cavities of such media possess "hard" walls. Therefore it becomes very important to choose a proper zeolite (with adequate free volume) to steer a reaction towards a particular product.

The above model leads one to conclude that "guests in hosts" are similar to balls in boxes. However, this analogy is inadequate. In addition to being "hard"

Table 3.2 Cation dependence of supercage free volume in MY zeolites.

Cation (M^+)	Ionic radius of the cation (Å)[a]	Vacant space[b] within the supercage (Å^3)	
		Y–Zeolite	X–Zeolite
Li	0.6	834	873
Na	0.95	827	852
K	1.33	807	800
Rb	1.48	796	770
Cs	1.69	781	732

[a] Ref. 133.
[b] Ref. 51.

or "soft", cavity walls must be characterized as "active" or "passive". When the interaction between a guest molecule and the cavity is attractive or repulsive, the cavity is termed "active" and when there is no significant interaction it is considered to be "passive". A zeolite reaction cavity has been characterized to be "active". Interactions may vary from weak van der Waals forces to hydrogen bonds to strong electrostatic forces between charged centers. Zeolite surfaces possess acidic silanol groups which may chemically interact with an adsorbed molecule. Furthermore, zeolite surfaces contain a large number of cations which can interact electrostatically with guests. Thus the cation–guest interactions are expected to play a significant role in controlling the fate of an excited molecule. Factors that determine the photochemical processes of a guest in a confined space include structural aspects of both the guest and the host zeolite and the nature of chemical and physical interactions between the two.

4. LOCATION OF GUESTS WITHIN ZEOLITES

There are several aspects to guest location within a zeolite. One should know its location before excitation and then during its stay on an excited surface. Information concerning the latter can only be obtained through photophysical probes. Because all guest molecules exert some motion within the channels/cages/cavities of zeolites, and because single crystals of zeolites are rare, routine X-ray structural characterization of a guest–zeolite complex is not common. However, based on other techniques considerable literature exists on the characterization of the location of guest molecules within zeolites. Most of these studies are concerned with small molecules such as benzene, substituted benzene, and pyridine. A brief summary of conclusions reached so far with these molecules is helpful to predict the possible location of a guest molecule.

The location of benzene on X- and Y-type zeolites alone has been probed by several techniques: IR spectroscopy[52–61], Raman spectroscopy[62], UV diffuse reflectance spectroscopy[63,64], NMR spectroscopy[65–72], neutron diffraction[73,74], small-angle neutron scattering[76–79], adsorption techniques[80], and quantum chemical calculations[81]. These studies have shown that at high loadings there are three distinct types of benzene molecules, located within the supercages—one at the cation site (site II or III), one at the 12-ring window site, and the other corresponding to benzene clusters within the cage. At low loading levels the clustering can be avoided and the distribution between the window and the cation sites can also be controlled by the loading level and by the nature of the cation.

Direct evidence for the location of benzene[82], xylenes[83,84], mesitylene[85], *meta*-dinitrobenzene[86] and pyridine[87] within NaY come from either powder neutron diffraction studies carried out at low temperature < 10 K or from X-ray powder study (*meta*-dinitrobenzene) at room temperature. At low loading levels, benzene, xylenes, and mesitylene prefer a cationic site. These aromatic molecules are stabilized by interactions with the Na ions. Location of *meta*-dinitrobenzene within NaY is slightly different. Interactions with Na ions are still dominant but they occur, as expected, with the nitro groups, the polar negative ends of the aromatic ring. The structure of pyridine in NaY in unique. In this case at low coverages preferential absorption occurs at the window site. It is not obvious why a non-cationic site is preferred in the case of pyridine.

The stability of the cationic site is believed to be due to a cation–π type electrostatic interaction. This is suggested to be augmented by dispersion interactions between the guest and framework. At the 12-ring window site, the interaction occurs through van der Walls forces and through acid–base interactions between the C—H bonds of benzene and the oxygens of the 12-ring window. The nature of interaction is believed to be C—H—O hydrogen bonding between aromatic hydrogen atoms and oxygen ions of the window.

Although the sites referred to above may be the preferred locations at < 10 K, at room temperature molecules have been established to be mobile. For example, NMR studies conducted above 77 K have revealed that benzene molecules are not stationary at the cation site[88–103], undergoing rotational motions at the site and also jumping from one site to another. ^{2}H-NMR line

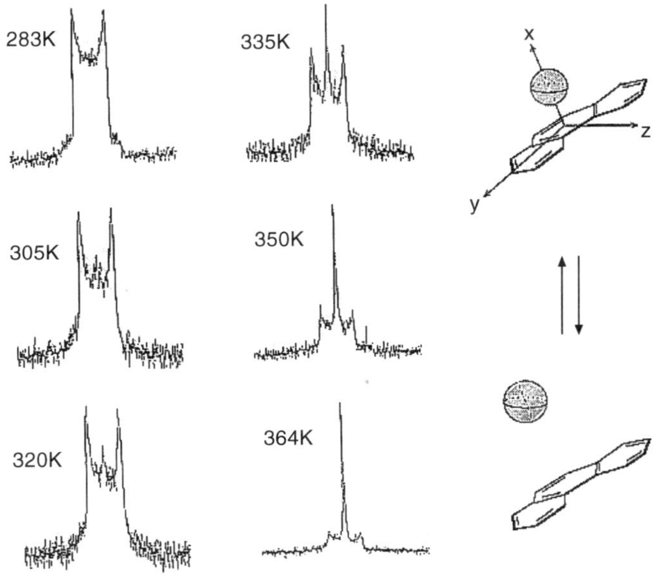

Figure 3.4 ^{2}H-NMR of perdeuterated phenanthrene included within $K^{+}X$ zeolites. Middle sharp peak is due to freely tumbling phenanthrene and the doublet is due to cation bound phenanthrene. (Reproduced by permission from Hepp *et al.*[104].)

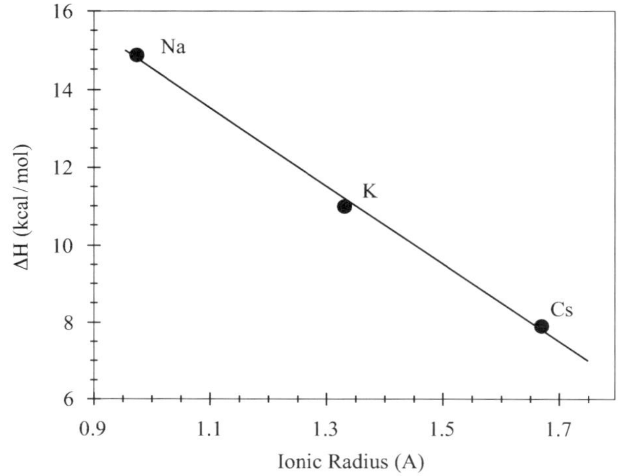

Figure 3.5 Relationship between the ionic radius of the zeolite cation and the binding energy of the cation-phenanthrene complex. (Reproduced by permission from Hepp *et al.*[104].)

shape analyses indicate that the mobility of benzene adsorbed on NaX and NaY is temperature dependent. While at 77 K only rapid rotation about the six-fold axis is occurring, at room temperature the molecule begins to jump between sites, resulting in nearly isotropic averaging of the electric field gradient.

In our laboratory, ^{2}H-NMR studies on larger aromatic molecules (naphthalene and phenanthrene) in MX and MY zeolites have been carried out[104]. As illustrated in Figure 3.4, the deuterium NMR spectra of phenanthrene-d_{10} consist of two components, narrow and static, with relative amounts of the two varying with temperature. On the assumption that the static component is due to the ion–phenanthrene bound state and the narrow component to a free state (unbound to ions) the heat of dissociation of the ion–phenanthrene complex has been calculated. A linear relationship between the heat of dissociation and the ionic radius of the cation has been observed (Figure 3.5). This is consistent with the general understanding that aromatic molecules interact with the cation present in a supercage. At room temperature they establish an equilibrium between bound and unbound states and the extent of binding is dependent on the field strength generated by the cation.

5. INHOMOGENEOUS DISTRIBUTION OF GUESTS WITHIN ZEOLITES

From the discussion above, it is clear that there is more than one site in which the guest molecule may be present in a zeolite. In homogenous solution, one rarely discusses the possibility of more than one site for the reacting solute. Fluid solutions allow the reacting molecules to experience an average microenvironment by virtue of fast relaxation time of the solvent and/or high mobility of reactant molecules. Important relaxation modes of low viscosity, non-polar solvents occur at rates near or greater than $10^{12}\,s^{-1}$ at room temperature, i.e. faster than the rates normally encountered for deactivation of S_1 and T_1 excited states[105–107]. Under such conditions, temporal fluctuations in microscopic solvent structure may permit all molecules in the solution to be exposed to very similar environments during their excited state lifetimes. In essence all reactant molecules experience a "single effective environment".

Unlike an isotropic liquid, the zeolite structure is time independent. Guest molecules may experience identical environments within zeolites only if they can establish an equilibrium between various sites through fast migration before they decay from the excited state. However, in zeolite media rates of guest diffusion are generally slower than the time period of a photoreaction[108]. These characteristics might lead to guest molecules undergoing photochemical and photophysical changes while residing at a number of different environments. Three parameters are generally utilized to describe a molecule within a zeolite[109]: occupancy; families of sites; and distribution of sites within a family. The term "occupancy" is used to indicate the aggregation status (monomer, dimer, microcrystals, etc.) of guest within zeolites. Even at low loading levels the supercages of X and Y zeolites may not be uniformly occupied, i.e. some may be singly occupied, some doubly occupied and some not occupied at all. The term average occupancy, represented by <S>, defined as the number of guest molecules over the number of available supercages, is used in the literature to indicate the loading levels in X and Y zeolites. The parameter <S> is only an average number and does not represent the heterogeneity of guest distribution within zeolites. The term "families of sites" indicates a group of sites which can be represented by a single *primary characteristic*. Instead of simply being called sites, the descriptor "family" is used to indicate that not all molecules within a family need be identical. This feature is described by the term "distribution of sites within a family". Thus, while the principal characteristics of these closely related sites may be defined by the primary site, the individual sites belonging to the same family might show properties which are slightly different from each other. For example, a guest molecule within a zeolite X can be present within a supercage or at the window, and these two sites are denoted as two families of sites. Within the supercage or at the window, not all molecules will be adsorbed at identical environments. Minor differences between molecules at these two sites may exist and this is represented by the term "distribution of sites".

Excitation of pyrene included within anhydrous NaY results in two emissions as shown in Figure 3.6, one from monomer and the other from excimer[110]. The so called excimer emission is definitely not an emission from a typical dynamic

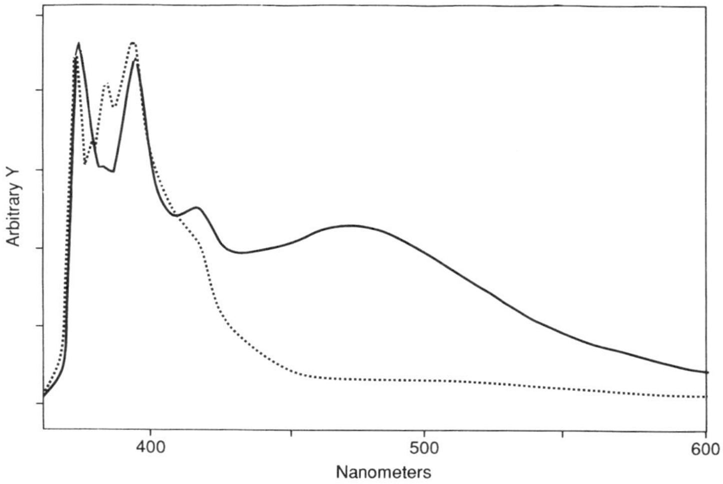

Figure 3.6 Fluorescence emission spectra of pyrene included within NaY. Dry sample shows both monomer and excimer emissions and the wet sample shows only monomer emission.

excimer since the excitation spectra for the monomer and the excimer emissions are different. Pre-aggregated dimers (static excimers) being responsible for the excimer emission is supported by the following observations: (i) no growing-in of the excimer in the nanosecond time scale was noticed when monomer and excimer decays were monitored by time resolved single photon counter; (ii) no negative pre-exponential term for excimer decay was obtained; (iii) the ratio of the intensities of excimer to monomer emission increased slightly upon lowering the temperature; and (iv) variation of loading level between 0.002 and 0.2 (average number of molecules per supercage) did not alter the pyrene monomer lifetime significantly. This example illustrates that the distribution of guest occupancy within zeolites may not be homogeneous.

Unlike in homogeneous solution media, fluorescence decay of pyrene monomer adsorbed onto NaY does not follow single exponential decay. More than 95% of the fluorescence decay can be analyzed satisfactorily on the basis of bi-exponential decay indicating that primarily there are two families of sites in which these aromatic molecules are located. Support for the independent nature of the two sites comes from three different observations: (i) with increasing pressure of xenon and oxygen fluorescence from pyrene is quenched. While xenon quenching of the excited singlet occurs through external heavy atom effect[111–116], oxygen quenches these states via an electronic energy transfer process. Surprisingly, only one of the two lifetimes dependend on the pressure of xenon and oxygen. This suggests that among the two sites in which the aromatic probe molecules are present only one is accessible to xenon and oxygen. At this stage on information on the nature of the two sites in available; (ii) it has been established that the singlet lifetime of pyrene is related to the polarity of the medium. The lifetime is fairly long in a non-polar medium (e.g. cyclohexane: 430 ns) and is relatively short in a polar medium (e.g. dimethylformamide: 280 ns)[117,118]. When pyrene is included within NaY two lifetimes, one short and one long, are measured. On the basis of the relationship between lifetime and polarity of the medium, it is likely that the one with shorter lifetime corresponds to pyrene closer to the cation (higher electric field/micropolarity; see next section) and the one with longer lifetime refers to the pyrene adsorbed onto the walls (lower electric field/micropolarity); (iii) spectral features in the emission spectra of pyrene are found to be time dependent in the time range 1 to 900 ns. The intensity of the first band at 373 nm is fairly strong at early times and weak at later times. In homogeneous solution, it has been established that the intensity of this band is dependent on the polarity of the

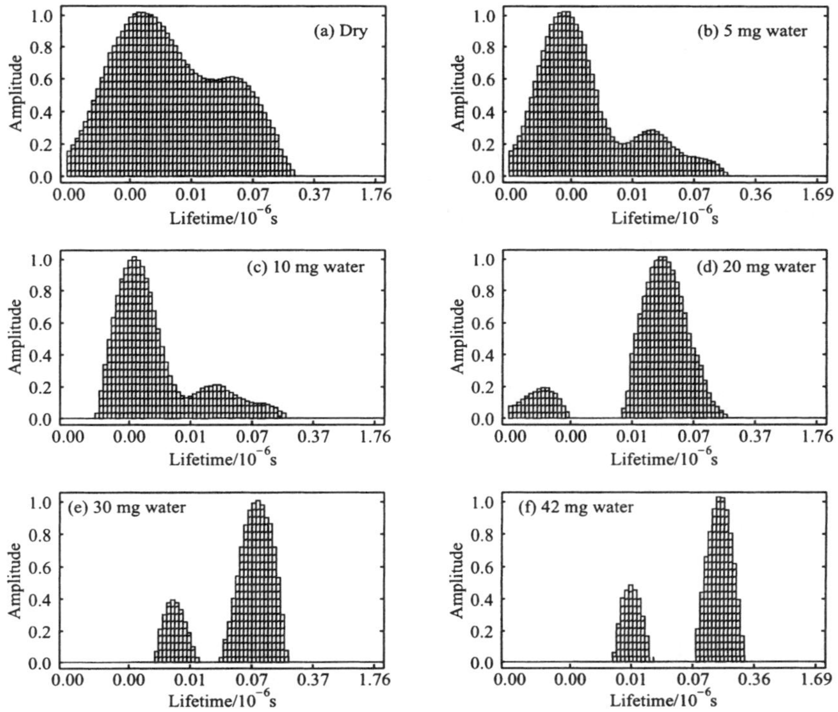

Figure 3.7 Distribution of excited singlet state lifetime of pyrene monomer included within NaY: distribution is dependent on the water content. (Reproduced by permission from Hepp *et al.*[104].)

medium[119–121]. The intensity variation observed within NaY is consistent with the proposal that pyrene molecules are present under two different micro-environments, one highly polar environment (the cation site) and the other moderately polar (the wall site).

Analysis of fluorescence decay from pyrene by a distribution analysis program provides an insight into the problem of distribution of sites within zeolites[122,123]. Pyrene provides an ideal example wherein "distribution of sites" exists. Distribution of lifetime obtained in the case of pyrene within anhydrous NaY is presented in Figure 3.7. The distribution is affected by co-adsorbents present within NaY along with the guest molecule. This is illustrated by the results obtained in the case of pyrene with water and other solvents as co-adsorbents. From Figure 3.7 it is evident that the lifetime distribution for pyrene within NaY co-included with water is fairly narrow when compared to that in anhydrous NaY. A similar decrease in distribution was obtained when organic solvents such as hexane, pentane, cyclohexane, acetonitrile, diethylether and methanol were used as co-adsorbents.

6. ELECTRIC FIELD

In dehydrated X and Y zeolites many, if not all, cations are located near the inner surface of supercages. As mentioned in an earlier section cations present in site I are hidden from the zeolite surface (Figure 3.3). Cations of site I enjoy an octahedral co-ordination of six oxygens and are therefore well hidden and well shielded within sodalite cages. On the other hand, site II cations are present on top of the hexagonal faces of the sodalite cages facing into the supercage and are co-ordinated in a pyramidal form to only three of the six oxygens of the ring. Site III cations are close to four oxygens rings on the square face of the sodalite cages. Both site II and III cations are present within supercages and are shielded only on one side; the side exposed to the center of the supercage being unshielded. Consequently, these partially unshielded cations create very large electric fields extending into the supercage[124]. The fields generated by these cations have been theoretically estimated for a number of alkali and alkali earth cations and are summarized in Table 3.3[125–127]. General conclusions from these studies are as follows: (i) the electric field of a cation at site III is several times stronger than that of the same cation at site II; (ii) within the same row, divalent cations exhibit higher field than monovalent cations; and (iii) for a fixed distance from a given cation, the average field increases with silicon/aluminum ratio. For example, cations in zeolite Y show higher field than in zeolite X.

As described in the paragraph above, cations present in supercages are only partially shielded, leading to partial shielding of the anionic framework of the zeolite as well. Thus, the zeolite surface can be viewed as a polyanion with counter ions scattered within supercages. The only difference between a

Table 3.3 Field variation as a function of cation and Si/Al ratio[a].

Cation; Si/Al	Field at indicated distance from site II cation (V/Å)			Field at indicated distance from site III cation (V/Å)		
	1 Å	1.75 Å	2.5 Å	1 Å	1.75 Å	2.5 Å
Na; 1.0	1.8	0.43	0.096	3.07	1.11	0.43
Na; 1.4	2.3	0.98	0.6	3.45	1.61	0.88
Ca; 1.0	6.09	2.68	1.34			
Ca; 1.4	6.41	3.04	1.63			

[a] Taken from Dempsey[125].

polymeric polyanion and a zeolite surface is that there is order and periodicity on a larger scale in the latter. A dry zeolite is expected to reduce its free energy by maximizing ionic interactions between the cations and anionic framework. Intuitively, it is obvious that the free energy of the system will be increased when a non-polar guest is placed in the cavity. In order to attain a lower energy, the zeolite will polarize the guest such that the negative end will be stabilized by the cation and the positive end by the anionic framework. The free energy of the guest–zeolite system will depend upon ease of polarization of the guest molecule by the zeolite supercage.

These qualitative arguments lead to two conclusions: (i) zeolite supercage can polarize the included guest molecule and thus distort the electron density of a guest molecule; and (ii) heterolytic bond cleavage leading to formation of ions and electron transfer leading to ions will easily occur within a zeolite. These conclusions have been supported by a number of experimental observations. For example, the origin of higher frequency band in carbon monoxide and carbon dioxide and pyridine adsorbed within Y zeolites is attributed to the field generated by cations[128–132]. As illustrated in Figures 3.8–3.10. there is a linear relationship between the field generated by the cation and the position of the high frequency band. A similar relationship has been established for zeolite adsorbed nitrogen and oxygen stretching frequencies. ESR techniques have also been used to probe the changes in the electron density of molecules adsorbed within zeolites[133–135]. An impressive linear relationship between the cation dependent electric field and the extent of hyperfine splitting has been observed for ClO_2 when adsorbed on alkali cation exchanged zeolites.

The most frequently observed ionization reaction in divalent cation exchanged zeolites is the ionization of water[136] which generates Brønsted acid sites within M^{2+} zeolites. Catalytic activity of zeolites, an outcome of the presence of Brønsted acid sites, is a direct result of the high field or electrolytic nature of the zeolite supercage. Clearly there is a direct relationship between the number of acidic sites and the field generated by the cation; a higher field and smaller cations favor a larger number of sites. Another interesting ionization

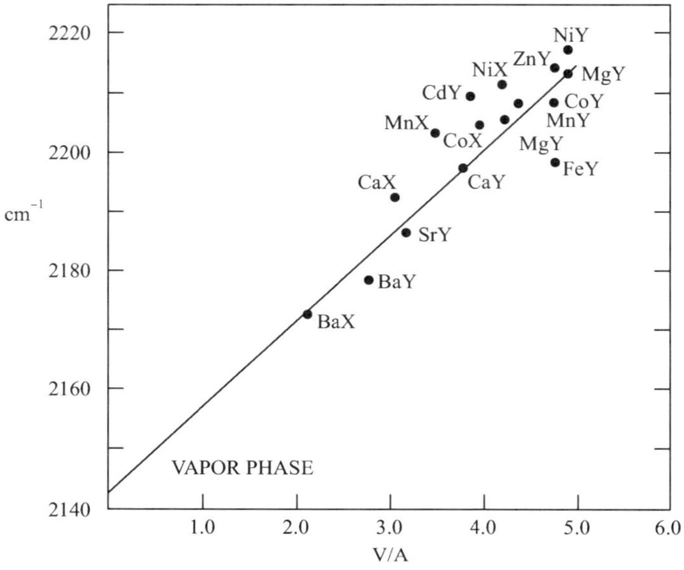

Figure 3.8 Dependence of the IR frequency of "CO" included within MY zeolite included on the cation. (Reproduced by Angell and Schaffer[130].)

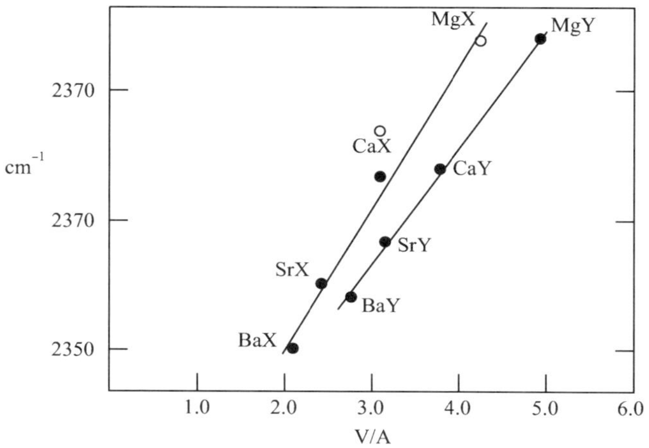

Figure 3.9 Dependence of the IR frequency of "CO_2" included within MY zeolite included on the cation. (Reproduced by permission from Ward and Habgood[128] and Angel[129].)

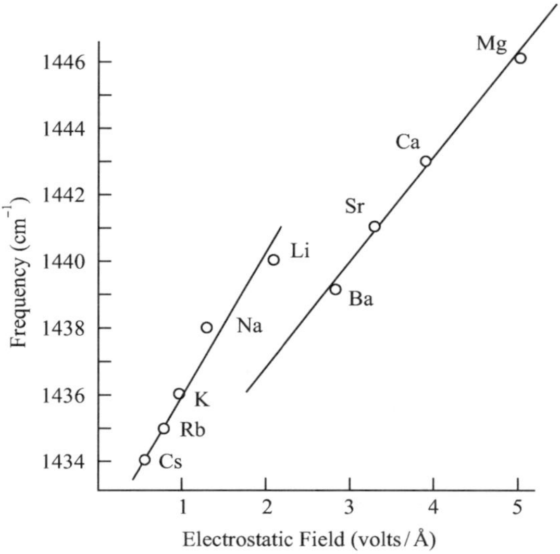

Figure 3.10 Dependence of the IR frequency of "pyridine" included within MY zeolite included on the cation. (Reproduced by permission from Ward[133].)

process is the interaction of NO and NO_2 radicals within zeolites[137,138]. When these radicals are included within NaY, they exist as NO^+ and NO_2^- as a result of spontaneous electron transfer. Another example of this type is the oxidation of Na metal within NaY where a spontaneous electron transfer occurs, Na becomes Na^+, and the electron is captured by a Na_4^{3+} cluster[139,140]. Both of these electron transfer reactions are highly endothermic in the gas phase (>100 kcal/mol). The ready occurrence of these processes within zeolites has been attributed to the zeolite electric field and to lowering of guest–zeolite energy when these ionized species are properly arranged within supercages.

From the above it is clear that the field generated within zeolites is dependent on the cation: (i) divalent cations generate higher field than monovalent cations; and (ii) smaller cations with lower atomic number (within the same group) generate higher field. Thus one would expect organic molecules to be polarized more readily within NaY than within CsY, and more easily within CaY than in NaY. Because, within a group, cations of lower atomic number play a significant role; we term this phenomenon the "Light Atom Effect".

Generally, organic chemists are not used to describing a medium in terms of its field. A term closer to the field is polarity, although they may not be exactly the same. In a qualitative sense the changes caused by the local fields described above are similar to those observed when the molecule is dissolved in a polar solvent, but with an important difference. In solvents, the dipole and quadrupole

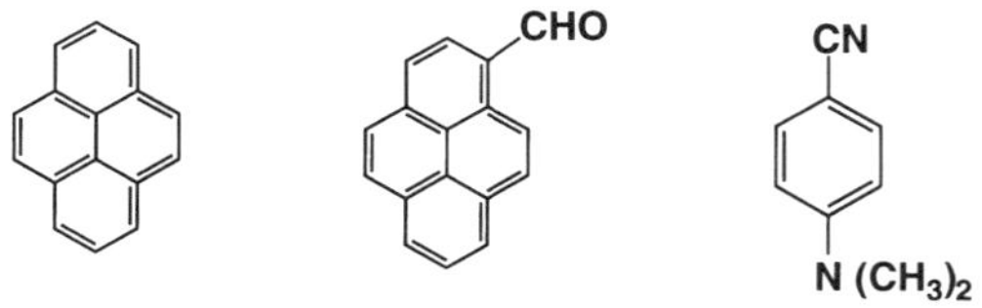

1 **2** **3**

Scheme 3.1

generated by solvent molecules fluctuate whereas in a zeolite these are stationary. A number of organic probes have been used to measure the micropolarity of a supercage[141]. For example, medium-sensitive photophysical properties of probes such as pyrene (**1**), pyrenealdehyde (**2**) and *para*-dimethylaminobenzonitrile (**3**) (Scheme 3.1) have been utilized to estimate the micropolarity of the zeolite interior[142,143]. Results of the photophysical measurements carried out in our laboratory are summarized in Table 3.4 (also listed in Table 3.4, for reference, are the measurements made for each probe in four solvents of varying polarity). Analysis of Table 3.4 reveals the following:

Table 3.4 Micropolarity of zeolite cages as probed by three conventional photochemical probes.

Zeolite	Pyrene[a] I_5/I_1	DMABN[b] emission λ_{max} (nm)	ε[c] (D)	$E_T(30)$[c] (kcal/mol	Pyrenealdehyde[d] emission λ_{max} (nm)
LiX	0.58	337 and 535	68	59	481
NaX	0.60	338 and 528	56	57	473
KX	0.49	336 and 510	42	55	460
RbX	0.53	339 and 477	16	49	454
CsX	0.60	342 and 471	11	48	450
LiY	0.81	334 and 523	55	57	467
NaY	0.73	334 and 512	44	55	463
KY	0.56	335 and 500	37	54	450
RbY	0.60	335 and 463	4	46	443
CsY	0.71	335 and 440[e]	–	–	440
Cyclohexane[f]	1.25	340 only	2.0	30.9	392
Acetonitrile	0.60	337 and 485	35.9	46.7	415
Methanol	0.69	352 and 500	32.6	55.4	450
Water	–	347 and 553	78.3	63.1	480

[a] Intensity of the monomer fluorescence at 392 nm versus 372 nm; 0.001 mg of pyrene in 250 mg of zeolite.
[b] λ_{max} of the normal emission and emission from TICT state; 0.01 mg of DMABN in 250 mg of zeolite.
[c] Solvent parameters, dipole moment (ε) and $E_T(30)$ were obtained from the literature.
[d] λ_{max} of the monomer emission; 0.01 mg of pyrenealdehyde in 250 mg of zeolite.
[e] Emission corresponds to phosphorescence.
[f] All solvent numbers correlate well with literature reports.

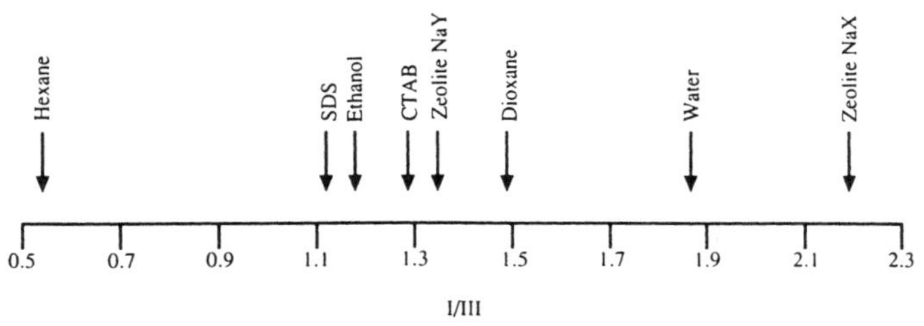

Figure 3.11 A comparison of the polarity of zeolites NaX and NaY with that of solvents. Micropolarity monitored with pyrene as a probe.

(i) supercages of cation exchanged X and Y zeolites are relatively polar; (ii) the supercage is very polar or possesses high electric field when the cation is Li or Na; (iii) the polarity/electric field of the cage decreases with the cation size (Li > Na > K > Rb > Cs); and (iv) the cage is more polar in the case of X zeolites than in Y zeolites (Figure 3.11). The above observations are in line with the results for electric field discussed above.

Solvent polarity is represented by more than 30 different polarity scales[144]. Of these, the E_T (30) scale takes into account both non-specific solute–solvent interactions as well Lewis acid–base type interactions making this scale particularly appealing for investigating guests within zeolites. A Lewis acid–base type interaction is expected between the cation and the probes described in this study. By utilizing the reported correlations of solvent parameters in the case of *para*-dimethylaminobenzonitrile one can roughly estimate the polarity within the supercages[145,146]. The dielectric constant and E_T (30) for Li, Na, K, Rb and Cs—X and —Y are provided in Table 3.4. Supercages of Li, Na and KX have polarities similar to a methanol–water mixture while those of Rb and CsX are close to that of propanol.

There have been other attempts, in addition to ours, to monitor the micropolarity of the supercages of NaX and NaY. The earliest such study is due to Baretz and Turro where pyrenealdehyde was used as the probe[147]. Although this investigation was complicated by excimer formation due to high loading levels, supercages were inferred to be polar. Yoon and Kochi, on the basis of charge transfer spectral maxima, have concluded that supercages of methyl viologen (MV^{2+}) exchanged NaY possess polarity comparable to that of 50% aqueous acetonitrile[148]. Dutta and Turbeville have suggested on the basis of salicylidenes as indicators that supercages of NaX and NaY zeolites exhibit polarity similar to that of polar hydroxylic solvents[149]. Iu and Thomas have suggested, on the basis of pyrene as the probe, that the supercages of X and Y zeolites are polar[150].

Micropolarity reported by organic probes is only an average number and may not be an accurate representation of the polarity of the supercage. As a result of non-uniform distribution of probe molecules within zeolites (see below), a distribution in micropolarity would be expected. Micropolarity of the environment monitored by the probe is expected to be dependent upon the distance between the cations and the probe as the electric-field generated by the cations, which is the major cause of the micropolarity, is expected to vary with distance (from cations) within zeolites. Therefore, one should view the knowledge gained through organic probes only as an approximate measure.

7. BRØNSTED AND LEWIS ACIDITY AND BASICITY

Sets of zeolites that are commonly used as photochemical reaction media are M^+X and M^+Y and $M^{2+}X$ and $M^{2+}Y$. The molecular formula as written for these X and Y zeolites, $M^+_x(AlO_2)_x(SiO_2)_y$, does not give any indication that they may possess reactive acid sites. However, it is well known to the zeolite community that activated $M^{2+}X$ and $M^{2+}Y$ zeolites contain Brønsted acid sites[151–159]. In addition, it has been noted recently that Brønsted acidity, of commonly used monovalent cation exchanged zeolites can vary significantly from source to source[160]. Even the presence of a very small number of acidic sites can afford catalytic activity and complicate the desired chemistry. We have found this to be the case in a number of oxidation reactions of olefins within zeolites, where a small number of acid sites can alter the chemistry significantly and result in a variety of unexpected products (resulting from the protonation of the olefin)[161]. Therefore, we believe that it is important to be aware to the acidic properties of zeolites. Considerable literature including reviews exist on this topic[151–159]. Below we provide a very brief introduction to the acidity of zeolites and for details readers should consult the original literature.

While discussing the acidity of zeolites one will have to consider the following: (i) the nature—Brønsted or Lewis; (ii) the location; (iii) the number; and (iv) the strength. Zeolites contain two functional groups (identified by MAS NMR) that could potentially show Brønsted acidity. These are aluminol (Al—OH) and silanol groups (Si—OH). The former is not acidic enough to be of any consequence, but the latter has been established to be acidic when adjacent to a tri co-ordinated Al (Lewis acidic site)[162–167]. When not adjacent to a tri co-ordinated Al site, these are not acidic enough to be significant. Therefore, the Brønsted acidic sites in zeolites are generally the bridging Si—O(H)—Al group (Figure 3.12).

Brønsted acid sites in alkali and alkaline earth cation exchanged X and Y zeolites are generated during the activation process. The role of the cation in this process has been established. According to one model, the electrostatic field associated with the charge balancing cation polarizes the adsorbed water molecule, resulting in the formation of acidic hydroxyl groups, as represented in Scheme 3.2. A linear relationship between the number of acidic sites and the

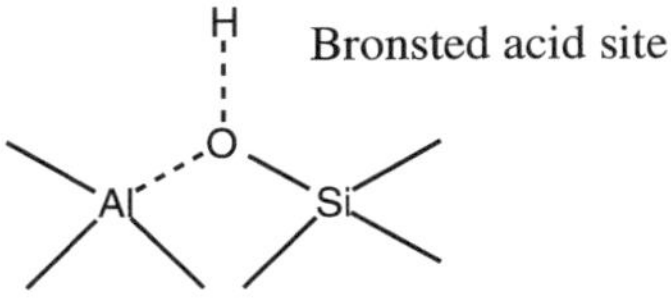

Figure 3.12 A representation of Brønsted acid sites in zeolites.

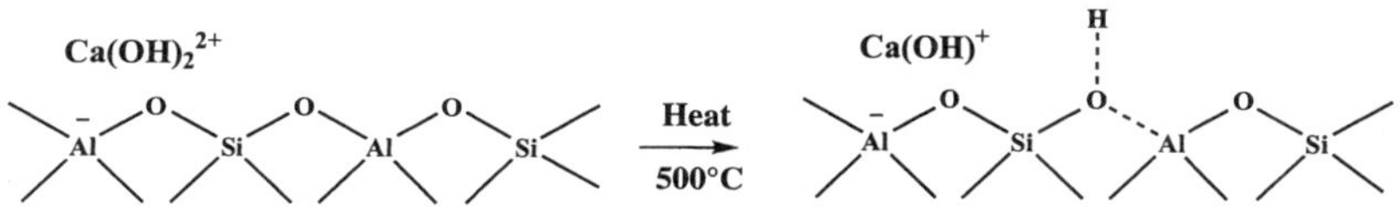

Scheme 3.2

polarizing power of the cation has been established (see Section 6). As expected, a larger number of acidic sites are generated within alkaline earth cation exchanged zeolites than in alkali cation exchanged zeolites. Even within the alkaline earth cation exchanged Y zeolites the number varies in the order expected Mg>Ca>Sr>>Ba. Although alkali exchanged (Li, Na, K, Rb and Cs) X and Y zeolites are thought to be non acidic we have found them to contain extremely small numbers of Brønsted acidic sites which could harm the expected chemistry by acting in a catalytic manner (see below)[160].

Brønsted acidity (both number and strength) in X and Y zeolites has been probed by a number of techniques[162–172]. UV-VIS using various color indicators, IR and Raman with probes such as pyridine, ammonia, and carbon monoxide, adsorption and temperature programmed desorption of probe molecules such as ammonia, MAS NMR, TRAPDOR NMR with alkyl amines and trialkyl phosphines as probes, photoelectron spectroscopy, and microcalorimetry. Readers should consult the literature for advantages and disadvantages as well as limitations of each technique. It is generally agreed that the acidic strength of a zeolite depends on a number of factors, the main ones being the Si/Al ratio of the zeolite framework[173] the Si—O(H)—Al bond angle[159], and the cation[154]. In effect, all the variables that alter the electron density of oxygen and the polarizability of the OH bond in Si—O(H)—Al will influence the acidic strength. Because not all Si—O(H)—Al bonds are expected to be under identical environments in a zeolite, a distribution of acidic strength has been noted.

We have recently utilized MAS NMR and color indicator techniques to quantify the number of Brønsted acidic sites in M^{2+} and M^{+} X and Y zeolites[174]. In the case of CaY, a concentration of ~16 Brønsted acid sites per unit cell (i.e two per supercage) was determined by adsorbing trimethylphosphine (TMP) molecules on the sample activated at 500°C. Only a small number (one per unit cell) of Brønsted acid sites was detected when samples were carefully calcined in a vacuum at 400°C and no Lewis acidity was observed. These results establish that the number of Brønsted acidic sites present in CaY is dependent on the activation conditions and that $M^{2+}Y$ zeolites must be viewed with caution when they are used as media for photoreactions. Prompted by the success with $Ca^{2+}Y$ zeolites, we proceeded to investigate $Na^{+}Y$ zeolites with solid state NMR. The basic probe molecules TMP, dimethylamine and methylamine were used to test both Brønsted and Lewis acidity in these materials. We could only infer from these studies that there must be less than one Brønsted acidic site per two unit cells. Although this technique failed to ascertain the presence of acid sites in M^{+} X and Y zeolites, the classical color indicator method (for structure of indicators see Scheme 3.3) showed that Y zeolites contain one acidic site per two unit cells (i.e. one per 16 supercages)[175]. These results clearly highlight the need to characterize a zeolite prior to use. Although extremely low levels of Brønsted acidity were detected, these levels are sufficient to alter the reaction pathways for a number of olefinic systems.

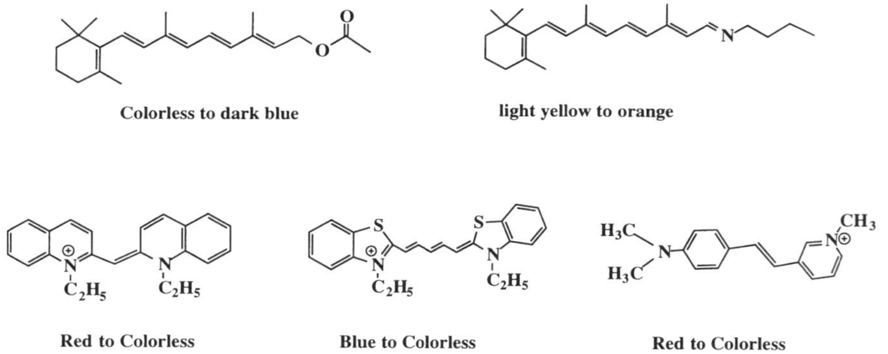

Scheme 3.3

Zeolites also contain Lewis acid sites, the site that can accept/interact with a pair of electrons[176–181]. These are the tri co-ordinated Al sites on the framework, the extra lattice (non framework) Al sites and the charge compensating cations. The former two sites are generated when a zeolite is activated at higher temperatures (>650°C). The nature, number and strength of Lewis acid sites are less well characterized, than are the Brønsted acid sites.

All the framework oxygens in a zeolite are potential Lewis basic sites[182–183]. Only the oxygens which acquire the highest negative charge have a true basic character. There are two types of oxygens: $O_3Si{-}O{-}SiO_3$ and $O_3Si{-}O{-}AlO_3$. Of these, the latter is most negative and therefore most basic. The oxygen acquires more negative charge in the latter because the Al to which it is bonded is negatively charged. Because the negatively charged framework AlO_4 interacts with a cation, the basicity of the oxygen would indirectly be influenced by the cation[183]. The Lewis basic sites (oxygen lone pair) are indirectly associated with Lewis acid sites (cations), forming acid–base pairs[184,185]. Thus, the stronger the acid (the cation), the weaker the conjugate base (oxygen). For a given Al content, whether the zeolite will behave as a Lewis acid or as a Lewis base will be controlled by the cation. With strongly acidic cations such as Li and Na the charge on the Si—O—Al oxygen is low and the zeolite behaves as a Lewis acid. When the cation is less acidic or has low electronegativity, (e.g., Rb, Cs), the zeolite shows basic properties.

From the above it should be clear that zeolite cavities are not inert and care must be taken in analyzing photochemical results by considering potential complications that might arise from acidic and basic sites.

8. EXPERIMENTAL PROTOCOL

Zeolites known as "molecular sieves" are routinely used to dry organic solvents. The feature that allows zeolites to be used as drying agents is also the feature that disqualifies them from serving as reaction vessels, and the water molecules must be desorbed to make room for organic guest molecules. Most commonly, zeolite samples are activated at ~500°C for several hours prior to exposure to the guest. In the case of relatively large organic compounds, complexation is carried out from volatile non-polar solvents with relatively low affinity for the zeolite such as hexane. Sometimes the solvent of choice is one with a kinetic diameter that is too large to compete with the substrate for complexation sites. For example, isopentane (2,2,4-trimethylpentane), with a kinetic diameter >7 Å, commonly used when a molecule is to be included within the channels of ZSM-5. Adsorption typically occurs over periods of several hours although for large guest molecules, this may continue for several days. From our experience we have found that it is important to stir the guest sample with "dry" zeolite in a non-polar hydrocarbon solvent for at least 24 h. When the size of a guest is close to the aperture size of a zeolite, it is important to stir the slurry at slightly higher temperatures (~60°C). After saturation, the solvent is filtered and the zeolite complex degassed under reduced pressure for several hours to eliminate traces of solvent. Residual water that might have been adsorbed during sample preparation can be satisfactorily desorbed from the zeolite only by degassing the sample at higher temperatures (10^{-4} torr at ~100°C). Zeolite complexes thus prepared may be characterized by diffraction analysis (X-ray or neutron diffraction), solid state CPMAS NMR and spectroscopic methods such as diffuse reflectance, electronic, IR and Raman.

Following drying, it is important to carry out further manipulations in a dry atmosphere. Although samples can be prepared within a dry box, such an operation can be cumbersome for photochemical studies. Most photochemical studies are carried out on hydrocarbon-guest–zeolite slurries and most likely all of these samples contain varying amounts of water. Therefore, when comparing data from different laboratories one needs to be aware that all samples may not be similar.

Photochemical reactions are normally carried out in continuously rotated (or stirred) cells maintained under reduced pressure (or a nitrogen atmosphere). After irradiation, retrieval of reaction mixtures is normally achieved by using a suitable polar solvent (e.g. diethyl ether, chloroform, methylene chloride, tetrahydrofuran) that will drive the products from the zeolite interior to the solvent. Soxhlet extraction and sonication are sometimes employed in situations where low recovery is observed for simple solvent extraction. Dissolution of the zeolite in 1 N HCl followed by organic solvent extraction can also give adequate results although this method is destructive and is limited to zeolites with high aluminum and to organic compounds stable in aqueous acid.

The exact protocol to be used for sample preparation and handing is still evolving. This is illustrated by the studies carried out on pyrene within NaY. Over a period of 12 years four groups have investigated the steady state fluorescence of pyrene included within NaY[110,186–190] and spectra obtained by each group are slightly different. Studies by these groups have provided important clues in terms of sample preparation and handling. Part of the problem has to do with the inherent tendency of zeolite to adsorb water. Figure 3.13 provides emission spectra of pyrene–NaY prepared under two different conditions. In one case pyrene was included within NaY by simply mixing for

Figure 3.13 Dependence of fluorescence emission of pyrene included within NaY on the method of preparation. (Right) Pyrene mixed with NaY by stirring on the rotatory evaporator. (Left) Pyrene mixed with NaY by stirring overnight as a hexane slurry. (Reproduced by permission from Suib and Kostapapas[187].)

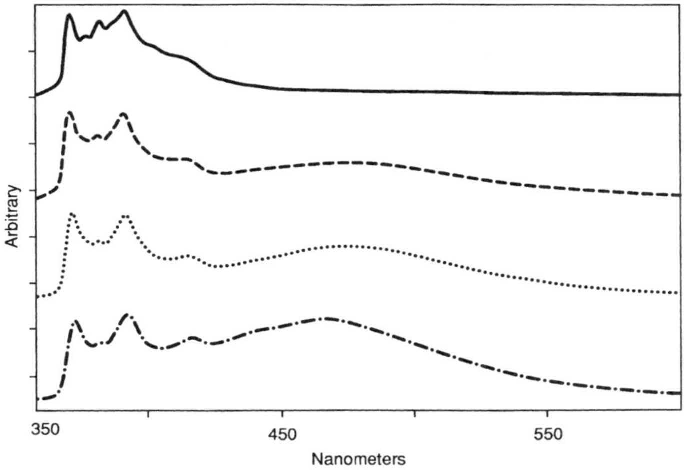

Figure 3.14 Dependence of fluorescence emission of pyrene included within NaY on water content. Sample prepared by stirring overnight as a hexane slurry and then dried on a vacuum line with mild heating.

15 min (in a rotary evaporator) a suspension of pyrene (in ethanol) with dry NaY, and in the other, the suspension was stirred overnight. Clearly the ratio of the excimer to monomer emission is dependent on the method adopted for sample preparation. In Figure 3.14 emission spectra of pyrene–NaY (low loading level) under wet and dry conditions are provided. The ratio of the monomer to excimer emission is dependent on the water content. This is an illustration of how spectra might differ between laboratories as a function on the manner in which the sample was dried. Figure 3.15 provides emission spectra of pyrene–NaY at a high loading level. The authors have shown that the nature of the spectra depends on the day the spectrum was recorded, and is attributed to the change in distribution of guests with time. Surprisingly, it has taken more

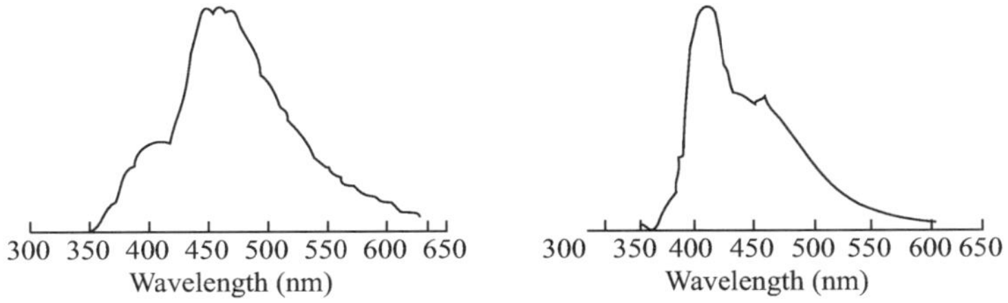

Figure 3.15 Dependence of fluorescence emission of pyrene included within NaY on the day it is recorded. (Left) Day 1. (Right) Day 16. (Reproduced by permission from Gzens *et al.*[190].)

than a month to attain an equilibrium distribution for a sample of pyrene within NaY at a loading level of one molecule in 10 supercages. The time it takes to attain an equilibrium distribution of guests within zeolites is expected to depend on the loading level and on the size (and mobility) of the guest and one cannot extrapolate experimental procedures that are described for low loadings to the ones for high loadings of guests within zeolites. The three examples provided here illustrate need to standardize the procedure adopted to prepare and handle samples.

9. CONSEQUENCES OF ELECTRIC FIELD ON PHOTOPROCESSESS: '*LIGHT ATOM EFFECT*'

In this section we discuss how a "small" (light) cation influences the photoprocesses of an organic guest molecule included within a zeolite. Small cations generate high electric field, polarize the electron distribution of a molecule by electrostatically interacting with the non-bonding and/or π-electrons of guest molecules and provide a high micropolarity. Although one might view these phenomena as independent, in a broad sense they are all the result of the field present within a zeolite. Therefore, we have combined these effects in the section below.

9.1 Perturbation of Electronic States

Changes in the absorption spectra of aromatic molecules such as benzene and anthracene adsorbed on zeolites have been noted as early as 1963[191]. For example, in benzene the symmetry forbidden 0–0 transition gains intensity when adsorbed within NaX and CaY[192,193]. Fluorescence emission from anthracene is red shifted (relative to isotropic solution medium) within NaY[194–197]. These are indications of changes in the electronic distribution and structural distortion of molecules when they are adsorbed on a zeolite surface. The perturbations of electronic states as well as the modifications of the vibrational modes are explained by a lowering of the symmetry of aromatic molecules in the adsorbed state. The presence of a strong field within supercages is expected to alter the distribution of π-electron density.

Our own observations with pyrene are consistent with the above suggestions[110]. Diffuse reflectance spectra recorded for pyrene, at a low loading level (Figure 3.16), within a number of cation exchanged X and Y zeolites suggest that the interaction between the cation and pyrene is strong and that the strength of interaction depends on the polarizability of the cation. As shown in Figure 3.16, the intensity of the 0–0 transition of the S_0 to S_1 band is dependent on the charge density of the cation present within supercages. Our working model is that the symmetry reduction of the aromatic molecule occurs

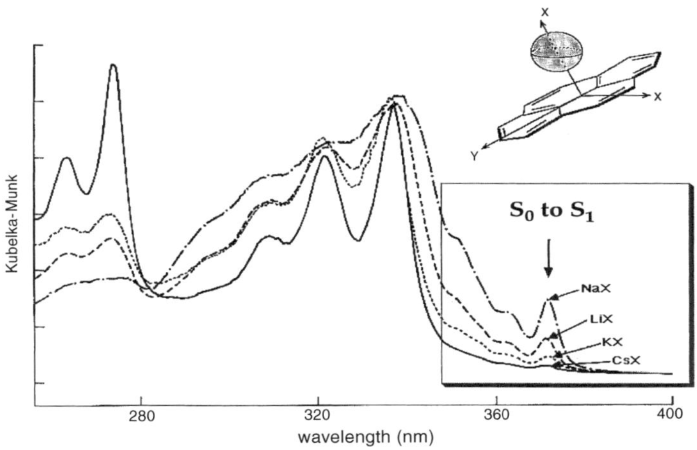

Figure 3.16 Diffuse reflectance absorption spectrum of pyrene included within M^+Y zeolites. The intensity of the 0–0 band depends on the cation.

due to strong interaction between the cation and the aromatic along the π face resulting in slight bending of the aromatic plane.

For ketones, the "light atom effect" manifests itself in altering the nature of the lowest triplet state[198]. There is a small difference in diffuse reflectance spetral maxima between LiX– and CsX–valerophenone complexes. Excitation (250–300 nm) of valerophenone included in zeolites at 77 K gives rise to phosphorescence in the region 375 to 550 nm. (The emission spectra of valerophenone in LiX and CsX are displayed in Figure 3.17). The triplet lifetimes for Na, K and RbX are all between 0.4–1.2 μs indicating little effect of the cation, at least within the limitation of the rather large variations in triplet lifetime from one sample to the other. However, there are significant changes when the cation is changed to either Li or Cs. For LiX samples, the triplet decay is much longer (4–5 μs) and does not give satisfactory fits to a single exponential decay. It is quite likely that the long lifetime observed in LiX is a consequence of the change in the character of the lowest triplet from $n\pi^*$ to $\pi\pi^*$. Such a proposal is supported by the phosphorescence spectra. As shown in Figure 3.17, the emission observed in LiX is distinctly different from that in CsX. While the emission in LiX is characteristic of $\pi\pi^*$, that in CsX is that of a $n\pi^*$ state[199]. Such a reversal in the character of the lowest triplet is known to occur in more polar solvents. This reversal is not unexpected[142,143] because the interior of LiX is more polar than other cation exchanged X zeolites.

Cation and zeolite dependent excited state chemistry of several carbonyl compounds reported in the literature may have its origin in the field within a zeolite. For example, photolysis of several alkylphenyl ketones in benzene (Scheme 3.4) and as complexes of zeolite M^+X and M^+Y (M = Li, Na, K, Rb,

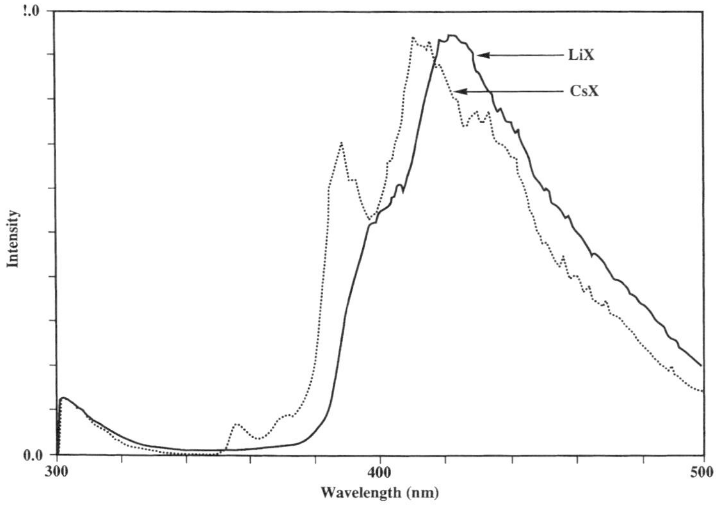

Figure 3.17 Emission spectra of valerophenone included within $M^{+}X$ zeolites.

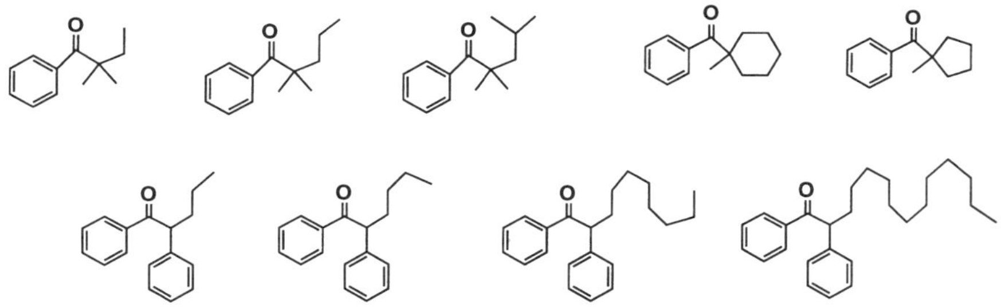

Scheme 3.4

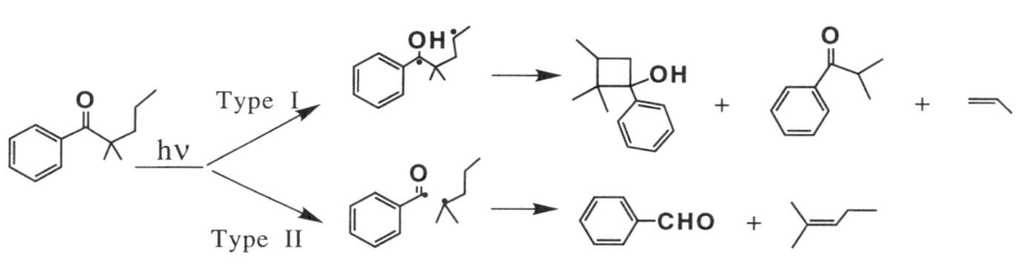

Scheme 3.5

and Cs) gave products resulting from both Norrish type I and type II processes (Scheme 3.5)[200,201]. Results with a few ketones are illustrated in Figure 3.18. It is clear from the figure that the yield of benzaldehyde a product of the Norrish type I process, is enhanced significantly within zeolites with respect to benzene

A Strategy to Enhance Type I Activity at the Expense of Type II

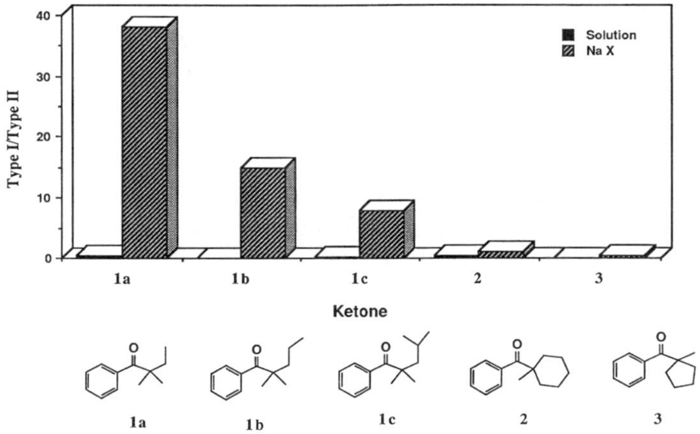

Figure 3.18 Enhancement of type 1 cleavage within zeolites.

solution. Cycloalkanones which normally do not undergo type I reaction in solution, yield products upon irradiation in M^+Y zeolites resulting from the type I process as major products. The relative yield of type I to type II products depends on the cation; smaller cations such as Li and Na have a larger influence. This alteration in chemistry can be understood on the basis that the cation interacts with the lone pair electrons of the carbonyl chromophore and thus influences the energetics of the $n\pi^*$ and $\pi\pi^*$ excited states. Greater contribution of $\pi\pi^*$ character to the reactive excited state will reduce the hydrogen abstraction rate, and under such conditions, the normally less dominant α-cleavage reaction is able to compete and give products from this process.

We have recently noticed that the reactivity of steroidal enones can be controlled within zeolites[202]. We believe that the observed effects could be the result of the field generated by cations present within zeolites. In isotropic solution, androstenedione has been established to react mainly from the cyclopentanone D-ring. As illustrated in Scheme 3.6, the epimerization to yield 13-α-androstenedione is the major reaction in most solvents. However, reduction of the cyclohexanone A-ring competes with the epimerization process in 2-propanol. Irradiation of androstenedione included in NaY gave only the reduction product (Scheme 3.6) and analysis at the initial stages of irradiation did not show the presence of the epimer. While this molecule reacts only from the cyclopentanone D-ring in hexane, no products due to reactions from the D-ring are seen when it is included within NaY. We believe that this is a reflection of the lowering of the energetics of the enone chromophore well below that of the cyclopentanone D-ring and that lowering of the $\pi\pi^*$ excited state of the A-ring is responsible for the changes in reactivity of androstenedione included within NaY. The observed β-selectivity in the reduction of the enone C=C bond can also be rationalized on the basis of changes in the characteristics of lowest excited state. Chan and Schuster have established in the case of 4a-methyl-4,4a,9,10-tetrahydro-2(3H)phenanthrone, a molecule closely analogous to the systems investigated here, that reduction occurs stereospecifically from $\pi\pi^*$ excited triplet to yield a *cis* fused bicyclic ketone[203], corresponding to β-addition in our examples. Based on this analogy, we suggest that the changes in the characteristics of the lowest excited triplet state of the enone chromophore discussed above are responsible for the observed selectivity.

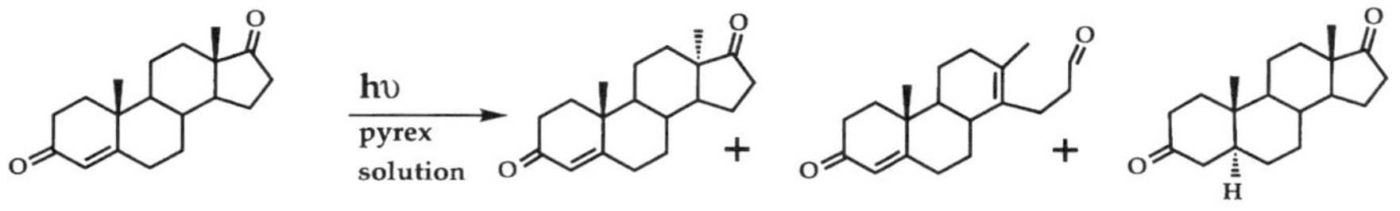

Hexane	40	4	–
Methylene Chloride	45	5	–
Methanol	42	4	–
Cyclohexane	36	3	–
2-Propanol	18	2	14
MCM-41 (Si/Al: >1000; Pure silica)	50 %		-
High Silica Y (Si/Al: >285)	30 %		-
Na X (Si/Al: 1. 2)	-		>90%
Na Y (Si/Al: 2. 5)	-		>85%

Scheme 3.6

During the investigation of the heavy cation effect on photoreactions of dibenzobarrelene and benzobarrelene we made a surprising observation which we believe has to do with the electric field within a supercage[204]. At this stage, our interpretation of the observation should be viewed with caution. For dibenzobarrelene (**4**), the Li^+ and Na^+ exchanged zeolites lead to surprisingly large amounts of triplet photoproduct **6** (67% and 62%, respectively), but the reaction is relatively slow (Scheme 3.7). This "light atom effect" disappears, however, when the complexes are equilibrated with water before use. Although the exact mechanism by which Li^+ and Na^+ bring about triplet chemistry in the case of **4** is not clear, we believe that it may result from the field generated by the light cations, Li and Na. We speculate that the field perturbs the symmetry characteristics of the guest (as discussed in the case of benzene and pyrene) and thus influences the intersystem crossing rates relative to an unperturbed molecule. Consistent with this suggestion, the effect falls off with Na^+ (62%) and K^+ (47%), which are expected to generate a weaker field. Further, where solvation of the cations by water disrupts their strong binding to dibenzobarrelene, the "light atom effect" of Li^+ is completely eliminated (20% triplet), and there is a normal progressive increase in perturbing power for the other cations (Na^+, 28%, K^+, 43%; Rb^+, 69%; Cs^+, 83%; Tl^+, >99%); the classical heavy atom effect becomes increasingly important as the atomic number of the perturbing ion increases.

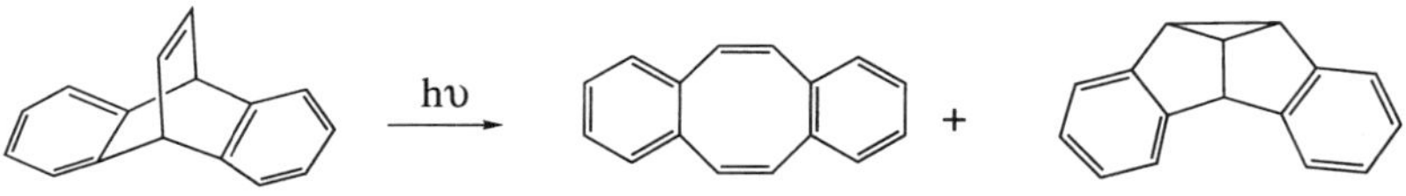

4 **5** **6**

Scheme 3.7

9.2 Stabilization of Hydrogen–Oxygen Contact Charge Transfer Complexes

Evans reported in 1953 that oxygen dissolved in aromatic solvents gives rise to a new absorption at longer wavelength than for the pure aromatic compound[205]. A few years later Munck and Scott reported that even saturated hydrocarbons, alcohols, and ethers, in presence of oxygen, show a longer wavelength absorption[206]. Evans attributed the longer wavelength absorption band to a S_0 to T_1 transition enhanced by dissolved oxygen, whereas Tsubomura and Mulliken attribute the band to a contact charge transfer complex between hydrocarbon and oxygen[207]. As a result of the efforts of a number of groups, it has become clear that the bands from the S_0 to T_1 transition and the transition from ground state solvent–oxygen contact charge transfer (CT) complex to a solvent–oxygen CT state are often in the same spectral region[208–211]. Ogilby *et al.* have established that the solvent–oxygen complex is weakly bound[212–215] and estimated the ΔG for the benzene–oxygen complex to be -0.9 kcal/mol. Thus it has been established that a number of organic molecules form a *weak* contact CT complex with oxygen.

Excitation of the contact CT complex via the CT band is expected to result in a reactive oxygen, most likely a superoxide anion and this has been established to be the case[216–227]. Chien demonstrated that oxidation of olefins, aromatics, and saturated hydrocarbons can be oxidized by exciting the solvent–oxygen CT band[218]. Since then, several Japanese research groups have established that oxidation of olefinic systems, amines, and aromatics can be achieved by irradiating at a longer wavelength region where the parent system does not absorb[216,217,219–225]. In none of these cases did the CT absorption extend beyond 370 nm, thus requiring oxidation to be carried out within the UV region. However, there is a report[228] in which oxidation of 2,3-dimethylbut-2-ene was conducted in an oxygen matrix at 10 K by exciting the CT band at 520 nm. Authors have recorded an absorption spectrum in which the CT band extends to about 550 nm.

Frei was the first to recognize the utility of a zeolite in the context of stabilization of a hydrocarbon–oxygen complex[229–235]. He reasoned that the high electric field present within a zeolite should stabilize the hydrocarbon–oxygen complex and should shift the absorption to longer wavelength. Indeed, Frei *et al.* have demonstrated for a number of cases that the absorption shifts to longer wavelength and the CT band can be reached with visible light. It is important to note NaY and BaY are preferable in this context, the choice being dictated by the field and acidity. Although Ca^{2+}, Mg^{2+}, Sr^{2+} zeolites possess higher electric fields, they are highly acidic. Among the monovalent cation zeolites, only Li^+ and Na^+ generate significant electric fields. Thus, the choice is limited to Li^+, Na^+ and Ba^{2+} exchanged zeolites. Once the CT spectrum is shifted to the visible region, Frei *et al.* utilized the band to carry out oxidation with visible light. (Systems, conditions and products of oxidation are provided in Schemes 3.8 and 3.9). The generality of this approach was demonstrated with several alkenes, alkanes, and aromatics as examples. Oxidation is conducted in zeolite solid matrix and products are monitored *in situ* by IR. The general

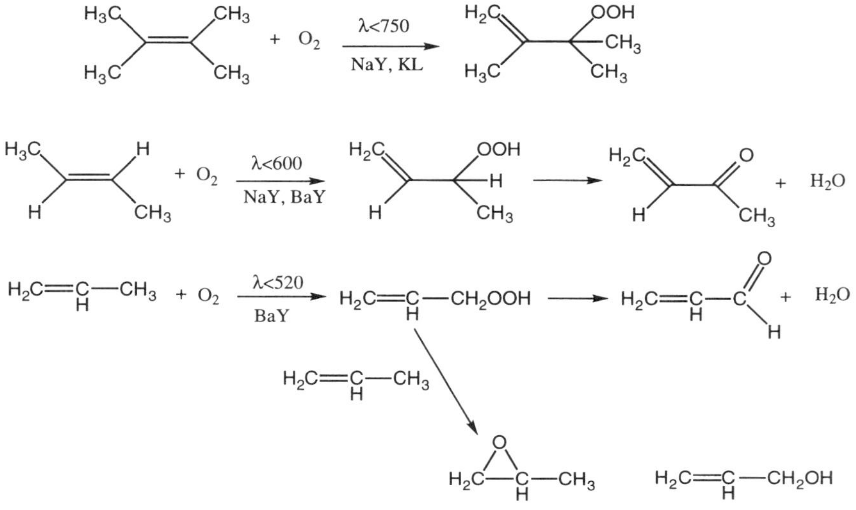

Scheme 3.8

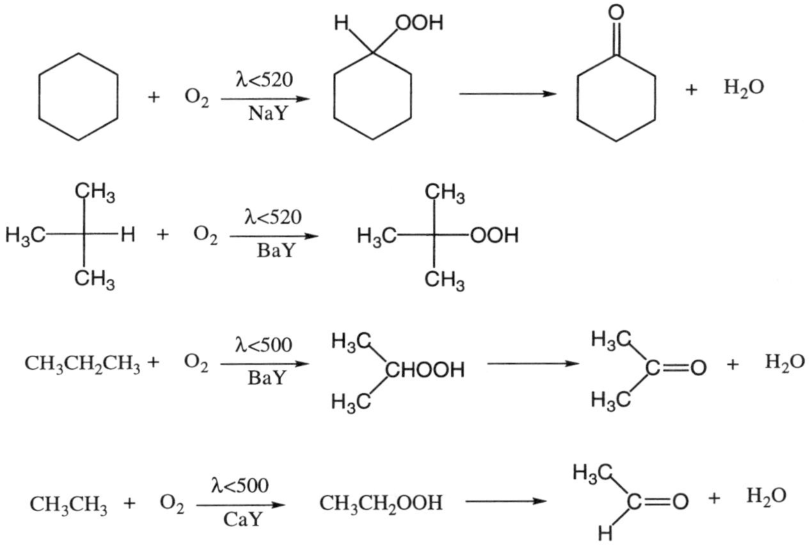

Scheme 3.9

pattern of the mechanism is the same in all cases (Scheme 3.10). The first step is an electron transfer from the organic molecule to oxygen, the second step is a proton transfer from organic radical cation to superoxide anion and the last step is the coupling of the radicals.

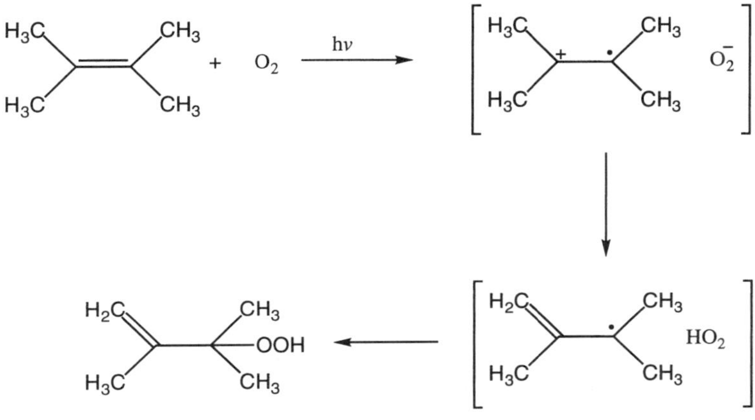

Scheme 3.10

9.3 Stabilization of Donor–Acceptor Complexes

Yoon *et al.* have established that donor–acceptor complexes of aromatics and pyridinium salts are stable within zeolites and that the radical ions generated by excitation of these complexes are longer lived (microsecond) by several orders of magnitude compared to that in solution (picosecond)[236–244]. The electric field present within a zeolite is believed to be the cause for such an unusually slow rate of back electron transfer. The primary step in generating intrazeolite CT complexes involves doping various zeolites (zeolites A, X, Y, ZSM-5, mordenite and L) with organic acceptors of various shapes (Scheme 3.11). These pyridinium salts N-methylacridinium (AC^+), N-methylquinolinium (Q^+), N-methylisoquinolinium (IQ^+), methyl viologen (MV^{2+}), diquat (DQ^{2+}), N-methyl-3-cyanopyridinium) (MCP^+), N-methyl-4-cyanopyridinium (PCP^+), and N-methyl-2-cyanopyridinium (OCP^+) are known acceptors in solution.

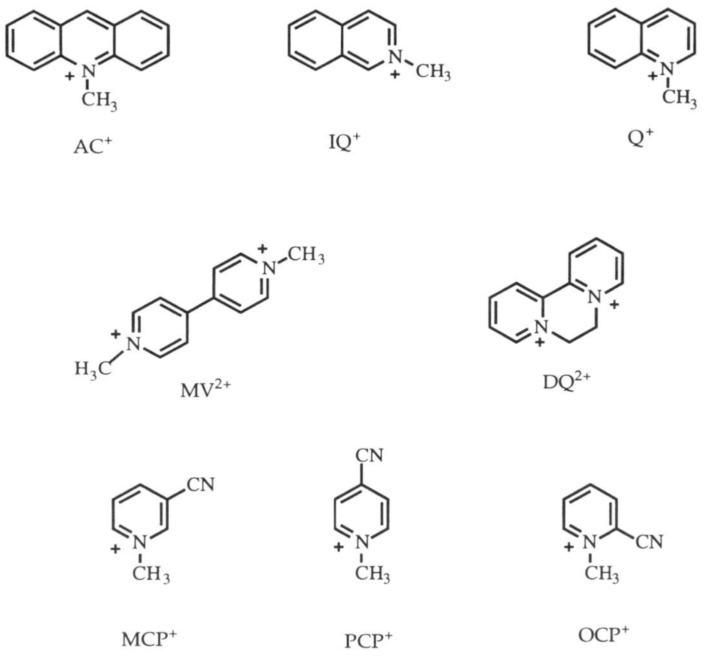

Scheme 3.11

Initially, both di- and tri-pyridinium salts were placed into zeolite Y via aqueous ion exchange, where the degree of doping could be easily controlled. The secondary step in generating the CT complexes within the zeolite supercage involved exposing the doped organic cation Y zeolite to the donor molecule chosen from a series of arenes (Scheme 3.12). First, the water molecules were removed from the doped zeolites by heating and evacuation and then the donors were introduced by exposing the dry acceptor–Y zeolites to hexane solutions containing the arene donors.

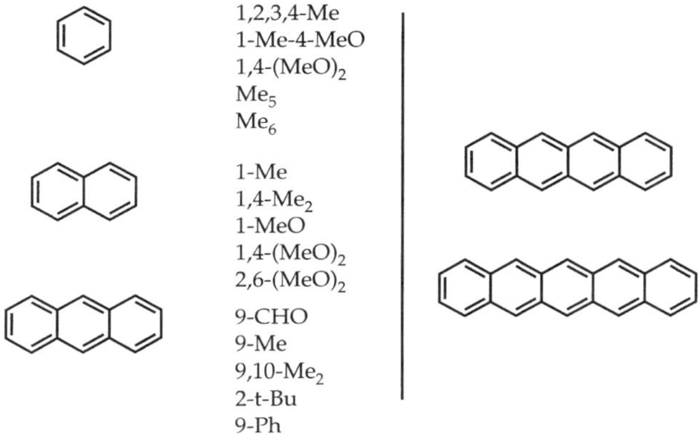

Scheme 3.12

The colorless zeolites immediately developed colors which included yellow, orange, purple, and green depending on the combination of donor–acceptor CT complexes and diffuse reflectance UV-VIS spectra of the colored samples showed new absorption bands. The CT nature of these complexes was assigned by comparison to authentic CT bands obtained in solution. (It is important to note that unlike hydrocarbon–oxygen complexes, the absorption spectra of the CT complexes of aromatics and pyridinium salts did not show bathochromic shift.) Further confirmation was made by the Mulliken relationship in the plot of the absorption maxima of the CT complex ($h\nu_{CT}$) with the ionization potential of the aromatic donor and the reduction potential of the acceptor pyridinium ions (E^{o}_{red}). The intrazeolite formation of CT complexes also serves as a unique visual probe to monitor zeolite shape selectivity. In order for the 1:1 CT complex to form, the zeolite supercage in Y must be able to accommodate both the bipyridinium acceptor and the arene donor simultaneously. Space limitations inherent in the doped zeolite supercage can be seen by the selective shapes of arene donors that can be accommodated. For example, addition of mesitylene, durene, and pentamethylbenzene results in the formation of bright yellow CT complexes with MV^{2+} doped zeolite Y yet addition of hexamethylbenzene causes no color change. All the monosubstituted naphthalene derivatives generate a bright orange colored zeolite. However, a noteworthy size selectivity is observed between the 1,4-dimethoxy naphthalene and the 2,6-dimethoxy naphthalene derivatives. Addition of the former leaves the doped zeolite uncolored whereas exposure to the latter generates a bright orange color. Shape selectivity is also observed for analogous 9-methyl anthracene and 9,10-dimethylanthracene. In this case, the monomethyl derivative is able to form a CT complex whereas the dimethyl homolog is not. Based on the size exclusions described above, a van der Waals "width" of 8 Å is judged to be sufficient to inhibit the complex formation with an immobilized MV^{2+} organic acceptor in Y zeolite.

An important observation was made by Kochi *et al.* during time resolved laser excitation of the immobilized pyridinium/arene/zeolite Y CT complexes by Nd^{3+}: YAG laser (10 ns pulse width) at 355 or 532 nm generated transient spectra[241]. In this investigation, the lifetimes of the CT ion pairs generated in the zeolites were in the microsecond range (0.11–9.2 ms), whereas the analogous irradiations carried out in solution gave transients with lifetimes in the picosecond time scale. Thus zeolite media prolong the lifetime of the transient radical ion pair generated by irradiation of the CT complex by a factor of 10^5–10^6.

The authors have proposed several factors to rationalize the lifetimes of these transients. It is thought that the electrostatic field within the zeolite supercage as well as the adsorption effects of ionic species at the polar aluminosilicate surface are factors that contribute to the slowing of the unproductive back electron transfer. Coulombic interactions between the positively charged transients and the negatively charged framework may explain the incomplete or retarded back electron transfer compared with the homologous solution chemistry. The positively charged transients may be pulled further apart from one another by interacting with the negatively charged framework. Furthermore, the negatively charged microenvironment of the zeolite may also influence the reduction and oxidation potential of the electron acceptor and electron donor respectively.

10. CONSEQUENCES OF LEWIS ACIDITY AND BASICITY: ZEOLITE AS A DONOR AND AS AN ACCEPTOR OF AN ELECTRON

The zeolite framework, as discussed above, consists of Brønsted acid sites (Al—O^+(H)—Si), Lewis acid sites (tri co-ordinated Al), and Lewis base sites (oxygen lone pair in Si—O—Si(Al),)[151–157]. In addition, there are exchangeable cations and depending on the method of activation, there may be extra lattice Al in these structures as well as often mentioned uncharacterized "defect sites". Every one of these sites has been suggested to participate in electron transfer processes within zeolites. In discussing electron transfer processes within zeolites, one should remember that the zeolite interior is highly polar (possesses high electric field) which can favor a charge separation process mediated by an electron transfer and, as a result, a second guest molecule which may not participate in an electron transfer process outside the zeolite may do so within a zeolite.

10.1 Ground State Chemistry

When certain organic molecules are included within zeolites they undergo spontaneous electron transfer to give long lived cation radicals. Stamires and Turkevich were the first to observe this phenomenon with 1,1-diphenyl ethylene, triphenyl amine, quinoline, perylene, aniline, and *para*-phenylene diamine as guests included within NH_4^+Y zeolites[246]. (The radical cations were characterized by their ESR signals.) Following this observation, numerous reports on the generation of cation radicals within mordenite, ZSM-5, X and Y zeolites have appeared. In spite of three decades of research, no clear picture on the mechanism of electron transfer has emerged.

Three types of zeolites, mordenite, ZSM-5, X and Y exchanged with H^+, alkali and alkaline earth cations have been extensively investigated[246–286]. The organic molecules investigated, the zeolites used, and the technique used to establish the formation of radical cations are summarized in Table 3.5. General conclusions from these studies are: (i) the method of activation determines the activity of a zeolite as an electron acceptor; (ii) aluminum on the framework is important; without the presence of aluminum on the framework, the zeolite does not serve as an electron acceptor; (iii) the number of radical cations generated (monitored by spin counts) varies depending on the organic molecule and the zeolite. For example, the original report by Stamires and Turkevich estimates this to be near 1.2×10^{19} spins per gram of NH_4Y in the case of triphenyl amine[246]. A much smaller number of cation radicals ($\sim 1.2 \times 10^{16}$ spins per gram of ZSM-5) was reported in the case of diphenyl polyenes. A recent report from Caldararu states that a maximum of $\sim 1.8 \times 10^{20}$ spins of phenothiazine cation radical was obtained within HY[263]. Although the number of long lived radical cations that can be generated within a zeolite is not high, it is sufficient to carry out spectroscopic studies; and (iv) estimations of the strength of acceptor sites have also been reported. For example, ZSM-5 is capable of oxidizing molecules whose oxidation potential is less than 1.62 eV versus SCE (ionization potential less than 7.0 eV). Similarly, HY and H-mordenite are reported to thermally oxidize systems whose ionization potential is less than ~ 8 eV.

The center that accepts an electron in a zeolite has not been unequivocally identified. There is a general consensus that Al is needed for the zeolite to act as an electron acceptor and that paramagnetic impurities such as Fe^{3+} are not responsible for the generation of radical cations. Although several groups have observed that adsorbed oxygen enhances cation radical formation, this observation has been interpreted differently by these groups. Shih believes that oxygen generates defect centers in zeolites and that these centers are the

Table 3.5 Radical cations investigated in zeolites.

Technique used	Substrate	Zeolite	Reference
ESR	propylene, isobutylene, 2,2,2,-trimethyl-1-pentene, cyclopentene	HZSM-5 ($SiO_2/Al_2O_3 = 69, 60, 38$)	269, 283
ESR	*o*-xylene, *p*-xylene, benzene, anthracene, mesitylene, 1,3,5-trideuterobenzene	HZSM-5	274, 279, 281
ESR	2,3-dimethyl-2-butene	HZSM-5	248
ESR	pentene-2, 2-methyl-2-butene, 4-methyl-2-pentene, pentamethylbenzene	CaY, CeY, Rare-earth-Y	274
ESR	triphenylamine, diphenyl ethylene, aniline, perylene	NH_4Y	246, 250
ESR	triphenylmethane	CaY, CaX, NH_4X	250
ESR	benzene, durene, hezamethylbenzene, pentamethylbenzene	HZeolon ($\approx$HMor)	264
ESR	benzene, biphenyl	NH_4Z ($\approx NH_4Mor$)	265
Diffuse reflectance and ESR	α,ω-diphenyl polyenes	NaZSM-5	280
Diffuse reflectance and ESR	thiophene oligomers	NaZSM-5, Naβ	287
ESR	diphenyldisulfide (thianthrenium)	NaZSM-5	275
ESR	propylanisole	NaZSM-5	284
ESR	acetyl and benzyloximes	ZSM-5	285, 286
ESR	2,5-dimethyl-2,4-hexadiene, 2-methylpropane	HZSM-5	288
Diffuse reflectance and ESR	xanthylium, dibenzotropylium and fluorenium cations	HY, Hβ	276
ESR	cycloalkenes	HMor	268
ESR	aniline	HMor, HY and HX	252
ESR	phenalenyl	M^+Y, M: Li, Na, K, Rb and Cs	258
ESR	9-octalin, piperylene, penta-1, 4-diene, hexa-1, 5-diene	HMor	266, 267
ESR	ethylbenzene	HMor	262
ESR	perrylene, napthalene, tetracyanoethylene, dinitrobenzene	HY	253
Diffuse reflectance	1,6-diphenyl-haxatriene, anthracene, pyrene, perylene	HY, HMor, HZSM-5	277
ESR	trimethylethylene	HMor	258

oxidizing sites[248]. (The structures of these centers are given in Scheme 3.13.) Chen and Fripiat[252] concur with an earlier suggestion by Dollish and Hall that adsorbed oxygen enhances the Lewis acidity of Al sites and that the Al bonded to molecular oxygen is a stronger electron acceptor than each of them individually with superoxide anion identified as a product. There have been suggestions that Lewis acid sites (aluminum) are the electron acceptor sites and this view has been reiterated by a number of groups over a period of three decades. On the other hand, the suggestion by Shih and Rohdes that defect sites generated by oxygen adsorption are the sites of action has not gained any support. Stamires and Turkevich were the first to suggest that Lewis acid sites generated during the activation process are the oxidation sites and this proposition has gained ground over a period of three decades. Hirschler hypothesized that Brønsted acid protons trap the electron from the donor[274] and this mechanistic model has been reinvoked by García and Corma recently[274–276].

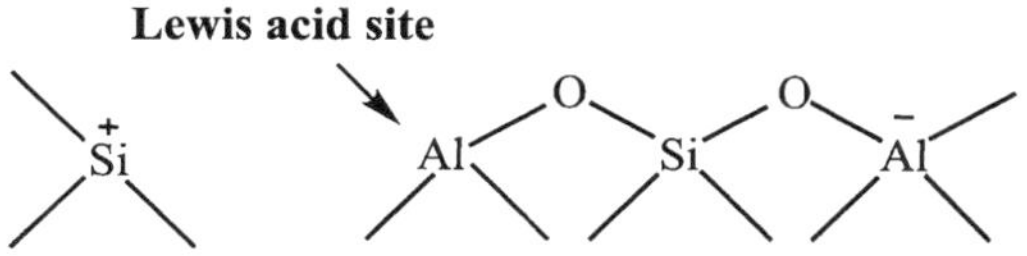

Scheme 3.13

Two recent and exhaustive studies carried out in the laboratories of Fripiat and Thomas identified the Lewis acid sites present in HY, HZSM-5 and H-mordenite as the centers that accept electrons from the adsorbed organic molecule[252,277]. Based on aluminum NMR, Thomas *et al.* suggest that extra lattice aluminum are the Lewis centers and that they play a crucial role in the electron transfer process. These authors have established a relationship between the extra lattice aluminum centers and the cation radical generation in HY, HZSM-5 and H-mordenite as acceptors and diphenyl polyenes as donors (the

$$L + R \longrightarrow L^{-} + {}^{\bullet}R^{+} \quad \text{(radical cation)}$$

$$L^{-} + O_2 \longrightarrow LO_2^{-} \quad (O_2^{-} \text{ in interaction with L})$$

$$L + O_2 + R \longrightarrow LO_2 + R \longrightarrow LO_2^{-} + R^{+\bullet}$$

L = Lewis Acid site
R = Reactant Molecule

Scheme 3.14

number of cation radicals generated in these cases was not measured.) Observations of Fripiat and Chen (on HY and H-mordenite) agree with those of Thomas *et al.*, and, in addition, they have shown from ^{27}Al NMR that extra lattice aluminium is important. Interestingly, they have identified superoxide anion by ESR as one of the products of the electron transfer process. The overall mechanism proposed by these workers is shown in Scheme 3.14.

Reports from the laboratories of García and Corma suggest that Brønsted acid sites are the acceptors of electrons from the adsorbed organic molecule (Scheme 3.15)[275,276]. (The zeolites investigated include HY, Hβ, H-mordenite and H-ZSM5). This disagrees with the generally believed mechanism of electron transfer within zeolites. These authors used xanthene and thiaanthrene as probes and characterized the Brønsted acid and Lewis acid sites by the pyridine adsorption method. They claim that removal of Lewis acid sites does not prevent radical cation formation. García *et al.* state that "oxidation must occur predominantly on the zeolite Brønsted acid sites". This view differs from that of Thomas *et al.* who believe that "Lewis acid sites associated with non framework aluminium atoms are responsible for the formation of cation radicals".

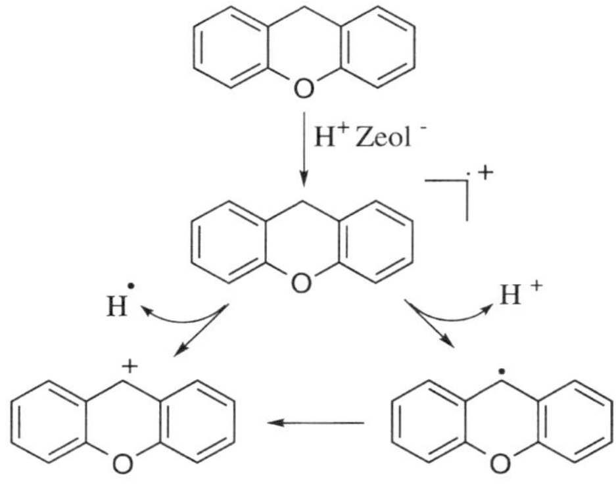

Scheme 3.15

In conclusion radical cations from organic molecules can be generated when they are included in activated divalent (Ca, Mg, Sr) cation, NH_4^+ and H^+ exchanged X and Y zeolites whereas monovalent exchanged X and Y zeolites rarely give radical ions. ZSM-5 zeolites containing aluminium on the framework undergo electron transfer reactions with a number of organic molecule and even Na^+ exchanged ZSM-5 act as electron acceptors. Activated NH_4^+ and H^+ mordenite is also active. While oxygen may not be essential for oxidation, its presence enhances the efficiency of oxidation.

10.2 Excited State Chemistry

Results of studies by a number of groups reveal that zeolites, depending on the nature of the electronically excited guest molecule, serve as an electron donor or as an acceptor[287]. While products of electron transfer-radical cation and radical anion—have been unequivocally characterized, the nature of the zeolitic sites that are responsible for the electron transfer process are less well identified. A final product of electron transfer in a number of cases is cation clusters (e.g. Na_4^{3+})[291].

The phenomenon of photoionization within NaX and NaY zeolites has been examined by a number of research groups with pyrene, anthracene, naphthalene, *trans*-stilbene, biphenyl, N-methyl carbazole and styrene as guests[292–301]. Radical cations, products of electron transfer, have been identified in each case from their transient absorption spectra. The electron that is ejected eventually is captured by Na^+ ions within a supercage and results in a Na_4^{3+} cluster that has been characterized[291] by its transient absorption centered at 550 nm. Although the nature of the ionic clusters has been spectrally characterized, it is not clear whether these are the primary electron acceptors within a zeolite. According to one model, the formation of clusters is considered to be a secondary process (Scheme 3.16). Although formation of clusters such as Na_3^{2+} and Na_2^+ is proposed during photoexcitation of arenes on the external surfaces of small pore size zeolites such as NaA, no detailed information is available on the photophysical behavior of these species. Although photoionization of arenes is observed in other alkali metal cation exchanged zeolites, the role of charge balancing cations in stabilizing the ejected electron is not known.

$$\mathrm{Py}_{\text{(active site)}} \xrightarrow{h\upsilon} \mathrm{Py}^{\cdot +} + e^-$$

$$4Na^+ + e^- \longrightarrow Na_4^{3+}$$

Scheme 3.16

Electron transfer properties of pyrene included in X and Y zeolites have been subjected to in-depth examination by the research groups of Thomas and Hashimoto[289–297]. Excitation of pyrene impregnated in X and Y zeolites led to the formation of transients such as $^1Py^*$, $^3PY^*$ and PY radical cations and anions. The co-existence of $PY^{\cdot +}$ and $PY^{-\cdot}$ was unambiguously proved by oxygen and amine quenching experiments[150]. Detailed analysis of pyrene photophysics suggests that pyrene molecules are present in at least two different zeolite sites, the active and passive[150,292,295]. It is speculated that upon excitation pyrene molecules at the active site eject an electron to form a radical cation whereas those at the non-active site either photoionize by absorbing a second photon or undergo inter-system crossing to give the triplets[292] (Scheme

3.17). Although the nature of the active site has not been identified, we speculate that these sites may be the Lewis acidic tri co-ordinated aluminium sites. It has been proposed that the active (acidic) sites form a ground state CT complex with pyrene[293] as a result of their electron accepting nature. It is the authors' contention that it is the CT complex that undergoes photoexcitation rather than the free pyrene molecules and that the excited CT complexes break up to give the arene radical cations and free electrons. Whether a free electron is formed or the electron is transferred to tri co-ordinated aluminum is not clear. The electron is eventually trapped by Na ions to form Na_4^{3+} clusters. In a few examples (anthracene and pyrene) the ejected electron is also trapped by the aromatic guest molecule to yield anion radicals.

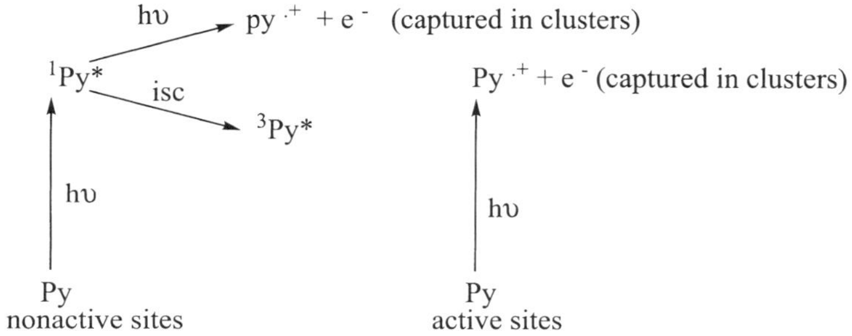

Scheme 3.17

The CT complex model is also supported by the behavior of benzene within ZSM-5. Photoionization of benzene to produce its radical cations in zeolite, ZSM-5 has been reported by Bolshov and Volodin[298] whereas, under normal circumstances, the photoionization of benzene is not possible due to its high ionization potential (9.25 eV), which cannot be achieved by UV light. However the formation of a strong CT complex of benzene with the strong acceptor site of ZSM-5 is proposed to reduce substantially the ionization energy.

The photoionization of stilbenes included within NaX zeolite has been investigated by three independent groups[299–301]. Photolysis of geometrical isomers such as *cis*- and *trans*-stilbenes in zeolites led to the generation of radical cations and trapped electrons. Photoionization is believed to be monophotonic in the case of *trans*-stilbone and biphotone in the case of *cis*-stilbene, yet both isomers yield the same transient signal corresponding to the *trans* isomer. It has been suggested that the formation of *trans*-St^{+} on photoexcitation of *cis*-Stilbene in NaX, proceeds via a rapid *cis–trans* isomerization followed by photoionization by absorption of a second photon. Guest molecules such as styrenes and 4-substituted styrenes produced radical cations in NaY on photoexcitation[302] and these radical cations were found to be relatively long lived compared to acetonitrile solutions.

It is generally accepted that a zeolite can accept an electron from a guest molecule although what part of the zeolite is the acceptor is yet to be conclusively identified. In a number of cases the electron that is lost from a guest is eventually trapped by Na^+ ions, but these sites have not been suggested as the primary acceptors of electrons. A strong contender for the electron acceptor is the tri co-ordinated aluminum site and it is likely that the electron may simply be ejected into the medium as a result of the highly polar nature of the zeolite. It is important to note that Na_4^{3+} clusters have not been identified during ground state electron transfer processes.

Zeolites have also been reported to be good electron donors to a number of acceptors: pyrene, methyl viologen, 1,2,4,5-tetracyanobenzene, pyromelleitic anhydride, dimethyl terephthalate, 1,4-dicyanobenzene and xanthenium cation[303–305].

Two detailed studies, one on pyrene[292] and the other on methyl vioToglen[304,305], have established the importance of the basicity of the oxygen lone pair on the electron transfer process. In the case of pyrene[292], factors such as Si/Al ratio, nature of the charge balancing cations, loading level, post hydration of zeolites and preactivation temperature influence the formation of $Py^{-\cdot}$. It has been found that the efficiency of $Py^{-\cdot}$ formation is more in X zeolites than the analogous Y series. This may be due to the increased basicity

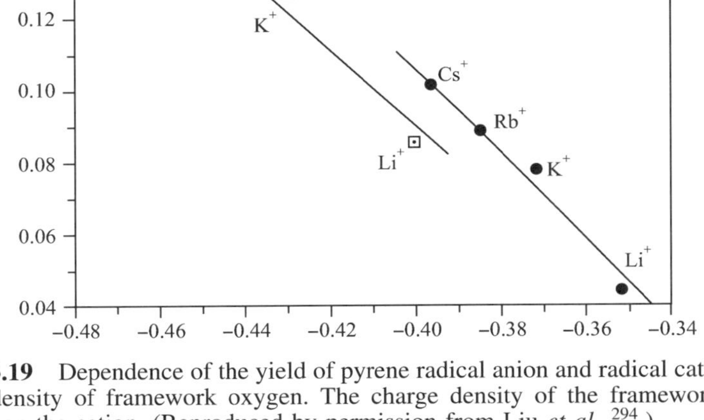

Figure 3.19 Dependence of the yield of pyrene radical anion and radical cation on the charge density of framework oxygen. The charge density of the framework oxygen depends on the cation. (Reproduced by permission from Liu *et al.* [294].)

of the former; charge on the framework oxygen is directly related to the basicity of the zeolites. Accordingly, the yield of the radical anion is expected to increase from Li^+ to Cs^+ within X or Y zeolites as a result of the increasing basicity of sites in the heavy cation exchanged zeolites and indeed, the yield of the radical anion increases linearly from Li^+ to Cs^+ (Figure 3.19).

Hashimoto has suggested that a new mechanism is weekly at higher loading levels of Pyrene as a guest molecule and that $Py^{-\cdot}$ is the result of direct quenching of the excited pyrene by a ground state pyrene molecule (Scheme 3.18). This process should be considered as an exception.

$$Py^* + Py \longrightarrow Py^{\cdot +} + Py^{\cdot -}$$

Scheme 3.18

11. CONSEQUENCES OF BRØNSTED ACIDITY AND BASICITY

Zeolites (H^+, NH_4^+ X and Y) are well established to be good proton donors and are known to protonate even saturated hydrocarbons. Monovalent cation exchanged X and Y zeolites are generally not acidic and (as discussed in Section 7) contain less than one proton per 16 supercages. On the other hand, divalent cation exchanged X and Y zeolites are acidic ($\sim$two protons per supercage) and one needs to exercise caution in using them as reaction media. Recently, these zeolites (M^{2+}, H^+, NH_4^+X and Y) have been used in the context of generating long lived carbocations that have extremely short lifetimes ($<\mu s$) in media other than zeolites. As a result, the photophysical properties of several carbocations have been investigated and these studies are briefly summarized here. A number of olefins (stilbenes, styrenes, 1,1-diarylethylenes, and indene) have been investigated within CaY zeolites[257,306,307] and all cases long lived carbocations are generated and characterized by diffuse reflectance absorption spectra. (The structures of olefins and persistent carbocations are provided in Scheme 3.19.) Diffuse reflectance spectra of the persistent carbocations from indene, vinyl anisole and 1,1-diphenyl ethylene are shown in Figure 3.20. Excited state properties of these carbocations have been investigated and emission spectrum of one of them is provided in Figure 3.21. The excited state singlet lifetime of these cations are in the order of nanoseconds.

It is important to note that when stilbenes, styrenes, 1,1-diarylethylenes, and indene are added to activated CaY, a bright color immediately develops and the color persists for months. For example, when CaY (250 mg) activated at 500°C under aerated conditions was added (after bringing to room temperature) to a hexane solution of indene (20 mg in 5 mL), a dark red color developed immediately and persisted for several months. The red color of the zeolite does

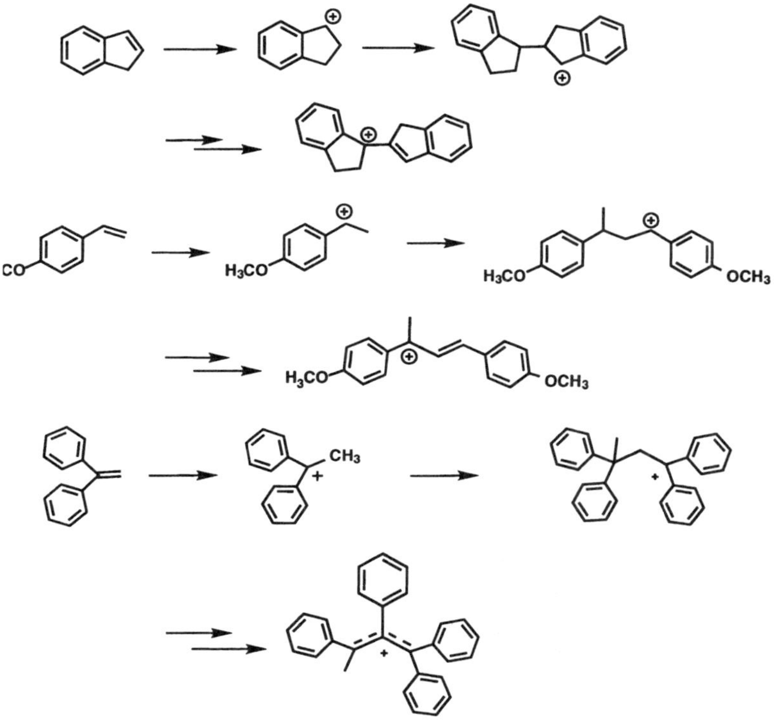

Scheme 3.19

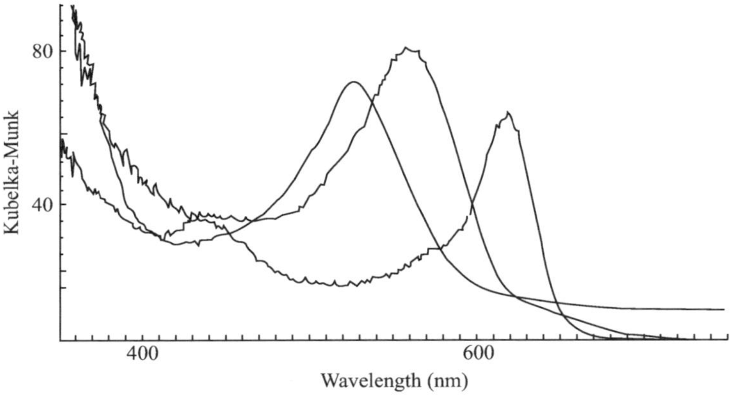

Figure 3.20 Diffuse reflectance spectra of the persistent cations from indene, vinyl anisole and 1,1′-diphenyl ethylene.

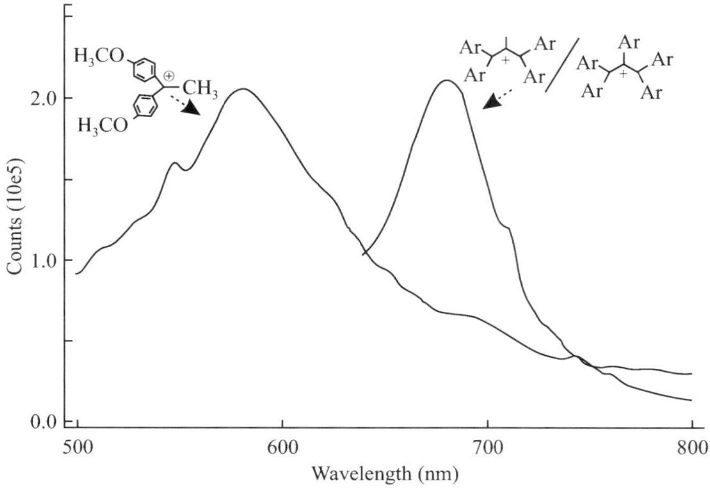

Figure 3.21 Emission from long lived cations from the monomer and dimer cations of 1,1′-diphenyl ethylene.

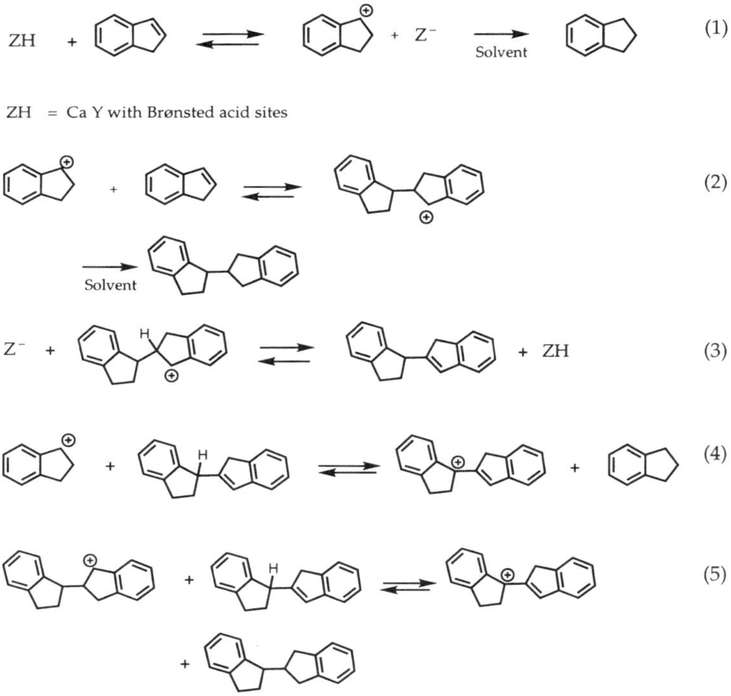

Scheme 3.20

not fade even after 3 months under laboratory conditions and remains colored after refluxing in methanol or water for 24 h. The persistent cation in each case is invariably a dimeric cation and is generated by a series of acid–base reactions listed in Scheme 3.20.

Stable carbocations can also be generated from precursors with good leaving groups. This approach has been utilized by García *et al.* to generate 1,3-diaryl allylic cations[308]. Carbocations have also been stabilized by a "ship in the bottle" synthesis approach[309]. The main theme of these studies is the investigation of the photophysical properties of the transient intermediates. The zeolite matrix helps to stabilize these intermediates and allows their photophysical properties to be investigated at leisure. The photophysics of protonated carbonyl systems and the photochemistry of azo compounds have been investigated within HY zeolites[310–312].

Although of interest, the generation of stable carbocations can interfere with other studies employing a zeolite as a reaction medium. The observation of bright colors upon including an organic molecule into a zeolite is a signal of potential complications from reactive intermediates.

12. HEAVY CATION EFFECT

The electrostatic interaction between exchangeable cations and guest molecules containing π-electrons is quite prevalent in zeolite chemistry. One of the consequences of such an interaction, as discussed above, is polarization of the electron cloud due to a high electric field. However, when the cation is large and has a high atomic number, the field generated by it is not high enough to be

Table 3.6 Estimated spin-orbit coupling constants for metal cations and the isoelectronic noble gas atoms[a].

Cation (isoelectronic noble gas)	Electronic configuration	Spin-orbit coupling constant ξ (cm^{-1})
Li^+ (He)	$1s^2$	0.7
Na^+ (Ne)	$1s^2\ 2s^2\ 2p^6$	520
K^+ (Ar)	$1s^2\ 2s^2\ 2p^6\ 3s^2\ 3p^6$	940
Rb^+ (Kr)	$1s^2\ 2s^2\ 2p^6\ 3s^2\ 3p^6\ 4s^2\ 4p^6$	3480
Cs^+ (Xe)	$1s^2\ 2s^2\ 2p^6\ 3s^2\ 3p^6\ 4s^2\ 4p^6\ 5s^2\ 5p^6$	6080
Li		0.23
Na		11.5
K		38
Rb		160
Cs		370

[a] For cations the values are taken from Anderson and Grissom[326]. For neutral atoms the values are taken from S. L. Murov, I. Carmichael, and G. L. Hug, *Handbook of Photochemistry*, Marcel Dekker: New York, 1993, p. 339.

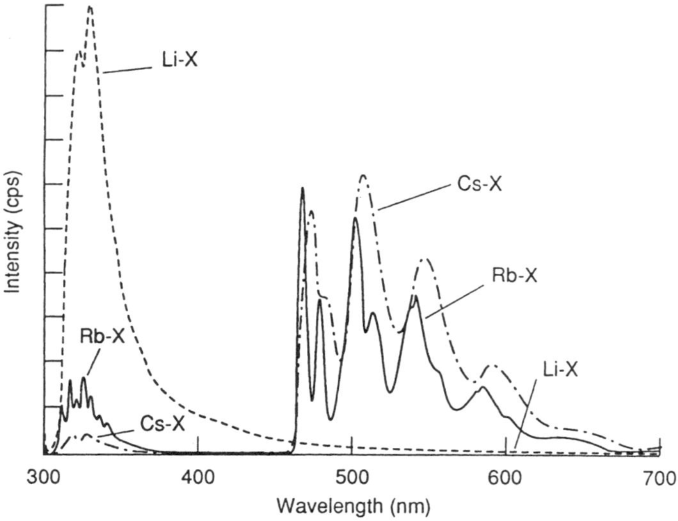

Figure 3.22 Cation dependent emission spectra of naphthalene included within M^+Y zeolites. (Reproduced by permission from Ramamurthyl *et al.*[314].)

important. On the other hand, larger cations such as Rb^+, Cs^+ and Tl^+ possessing larger spin-orbit coupling parameters (Table 3.6), alter the photo-processes by providing an additional mechanism for spin interconversion between excited states. Such an effect is termed the "heavy cation effect".

As shown in Figure 3.22, the emission spectrum of napthalene is profoundly affected by inclusion in X and Y zeolites[314–317]. For low-mass cations such as Li^+ and Na^+, the emission spectra show the typical naphthalene blue fluorescence. However, as the mass of the cation increases (e.g. from Rb^+ to Cs^+ to Tl^+), there is a dramatic decrease in fluorescence intensity and a simultaneous appearance of a new vibronically structured low-energy emission band that is readily identified as the phosphorescence of naphthalene. Table 3.7 lists excited singlet (at 298K) and triplet lifetimes (at 77K) of naphthalene included within various cation exchanged zeolites and it is clear that both these lifetimes are cation dependent. The following observations have been used to conclude that the heavy cation effect is responsible for the enhanced phosphorescence and decreased singlet and triplet lifetimes for naphthalene within K, Rb, Cs and Tl cation exchanged faujasites. It is well-known that the effect of external heavy atom perturbation scales with the square of the perturber's spin-orbit coupling constant, ξ^2 and that a log–log plot of τ_T^{-1} versus ξ^2 should be linear with a maximum predicted slope of unity and as shown in Figure 3.23, the expected dependence is observed. For comparison, the linear relationship observed in two systems, namely 1-halonaphthalenes and 1,5-naphtho-22-crown-6, where the external and internal heavy atom effects,

Table 3.7 Photophysical parameters for naphthalene included within zeolites.

Zeolite host	Triplet lifetime[a] (S)	Singlet lifetime[b] (ns)	P/F[c]
LiX	–	33.0	1.0×10^{-4}
NaX	–	35.4	7.3×10^{-2}
KX	1.72	19.4	0.16
RbX	0.72	2.22	8.1
CsX	0.20	0.23 (87%), 1.87 (13%)	45
TlX	0.0012	–	only P
LiY	–	31.8	1.2×10^{-3}
NaY	–	25.1	1.0×10^{-3}
KY	–	13.8	0.1
RbY	–	3.8	9.0
CsY	–	0.7	60

[a] Lifetime measured at 77 K.
[b] Lifetime measured at 298 K.
[c] Phosphorescence to fluorescence intensity ratio estimated at 77 K; the number is independent of the wavelength of excitation.

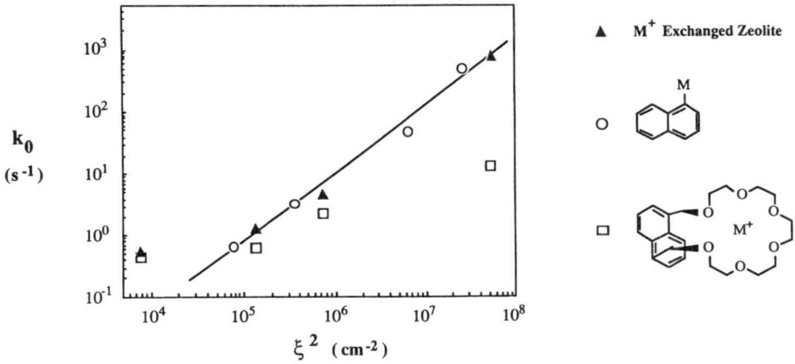

Figure 3.23 Dependence of excited singlet state (S^1) lifetime on the spin-orbit coupling parameter of the cation.

respectively, are presumed to operate is also provided. The magnitude of the heavy atom effect observed in zeolites is significantly larger than that observed for the 1,5-naphtho-22-crown-6 exchanged with heavy atom cations where the cation is rigidly held over the naphthalene π-face. In fact, the zeolite samples show heavy atom effects nearly as large as for a series of 1-halonaphthalenes where the perturbers are covalently attached to the chromophore. This is attributable both to the close approach between naphthalene and the heavy atom which is enforced by the zeolite supercage and to the presence of more than one heavy atom cation per supercage which leads to highly effective concentrations of the heavy-atom cation in the vicinity of the naphthalene molecule. If the heavy cation effect is indeed responsible for the variations in singlet and triplet lifetimes, one would expect a linear relationship between singlet and triplet decays with cation variation which is observed. While the mechanism involved in triplet production with Cs^+ ion has been accepted to be a heavy cation effect, that in the case of Tl^+ ion is still being debated[317].

Heavy atom induced phosphorescence which allows the use of optical detection of magnetic resonance (ODMR) in zero applied magnetic field[318] has provided information concerning the geometry of the cation-aromatic (naphthalene) interaction in X- and Y-type faujasites. The sublevel specific dynamics for adsorbed naphthalene show a distinct increase in relative radiative character and total rate constant of the out-of-plane x-sublevel with increasing mass of the cation perturber. Enhancement from only triplet sublevel T_x is expected for a heavy atom approach along the x-axis. ODMR kinetic results suggest that the naphthalene is adsorbed through its π-cloud at a cation site (Figure 3.24).

The above heavy cation effect within zeolites is found to be general and enhanced phosphorescence is observed for a wide range of different organic guests such as anthracene, acenaphthene, phenanthrene, chrysene, fluoranthene, pyrene, and 1,2,3,6,7,8-hexahydropyrene when included in Tl^+-exchanged faujasites. It is easy to appreciate the potential of the unusual environment of the zeolite when one realizes that even olefins, systems that under normal conditions do not show phosphorescence, emit from their triplet states when included in Tl^+-exchanged zeolites. All-*trans* α,ω-diphenylpolyenes exhibit very low intersystem crossing efficiencies and efficient fluorescence and no

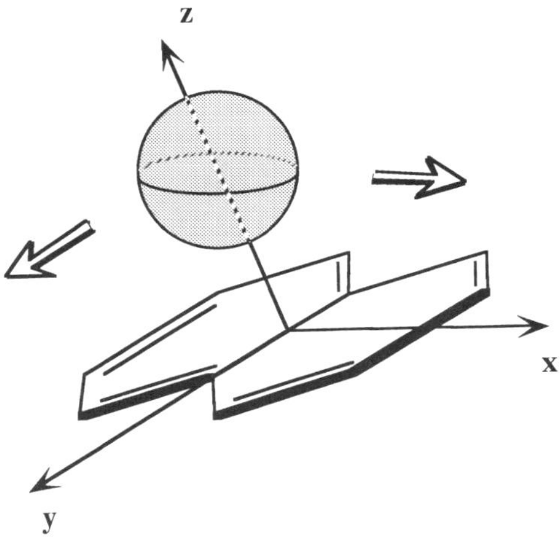

Figure 3.24 Geometry of cation–aromatic interaction on the basis of ODMR results.

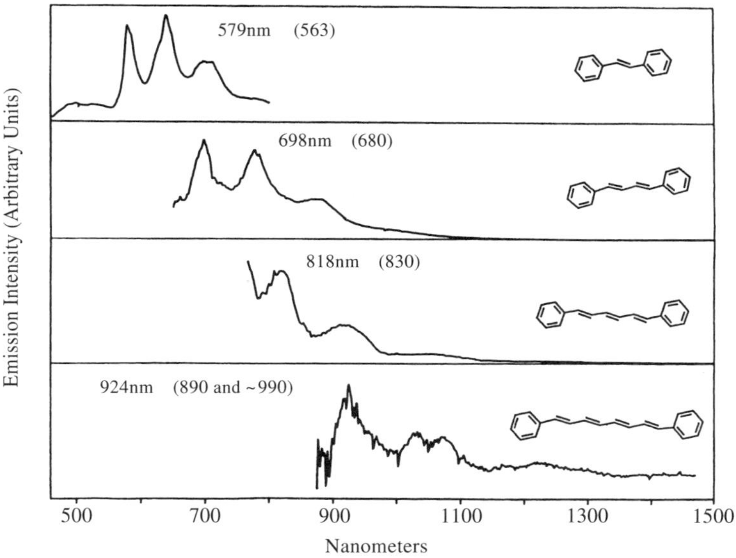

Figure 3.25 Phosphorescence emission from polyenes included within Tl^+ exchanged ZSM-5 zeolite. (Reproduced by permission from Ramamurthy *et al.*[314].)

authentic phosphorescence spectra from these have ever been reported in any conventional media. Figure 3.25 shows the observed phosphorescence of the α,ω-diphenylpolyenes included in TlX. At 77K, a well-resolved structured emission for each of the polyenes with a prominent vibronic spacing of 1200–1400 cm^{-1} as expected for triplet phosphorescence is observed. The singlet–triplet energy gaps ($\Delta_{T_1 \rightarrow S_0}$) obtained from the observed 0–0 lines are in excellent agreement with literature predictions[319]. In addition to the above systems, enhanced cation dependent phosphorescence from xanthone[320,321] and *para*-dimethylaminobenzonitrile[320] included within ZSM-5 and Y zeolites has been observed recently.

The heavy cation effect has also been utilized to control product distribution in both unimolecular and bimolecular photochemical reactions. The photo-behavior of acenaphthylene is unique in that it has been extensively studied in various constrained media and has been subjected to one of the largest heavy atom effects on its dimerization[322]. The irradiation of acenaphthylene in solution yields the *cis* and the *trans* dimers; the singlet gives predominantly *cis* dimer, whereas the triplet gives both *cis* and *trans* dimers in comparable amounts (Scheme 3.21). Photolyses of dry solid inclusion complexes of acenaphthylene in various cation (Li, Na, K, Rb)-exchanged Y zeolites gave the *cis* and *trans* dimers[323] (*cis* to *trans* dimer ratio, relative efficiency of dimerization, relative triplet yields, and triplet lifetimes of acenaphthylene are

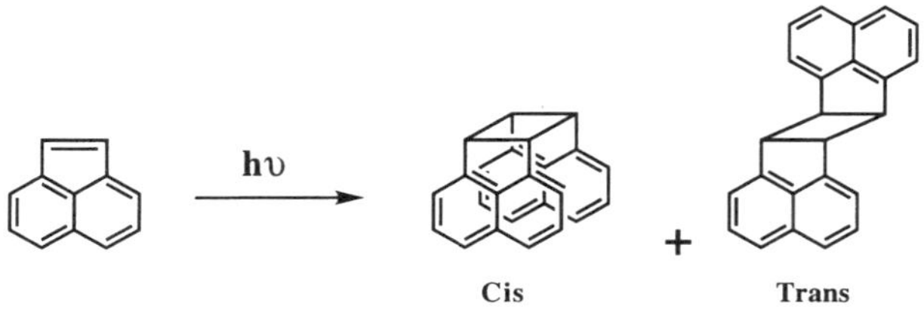

Scheme 3.21

Table 3.8 Cation dependent photodimerization of acenapthylene included in M^+Y type zeolites ($\langle S \rangle = 0.5$).

Zeolite	*Cis/Trans* dimer	Relative efficiency of dimerization	Relative triplet yield	Triplet lifetime (μs)
LiY	25	0.2	–	–
NaY	25	0.2	–	–
KY	2.3	0.4	0.2	9.6
RbY	1.5	1.0	0.5	5.7
CsY	4.2	0.8	0.7	2.1

dependent on the cation as summarized in Table 3.8). The absence of triplet formation in LiY and NaY is consistent with the solution behavior in which the intersystem crossing yield from S_1 to T_1 is reported to be near zero. This observation, as well as the exclusive formation of the *cis* dimer, support the conclusion that the dimerization in the supercages of LiY and NaY is from the excited singlet state. The high triplet yield in K and RbY is thought to be a consequence of the heavy atom effect caused by the cations present within the supercage. The trends observed in the variation of the triplet yield and the triplet lifetime with the increasing mass of the cation is consistent with the expected spin-orbit-coupling-induced triplet formation. Formation of the *trans* dimer (the triplet-derived product) in the cages of K and RbY is in agreement with triplet generation.

Both dibenzobarrelene and benzobarrelene also react differently from their triplet and excited singlet states (Scheme 3.7)[324,325] and are thus ideal examples to test the cation effect on chemical reactions. As illustrated in Table 3.9, the heavy atom response of dibenzobarrelene and benzobarrelene is strong and essentially 100% triplet state behavior is observed with Tl^+ as the heavy atom. There is a clear trend in the triplet product contribution with the spin orbit coupling parameter of the cation (Tl > Cs > Rb > K).

The final example was reported by Anderson and Grissom[326] in which the heavy cation effect was observed in the photofragmentation of 2,3-diaza-

Table 3.9 Photoproduct distribution from irradiation of compounds dibenzobarrelene and benzobarrelene in M^+X and M^+Y zeolites[a].

			Slurry[b]	
Medium	Compound	% Conversion	% COT	%SBV
CH_3CN	Dibenzobarrelene		77	23
Acetone	Dibenzobarrelene		0	100
LiX	Dibenzobarrelene	4 (23)[c]	33 [80][d]	67 [20]
NaX	Dibenzobarrelene	6 (20)	38 [72]	62 [28]
KX	Dibenzobarrelene	17 (30)	53 [57]	47 [43]
RbX	Dibenzobarrelene	16 (26)	25 [31]	75 [69]
CsX	Dibenzobarrelene	24 (41)	13 [17]	87 [83]
TlX	Dibenzobarrelene	98 (88)	<1 [<1]	>99[>99]
Hexane	Benzobarrelene		96	4
Sens[e]	Benzobarrelene		0	100
NaX	Benzobarrelene	77	95	5
KX	Benzobarrelene	59	92	8
RbX	Benzobarrelene	63	86	14
CsX	Benzobarrelene	84	88	12
TiX	Benzobarrelene	90	8	92

COT cyclooctatetraene (single product), SBV, semibulvalene (triplet product).
[a] The product ratios were independent of % conversion within the estimated error limits of ± 2% and represent an average of a least five independent runs.
[b] Slurry irradiations were conducted in hexane for 2 h. Solid state irradiations were carried out for 20 h. Conversions are comparable, since all irradiations were conducted under identical conditions.
[c] Numbers in parentheses are for solid state irradiations.
[d] Numbers in brackets are for zeolites saturated with water.
[e] Acetophenone sensitization in hexane solution.

bicyclo[2.2.2]oct-2-ene (Scheme 3.22). The ratio of the singlet product, bicyclo[2.2.0]hexane, to the triplet product, 1,5-hexadiene, was shown to be dependent on the cation present in Y zeolites. As with other examples detailed above, a clear trend in the increased formation of triplet product is observed when going from Na to Cs (Table 3.10). The effect of the heavy cation, according to the authors, is to enhance the intersystem crossing at the diradical stage rather than at the $^1n\pi^*$ to $^3n\pi^*$ stage.

13. TWO COMPONENT SYSTEMS WITHIN ZEOLITES

13.1 Energy Transfer

Several investigators have explored energy transfer between molecules to expand the range of photochemical processes carried out in zeolite media. Most of these studies were directed towards placing molecules in their excited triplet

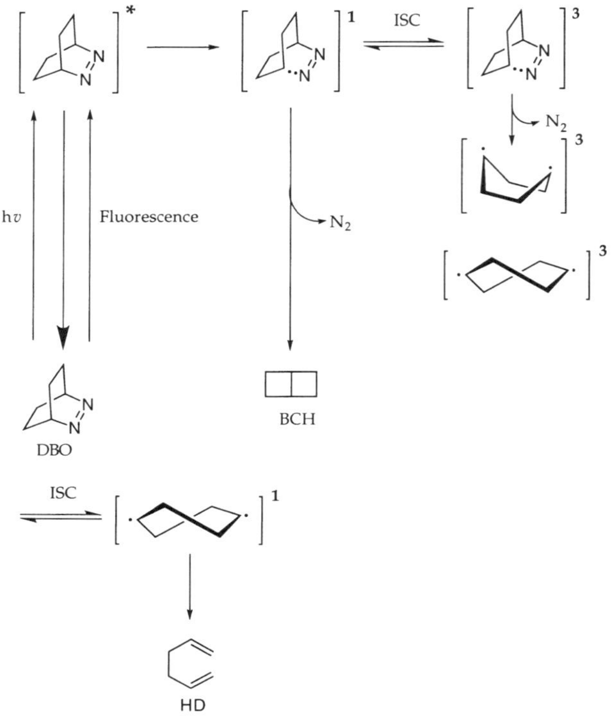

Scheme 3.22

Table 3.10 Heavy cation effect during the decomposition of 2,3-diazabicyclo [2.2.2]oct-2-ene[326].

Zeolite	% of 1,5-hexadiene	% of bicyclo[2.2.2]hexane
LiY	62.6	38.4
NaY	64.8	35.2
KY	65.6	34.4
RbY	72.5	27.5
CsY	74.8	25.2

surface. A general strategy has been to place a well known triplet sensitizer within a zeolite and demonstrate the energy transfer by observing triplet products in a chosen reaction or follow the photophysical properties of the acceptor molecule.

Pettit and Fox were the first to demonstrate that energy transfer can be performed within a NaY zeolite[327]. They used $Ru(bpy)_3^{2+}$ exchanged zeolite Y and showed that it acts as an effective, heterogeneous singlet oxygen sensitizer.

Singlet oxygen thus generated freely diffuses and the photocatalyst is reusable and has a high catalytic turnover. Tetramethylethylene (TME) and 1-methyl-1-cyclohexene (MCH) were used as probes for the generation of singlet oxygen.

Scaiano *et al.* demonstrated the use of zeolite sensitization in one-way isomerization of stilbenes[328]. The concept has been expanded and modified by García *et al.*[329,330]. Scaiano *et al.* used zeolite, silicate (>99% SiO_2), to investigate the *cis–trans* isomerization of stilbene using 3,4-dimethylbenzophenone as sensitizer. Of the reagents, *cis*-stilbene did not fit into the silicalite channels, whereas the *trans* isomer did. Irradiation of 3,4-dimethylbenzophenone and the *cis* isomer in a slurry with silicate generated a higher percentage conversion to *trans* than did the solution analog. It was proposed that energy transfer from DMBP generates a mixture of *cis* and *trans* isomers where the former remains in solution and accessible for future energy transfer events but the latter is incorporated into the channel system of the silicalite away from the sensitizer. In this process, sensitization occurs outside the zeolite and the zeolite, silicate, traps the product *trans*-stilbene as formed making unavailable for further reaction.

García *et al.* have explored and demonstrated the general utility of anchored sensitizers[329,330]. In these studies, 4-aminobenzophenone is anchored within the H^+ zeolites HY, Hβ, HMor and HZSM-5 via protonation of the amino group. The resulting composites were characterized by diffuse reflectance, thermogravimetric analysis, differential scanning calorimetry and FT-IR. Triplet formation and sensitization was demonstrated via time-resolved diffuse reflectance studies. In a chemical demonstration, 1,3-cyclohexadiene was used to probe photosensitization behavior of 4-aminobenzophenone where energy transfer results in $[2+2]$ dimers whereas the electron transfer results in $[4+2]$ dimers. These results indicated that energy transfer is the main mode of action in most of these composites. Further exploration of these 4-aminobenzophenone composite systems by García *et al.* resulted in a shape selective photosensitization of stilbene[330]. *Trans*-stilbene was selectively photochemically isomerized to the *cis* isomer in the presence of 4-aminobenzophenone–HZSM-5 while the *cis* to *trans* isomerization failed. In the cases of the 4-aminobenzophenone–(HY, Hβ, and HMor) composites, no such differentiation in isomerization efficiencies was observed. While there is no question that anchored 4-aminobenzophenone is an excellent triplet energy sensitizer, one has to be cautious in using H^+ zeolites as the proton can initiate unwanted side reactions.

Exploration of the dependence of triplet energy transfer on distance in NaY zeolite was examined by Scaiano *et al.* for the quenching of the triplet state of xanthone by 1-methylnaphthalene[331]. Energy transfer was shown to take place within 100 ns and importantly to occur between nearest neighbor pairs. When the donor–acceptor pair is separated by more than a single cage the energy transfer is believed to be slow.

In our laboratory, energy transfer studies have utilized non-acidic zeolites. Most systems were found to be thermally reactive within H^+ zeolites. In one set of studies, cationic dyes were exchanged into Na^+ zeolites[332–334] and in another, conventional sensitizers were physisorbed within zeolites[335,336]. To investigate the triplet–triplet energy transfer in zeolitic environment, we chose known triplet energy donors (acetophenone, *para*-methoxyacetophenone, and α-aminoacetophenone) and several acceptors that give different products from excited singlet and triplet states (Scheme 3.23)[335,336]. Results from one acceptor, dibenzobarrelene **4**, are provided in Table 3.11.

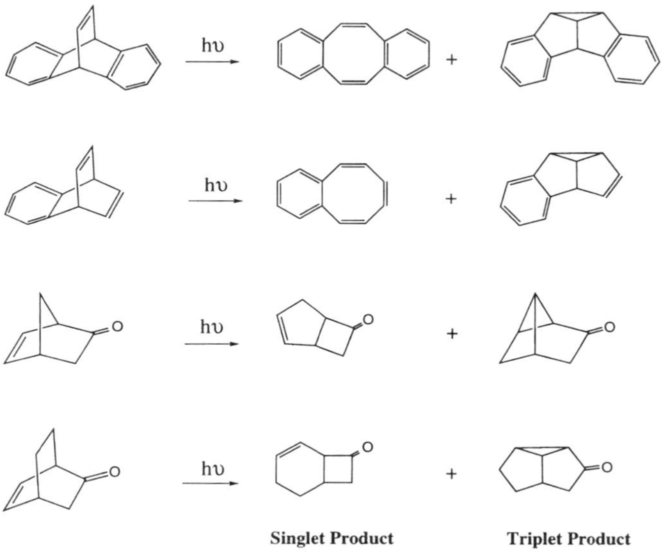

Scheme 3.23

These results shown that T–T energy transfer does indeed occur in the zeolite interior and *para*-methoxyacetophenone is the sensitizer of choice. Loading levels were varied and the extent of T–T energy transfer was shown to increase as the number of molecules per supercage increased. The zeolite KY exchanged with α-aminoacetophenone hydrochloride was reusable. Since the sensitizer is anchored to the surface of the zeolite higher loading levels of the sensitizer (one in 25 cages) was required to achieve efficient sensitization. Sensitizer *para*-methoxyacetophenone had a different with a long triplet lifetime that allowed efficient sensitization even at low loading levels (one in 85 cages) corresponding to a separation of four cages between the donor and the acceptor.

Following the establishment of feasibility of the sensitization, we have explored the possibility of product control in singlet oxygen mediated reactions

Table 3.11 Photoproduct distribution from irradiation of dibenzobarelene and benzobarrelene in $K^{+}Y$ zeolite containing various triplet energy sensitizers[a].

Sensitizer	Compound	Loading level[b]	Slurry[c]		Solid	
			%COT	%SBV	%COT	%SBV
4-Methaoxyacetophenone	A	25	<1	>99	<1	>99
	A	40	4	96	<1	>99
	A	85	13	87	<1	>99
α-Aminoacetophenone HCl	A	25	8	92	5	95
	A	40	16	84	10	90
4-Methoxyacetophenone	B	13	10	90	1	99
	B	25	34	66	2	98
	B	40	51	49	20	80
α-Aminoacetophenone HCl[d]	B	13	39	61	18	82
	B	25	54	46	63	37

A, dibebzobarrelene; b, benzobarrelene; COT, cyclooctatetraene (singlet product); SBV, semibulvalene (triplet product).
[a]The product ratios were independent of % conversion within the estimated error limits of ±2% and represent an average of at least five independent runs.
[b]Loading level refers to the average number of supercages per guest molecule. For example, a value of 25 indicates one probe molecule and one sensitizer molecule per 25 supercages.
[c]Slurry irradiations were conducted in hexane for 2 h. Solid state irradiations were carried out for 20 h. Conversions are comparable, since all irradiations were conducted under identical conditions
[d]In this case, the sensitizer is anchored to the zeolite via an ionic bond, which allows the photoproducts to be removed selectively and the complex reused. No decrease in zeolite efficiency was seen after as many as six runs.

of olefins[333,334]. In these systems, generation of singlet oxygen resulted from the irradiation of a thiazine dye located within the supercages of zeolite Y and X. These dyes were placed in the zeolites via a cation-exchanged process. Further exploration of these photosensitizers showed that highly regioselective hydroperoxidations were possible and that the NaY zeolite supercages acted as "active reaction sites".

In isotropic media, there is little regio- and stereo-control of ene reaction of singlet oxygen with an alkene. By contrast, a supramolecular assembly made up of dye, olefin, oxygen and zeolite provided an unprecedented opportunity to conduct selective oxidations of olefins[337]. A series of olefins, containing several distinct features were loaded into the dye exchanged zeolites prior to irradiation, and after irradiation, hydroperoxides were quenched with triphenylphosphine and extracted with tetrahydrofuran. Examples of the selectivity achieved are shown in Scheme 3.24 from which it is clear that the medium plays an important role in controlling selectivity. The cation plays a critical role in deciding the extent of selectivity with smaller and lighter cations such as Li and Na providing the highest selectivity. At this stage the origin of selectivity is unclear but it is possible that the field provided by the cation and the steric features offered by the zeolite surface are important.

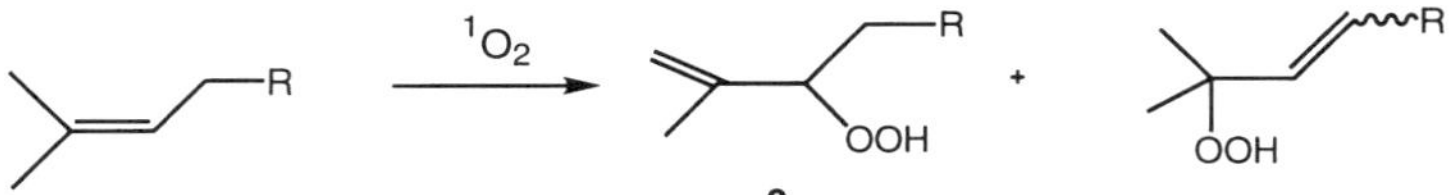

	1	2	3
R = CH$_3$			
Thionin/ CH$_3$CN		40	60
NaY/Thionin		99	1
R = CH$_2$CH$_2$CH$_3$			
Thionin/ CH$_3$CN		40	60
NaY/Thionin		99	1
R = CH$_2$(CH$_2$)$_3$CH$_3$			
Thionin/ CH$_3$CN		40	60
NaY/Thionin		99	1

Scheme 3.24

Thus, there have been three approaches to carry out sensitization within zeolites: (i) anchor the sensitizer within H^+ zeolites; (ii) exchange cationic sensitizers within M^+ zeolites; and (iii) adsorb the sensitizer within a zeolite. The first two approaches provide a reusable sensitizer–zeolite composite whereas the last one will allow only one time use. In using the first strategy it is important to ensure that there are no thermal Brønsted acid catalyzed reactions.

13.2 Electron Transfer

Three research groups (Dutta, Mallouk and Kincaid)[338–347] have exploited the zeolite framework to control the back electron transfer process in the donor-acceptor system tris (2,2′-bipyridine) ruthenium II ($Ru(bpy)_3^{2+}$) and methyl viologen (2^+). They have established conditions under which the radical ions generated by the photoelectron transfer process live for several hours. In preliminary studies, Dutta *et al.* explored the use of the zeolitic, three-dimensional framework to assemble the redox system[338]. Resonance Raman spectroscopy was used to explore the photoelectron transfer initiated by visible light in the $Ru(bpy)_3^{2+}$-MV^{2+}-hydrated NaY zeolite system examined. The electron donor ($Ru(bpy)_3^{2+}$), was synthesized inside the supercage of faujasite Y then MV^{2+} was ion exchanged. Although $Ru(bpy)_3^{2+}$ is trapped inside the supercages, it still interacts with the acceptor through the 7 Å windows. On the other hand, MV^{2+} is able to migrate through the zeolite channels and cages.

Upon photoexcitation of $Ru(bpy)_3^{2+}$ in this system, the $MV^{+\cdot}$ radicals are generated and are formed in neighboring cages. The $MV^{+\cdot}$ was found to be stable on the hour-long scale and decayed via the back electron transfer to $Ru(bpy)_3^{2+}$. The resonance Raman spectra of the $MV^{+\cdot}$ species demonstrated that intercage electron transfer occurred in zeolites. These results showed that for $Ru(bpy)_3^{2+}$ trapped inside the supercage, the closest electron acceptor cation must be in a neighboring supercage separated from the Ru^{2+} center by 7–10 Å.

Mallouk and Kim also explored the $Ru(bpy)_3^{2+}/MV^{2+}$ in zeolite Y system using steady-state and time-resolved spectroscopic techniques[339]. They found that the average charge separated lifetime was on the order of tens of microseconds, and that the quantum yield for charge separation was in the range of 7–9%. Further investigations by Dutta *et al.* for the same system monitored by time-resolved Raman spectroscopy confirmed the above findings and demonstrated that there are two pathways for electron transfer[340]: back electron transfer from $MV^{+\cdot}$ to $Ru(bpy)_3^{3+}$ (observed to be on the microsecond time scale) and a slower process that leads to charge separation and stabilization of the redox pairs for longer periods of time. Because MV^{2+} can diffuse thoughout the cages, it is believed that these acceptors orient themselves in order to facilitate electron hopping along adjacent molecules leading to charge separation. Along with the electron migration, positive charges reposition themselves in order to maintain charge neutrality. The alkali cations, such as Na^+ can accomplish this feat. The importance of migration can be seen when one compares the behavior of $MV^{+\cdot}$ in the Li^+ exchanged zeolite with that in the Na^+ zeolite. The rate of generation of $MV^{+\cdot}$ is slower in Li^+ exchanged zeolite than it is in the Na^+ exchanged zeolite as a result of the increased coulombic interaction between the Li^+ cation and negative three-dimensional framework. Work by Dutta and Borja exploited the ion exchange capabilities of zeolites to move the photogenerated $MV^{+\cdot}$ outside the zeolite supercage and into supernatant solution thereby further separating the charge[341,343]. In order to achieve this, the photolysis of $Ru(bpy)_3^{2+}$-Y zeolite system was carried out in a 0.1 mol/dm^3 solution of MV^{2+} monitored by UV-VIS spectroscopy for the $MV^{+\cdot}$ (Scheme 3.25).

Because of the importance of zeolite as ion exchanger, the thermodynamics of the exchanges was investigated by Mallouk *et al.* using electron and energy transfer as probes[342]. Photochemically and electrochemically induced reactions were studied by emission spectroscopy. Electron (MV^{2+}) and energy ($Os(bpy)_3^{2+}$) transfer quenching of the MLCT excited state of $Ru(bpy)_3^{2+}$ provide a convenient method to probe the kinetics of interparticle exchange. $Ru(bpy)_3^{2+}$ was synthesized in the zeolite whereas MV^{2+} and $Os(bpy)_3^{2+}$ were mobile ions. Under conditions of low ionic strength (0.1–0.3 mM electrolyte), the time constants for interparticle exchange were found to be significantly longer than those typically associated with either photoinduced charge

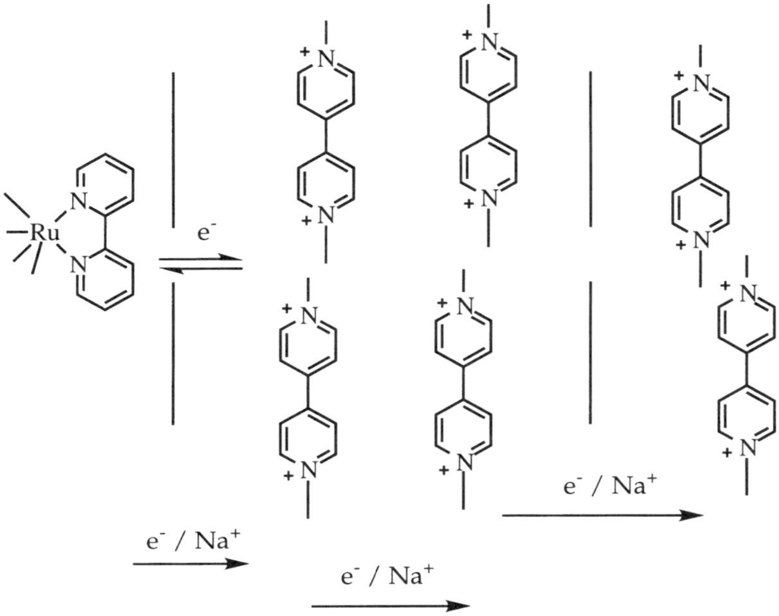

Scheme 3.25

separation or cyclic voltammetry (10^3–10^5 s versus 10^{-1}–10^1 s). These results are consistent with a direct intrazeolite electron transfer model but not with mechanisms that invoke exchange of electroactive species into solution prior to electron transfer.

Dutta and Borja also achieved efficient donor–acceptor electron transfer with a slow back electron transfer and easy separation of products employing a trimolecular/zeolitic redox system[343]. Their experiment consisted of the photosensitized donor $Ru(bpy)_3^{2+}$ trapped in the zeolite supercage and the neutral zwitterionic acceptor (propylviologen sulfonate (PVS)) placed in solution with the electron transfer mediated by DQ^{2+} (N,N′-dimethyl-2,2′ bypyridinium) loaded via ion exchange. Although results indicated that there was a low quantum yield for this system (5×10^{-4} for $PVS^{-\cdot}$), there are advantages to this relayed electron transfer because the ultimate electron acceptor (DQ^{2+}) is free in solution and physically separated from the donor.

Sykora and Kincaid expanded on this ternary redox system ($Ru(bpy)_3^{2+}$/ DQ^{2+}/PVS)[344], focusing on the development of synthetic methods to elaborate the organizational possibilities inside the three-dimensional zeolite framework. They believed that they could improve the quantum efficiency of the electron transfer by organizing and arranging the active components (donor, acceptor, and sensitizing intermediate molecule) so that they would occupy adjacent cages. Two complexes were synthesized and entrapped in adjacent cages; $Ru(bpy)_2(bpz)^{2+}$ and $Ru(mmb)_3^{2+}$ where bpz is 2,2′-bipyrazine and mmb is 5-monomethyl-2,2′-bipyridine. Their strategy illustrated in Scheme 3.25 succeeded in separating the ion pair. Reductive quenching of the oxidized sensitizer by adjacent $Ru(mmb)_3^{3+}$ is efficient and results in the ion pair $Ru(mmb)_3^{3+}/DQ^+$ separated by the insulating reduced sensitizer $Ru(bpy)_2(bpz)^{2+}$.

Finally, exploitation of the photosynthetic assembly mimics have led Ledney and Dutta to use the charge separated states generated with the redox systems to do useful chemistry[345] in the oxidation of water to dioxygen by $Ru(bpy)_3^{2+}$ encapsulated inside Y Zeolite. The mechanism of this reaction has been investigated with the use of resonance Raman, diffuse reflectance and electron paramagnetic resonance spectroscopies.

Mallouk *et al.* investigated zeolite (Y, L, mordenite) based photoelectron transfer reactions[347] using a covalent aliphatic linkage between the donor and acceptor molecules [D-$(CH_2)_n$-A] ($n = 2$–5, 7, 8) where D and A are (2,2′-bipyridine)$_2$Ru(4-Ch$_3$-2, 2′-bypyridine-4′) and (4, 4′-bipyridinium-CH_3)$^{4+}$ respectively (Scheme 3.26). From solid state CPMAS spectra of ^{13}C labeled compounds, it was established that the D–A molecules occupy surface sites where A is in the zeolite channels but size excluded D remains exposed in solution. The rate of electron transfer from D to A decreased with increasing spacer length and the back electron transfer was 10^5 times slower for the D–A diad on the surface versus the D–A in solution. Expansion of the investigation involved addition of size excluded electron donor (D′) to the suspension to give triad (D′–D–A). Long lived charge transfer separated states were formed between D′ and A (where one of the D′ was the promethazine cation (PMZ^+)).

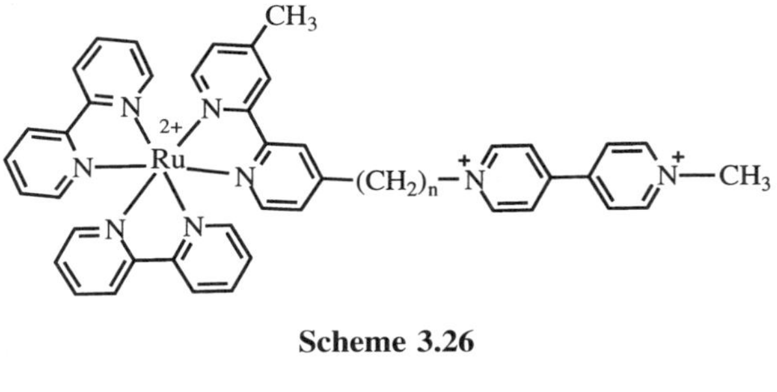

Scheme 3.26

Not only have zeolites been used to investigate charge separation in redox systems but they are how being employed as a methodology to modify and control the photophysical and photochemical behavior of adsorbed organic guests. Recent examples demonstrate the exploration of this area.

García *et al.* have used the strategy of synthesizing the photosensitizer molecule within the zeolite[348,349]. They have encapsulated in zeolite Y the 2,4,6-triphenylpyrylium cation (TPP^+) as well as a number of trityl cations within H^+ zeolites (Scheme 3.27) where they act as heterogeneous photosensitizers to promote the sensitized dimerization of 1,3-cyclohexadiene (CHD) and isomerization of *cis*-stilbene[350,351]. Because all zeolites incorporated with electron transfer sensitizers contain Brønsted acid sites, one needs to be careful to avoid proton transfer chemistry in these systems.

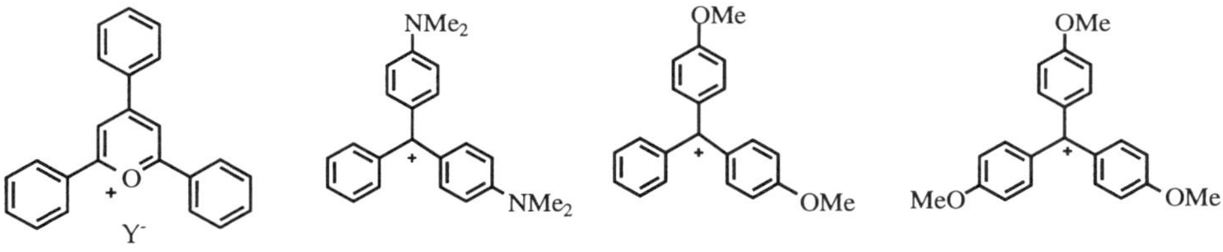

Scheme 3.27

Other intrazeolite photochemical reactions involved photosensitization of amines via acetophenone in the zeolite NaY[352]. Laser flash photolysis of acetophenone (ACP)/aliphatic amine (AA) mixtures in NaY cavities generates triplets, the ketyl radical of ACP and dimers of the amine radical cations. Irradiation of the same mixture in solution (acetonitrile/water) generates only ACP triplet and $ACP^{-\cdot}$. It is thought that the amine cation radical interacts with a second neutral amine to give the dimeric species (480–580 nm) that has never been seen in solution except for cyclic amines where the nitrogen centers are in close proximity. Product studies from the NaY zeolite system reveal the formation of hydrazine products.

Other photochemically induced electron transfer reactions involve the oxidation of stilbenes (*trans*-stilbene) (TS) and 4,4′-dimethyoxy stilbene (DMS)) included in zeolites[353] in which the alkali cations present in NaY and NaX had been exchanged with organic cations (methylene blue, thionin and methylene green). These thiazine dyes served as sensitizers for the oxidation reaction (< one dye molecule per ~25 cages). The reaction yield the corresponding aldehyde as the main product. The authors believe that an electron transfer from either TS or DMS to the singlet excited thiazine dye initiates a non-chain oxidation process. The oxidation was shown to be more efficient in the zeolite than in homologous media as a result of "localizing and confining" the sensitizer and oxidant. The efficiency of the oxidation was also found to be dependent on the cation (more efficient in NaY and inefficient in CsY and RbY).

Brancaleon *et al.* investigated the lifetimes and reaction pathways of radical cations of aryl alkenes (*trans*-anethole, 4-vinylanisole, phenyl vinyl ether, and two indenes) generated by photosensitized electron transfer (Scheme 3.28)[354]. Product studies, fluorescent spectroscopy and diffuse reflectance laser flash photolysis were used to monitor these radical cations inside NaX.

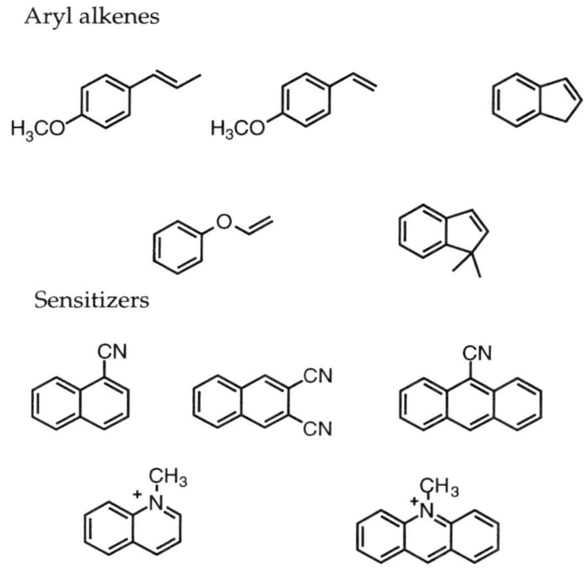

Scheme 3.28

Using ionic and cyanoaromatic sensitizers, dimeric cyclobutane products were obtained in both solution and zeolite media. However the distribution of dimers were different and the *cis/syn* cyclobutane was formed preferentially in the zeolite. The zeolite supercage is more conducive to a *cis/syn* isomer which has a spherical shape similar to the geometry of the cage rather than the linear, elliptical *trans/anti* isomer.

13.3 Chirality Transfer

In the past, chiral solvents, chiral auxiliaries, circularly polarized light, and chiral sensitizers have been utilized to conduct enantioselective photoreactions. Of the many organized media, the crystalline state and solid host–guest assemblies have provided the most encouraging results. In the context of asymmetric catalysis, zeolites offer advantages over other organized assemblies.

Two approaches have been used to achieve chiral induction in the crystalline state. In one, due to the Weizmann Institute Group, the achiral reactant is crystallized into a chiral space group[355], although possibility of this happening is limited and therefore this approach is less general. Alternatively, Scheffer *et al.* used an ionic chiral auxiliary to effect a chiral environment[356]. This approach is limited to those acidic molecules that form crystallizable salts with chiral amines or vice versa. Another moderately successful approach used organic hosts that contain chiral centers (deoxycholic acid, cyclodextrin, 1,6-bis(*o*-chlorophenyl)-1,6-diphenyl-2,4-diyne-1,6-diol) or crystallize in a chiral space group (urea)[357]. The success of this approach is limited to guests that can form solid solutions with the host without disturbing the host's macrostructure. Zeolites, on the other hand, can include a large number of different types of molecules, with the only limitation being that the dimensions of the guest must be less than the pore dimensions of the zeolite.

The reactivity of molecules in the crystalline state and in solid host–guest assemblies is controlled by the details of molecular packing. Consequently, chemical reactivity in the crystalline state continues to evade prediction. On the other hand, guest molecules present within zeolites possess greater freedom than in crystals/solid host–guest assemblies and less freedom than in isotropic solvents. This permits molecules within zeolites to react more readily than in crystals/solid host–guest assemblies and more selectively than in solution.

The ideal approach to achieve chiral induction within zeolites is to make use of "chiral zeolites" as the medium for a photoreaction. There are currently no zeolites that exist in a stable chiral form. Only zeolite beta and titanosilicate ETS-10 have unstable chiral polymorphs, but they have not been isolated in pure enantiomorphous forms. Although many other zeolites can, in theory, exist in chiral forms (e.g. ZSM-5 and ZSM-11), as far as we know, none has been isolated in such a state. Thus *at the present time*, the use of chiral zeolites to achieve asymmetric induction is not an option. In the absence of readily available chiral zeolites, an asymmetric environment within zeolites has been created by the adsorption of chiral organic molecules. Thus one will have to resort to the two component approach to achieve enantioslectivity in photochemical reactions within zeolites.

The use of functionalized zeolites (chiral inductors anchored through either covalent or ionic bonds) may not be limited to a single use. The effectiveness of zeolites as reusable chiral catalysts will be determined by the stability of the chiral inductor to light and to leaching by solvents. Current developments in the area of zeolite membranes and zeolite thin films provide considerable opportunity for the development of zeolite based chiral catalysts. In summary, while crystalline and host–guest assemblies have been very useful to conduct enantioselective photoreactions, their general applicability has been limited. We believe that zeolites offer a solution to this limitation. With this rationalization in mind, we have been conducting studies aimed at achieving chiral induction during a photochemical reaction.

The feasibility of asymmetric induction within zeolites has been established with two ketones–*trans*-4-*tert*-butyl-1-methylcyclohexyl aryl ketone and 1-adamantyl-*p*-carbomethoxyacetophenone[358,359]. Results on these two systems are briefly summarized in Schemes 3.29 and 3.30. The fact that non-zero enantioselectivity has been obtained with two systems suggests that there is control on the chirality of the products of the type II reactions. In addition the observation that the chiral induction can be obtained within zeolites but not in isotropic solution highlights the uniqueness of zeolites as media for conducting enantioselective reactions.

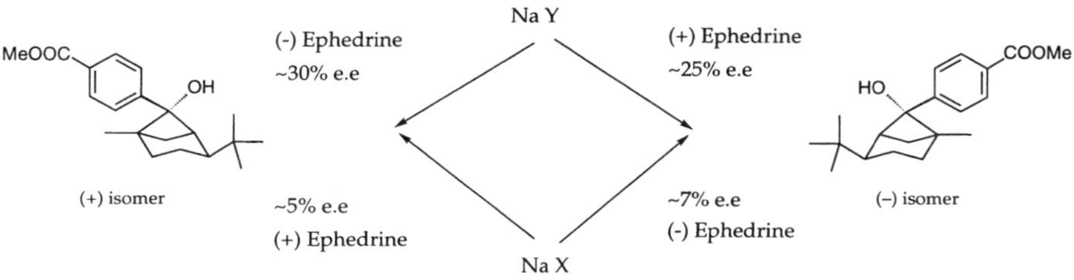

Scheme 3.29

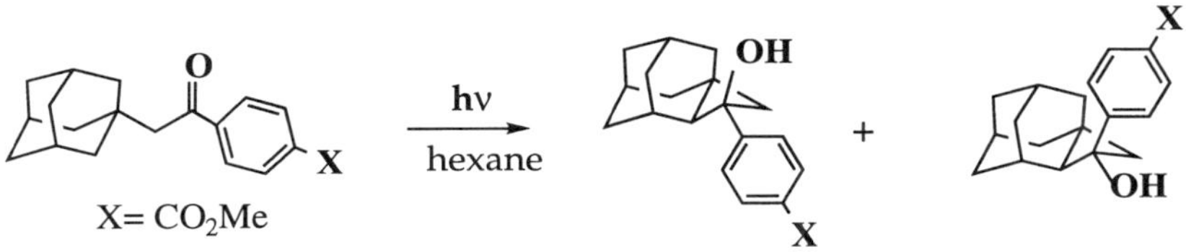

Hexane	0.4 (e.e.: 0)	1 (e.e.: 0)
Na Y	1 (e.e.: 0)	1 (e.e.: 0)
Na Y / (-) ephedrine	0.7 (e.e.: 35%)	1 (e.e. 3%)

Scheme 3.30

Another class of reactions that has been investigated in zeolites is the electrocylization of cyclic polyenes. Preliminary results reported from our laboratory on tropolone methyl ether have provided promising results (Scheme 3.31)[360]. Once again, in solution the e.e. is zero, whereas the extent of e.e. within zeolites depends on the nature of the chiral inductor, the nature of the zeolite and the temperature of irradiation (Scheme 3.32).

Strategies of heavy cation effect and chiral induction have been combined to achieve enantioselectivity during photoisomerization of benzonorbornadiene[361]. Benzonorbornadiene yields benzonorbornene via the triplet excited state and as expected, irradiation of benzonorbornene within NaY did not yield any products. The heavy cation strategy led the authors to use TlY and this resulted in the formation of benzonorbornene. To effect chiral induction in this process a number of chiral agents were used. Of these ephedrine hydrochloride gave the highest e.e. (15%; Scheme 3.33).

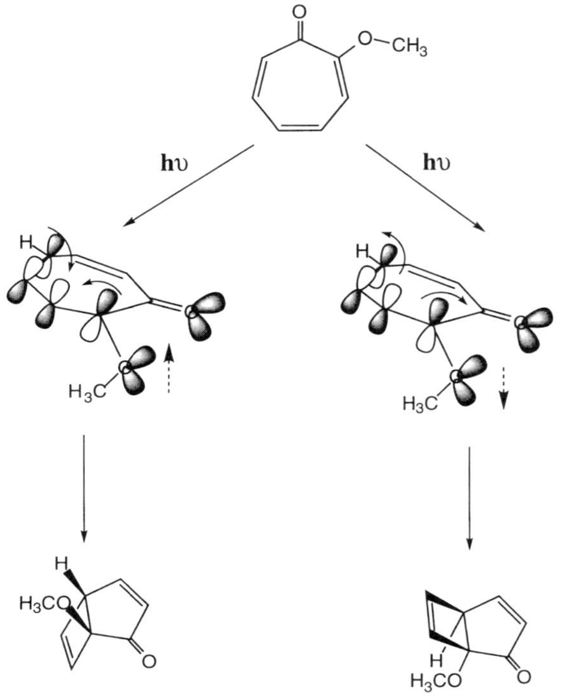

Scheme 3.31

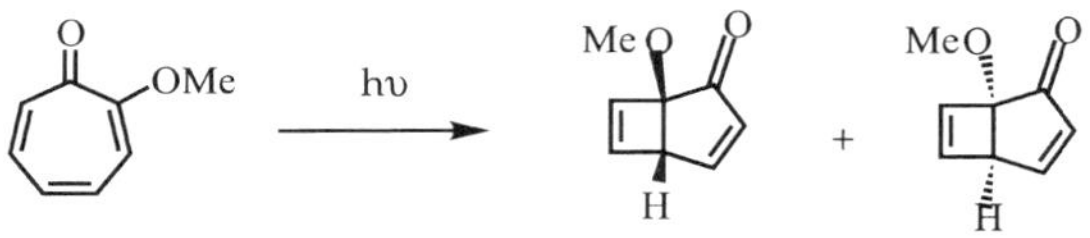

Zeolite/(-) norephedrine	Percent ee
LiY	15
NaY	35
KY	31
RbY	40
CsY	20

Scheme 3.32

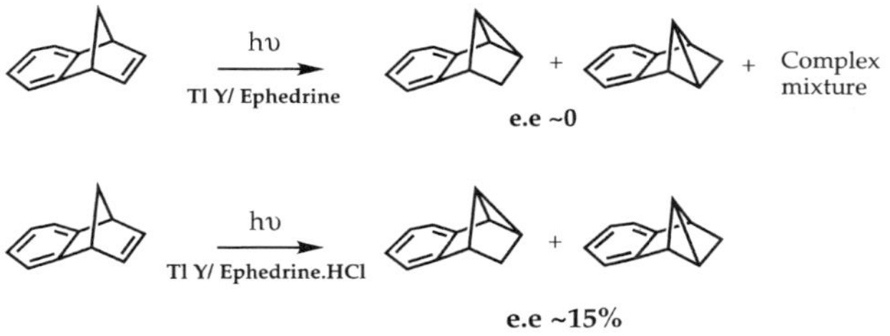

Scheme 3.33

14. CONCLUDING REMARKS

In this chapter we have highlighted the internal structure and characteristics of zeolites, especially X and Y. Cations control the Lewis and Brønsted acidities and basicities and field present within a supercage. Cations behave as Lewis acids and interact with π-electrons of included organic molecules influencing their location, adsorption mode and conformation. Such controls can be translated into products, and examples of these are presented. Proton transfer and electron transfer are the two fundamental reactions that take place within a zeolite and most often the zeolite matrices themselves participate in these reactions. The ability to include two molecules within a supercage provides an opportunity to establish a communication between two molecules in an excited state, resulting in energy transfer, electron transfer, and chiral induction. As yet, techniques to place two molecules in a specified arrangement within a zeolite are lacking. Techniques to specifically functionalize the supercage with active organic molecules need to be developed. "Naked" cations within a zeolite provide an unprecedented opportunity to explore reactions catalyzed by cations as these cations are capable of polarizing π-electrons and thus may bring about reactions that are normally considered less likely in solution. This aspect has received less attention.

15. ACKNOWLEDGMENTS

VR thanks the Division of Chemical Sciences, Office of Basic Energy Sciences, Office of Energy Research, US Department of Energy for support of the zeolite program in his laboratory. VR has benefited immensely through periodic discussions with Professors C. P. Grey (SUNY, Stonybrook), J. R. Scheffer (UBS, Vancouver), N. J. Turro (CU, New York) and R. G. Weiss (GTU, Washington, DC) and Drs D. R. Corbin (Du Pont, Wilmington), W. E. Farneth (Du Pont, Wilmington), H. Frei (UC, Berkeley) and L. J. Johnston (NRC, Ottawa). VR appreciates their generosity in sharing their knowledge with him.

16. REFERENCES

1. G. M. J. Schmidt *et al.*, *Solid State Photochemistry*, ed. by D. Guinsburg. Verlag Chemie, Weinheim (1976).
2. J. K. Thomas, *Chemistry of Excitation at the Interfaces*. American Chemical Society, Washington, DC (1984).
3. M. A. Fox (ed.), *Organic Phototransformations in Non-homogeneous Media*. American Chemical Society, Washington, DC (1985).
4. V. Ramamurthy, J. R. Scheffer and N. J. Turro (eds), *Tetrahedron* **43**, 1197 (1987).
5. K. Kalyanasundaram, *Photochemistry in Microheterogeneous Systems*. Academic Press, New York (1987).
6. T. Matsuura and M. Anpo (eds), *Photochemistry on Solid Surfaces*. Elsevier, Amsterdam (1989).
7. V. Balzani and F. Scandola, *Supramolecular Photochemistry*. Ellis Horwood, New York (1991).
8. H. J. Schneider and H. Durr (eds), *Frontiers in Supramolecular Organic Chemistry and Photochemistry*. VCH, Weinheim, (1991).
9. V. Ramamurthy (ed.), *Photochemistry in Organized and Constrained Media*. VCH, New York, (1991).
10. K. Honda (ed.), *Photochemical Processes in Organized Molecular Systems*. North-Holland, Amsterdam, (1991).
11. J. M. Lehn, *Supramolecular Chemistry*. VCH, Weinham, (1995).
12. H. Van-Bekkum, E. M. Flanigen and J. C. Jansen, *Introduction to Zeolite Science and Practice*. Elsevier, Amsterdam (1991).
13. A. Dyer, *An Introduction to Zeolite Molecular Sieves*. John Wiley & Sons, New York (1988).
14. D. W. Breck, *Zeolite Molecular Sieves: Structure, Chemistry and Use*. John Wiley & Sons, New York (1974).
15. J. C. Jansen, M. Stocker, H. G. Karge and J. Weitkamp (eds), *Advanced Zeolite Science and Applications*. Elsevier, Amsterdam (1994).
16. N. Herron and D. R. Corbin (eds), *Inclusion Chemistry with Zeolites: Nanoscopic Materials by Design*. Kluwer Academic Publishers, Dordrecht (1995).
17. V. R. Choudhary and D. B. Akolekar, *J. Catal.* **117**, 542 (1989).
18. S. Csicsery and P. Laszlo (eds), *Shape-Selective Catalysis*. Academic Press, San Diego (1987).
19. R. M. Barrer and H. Villiger, *Z. Kristallogr.* **128**, 352 (1969).
20. W. M. Meier, *Z. Kristallogr.* **115**, 439 (1961).
21. J. A. Gard and J. M. Tait, *Acta Cryst. B* **28**, 825 (1972).
22. E. Galli, *Rend. Soc. Ital. Mineral. Petrol.* **31**, 599 (1975).
23. J. M. Newsam, M. M. J. Treacy, W. T. Koetsier and C. B. de-Gruyter, *Proc. R. Lond. A* **420**, 375 (1988).
24. S. Qiu, W. Pang, H. Kessler and J.-L. Guth, *Zeolites* **9**, 440 (1989).
25. M. E. Davis, C. Montes, P. E. Hathaway, J. P. Arhancet, D. L. Hasha and J. Garces, *J. Am. Chem. Soc.* **111**, 3919 (1989).
26. J. A. Martens, E. Feijen, J. L. Lievens, P. J. Grobet and P. A. Jacobs, *J. Phys. Chem.* **95**, 10025 (1991).
27. M. Estermann, L. B. McCusker, C. Baeriocher, A. Merrouche and H. Kessler, *Nature* **352**, 320 (1991).
28. J. S. Beck, J. C. Vartuli, W. J. Roth, M. E. Leonowicz, C. T. Kresge, K. D. Schmidt, C. T. W. Chu, D. H. Olson, E. W. Sheppard, S. B. McCullen, J. B. Higgins and J. L. Schlenker, *J. Am. Chem. Soc.* **114**, 10834 (1992).

29. Y. F. Shepelev, A. A. Anderson and Y. I. Smolin, *Zeolites* **10**, 61 (1990).
30. R, Schöllner, R. Broddack, B. Kuhlmann, P. Nözel and H. Herden, *Z. Phys. Chem.* **262**, 17 (1981).
31. T. S. Egerton and F. S. Stone. *J. Chem. Soc.* **66**, 2364 (1970).
32. H. Herden, W.-D. Einicke, R. Schöllner and A. J. Dyer, *Inorg. Nucl. Chem.* **43**, 2533 (1981).
33. V. M. Radak, T. S. Germacic and I. M. Zivadinovic, *Naturforsch.* **336**, 1116 (1978).
34. C. Peucker and D. Kunate, *J. Chem. Soc., Faraday Trans. I* **77**, 2079 (1981).
35. H. Herden, W.-D. Einicke, R. Schöllner, W. J. Mortier, L. R. Gellens and J. B. Uytterhoeven, *Zeolites* **2**, 131 (1982).
36. Y. H. Yeom, S. B. Jang, Y. Kim, S. H. Song, and K. Seff, *J. Phys. Chem. B* **101**, 6914 (1997).
37. C. Forano, R. C. T. Slade, E. K. Andersen, I. G. K. Andersen and E. Prince, *J. Solid State Chem.* **82**, 95 (1989).
38. Y. F. Shepelev, I. K. Butikova and Y. I. Smolin, *Zeolites* **11**, 287 (1991).
39. W. J. Mortier, *Compilation of Extra Frame Work Sites in Zeolites.* Buttworth Scientific Ltd, Guildford (1982).
40. H. Herden, W.-D. Einicke, R. Schöllner and A. J. Dyer, *Inorg. Nucl. Chem.* **43**, 2533 (1981).
41. Y. F. Shepelev, A. A. Anderson and Y. I. Smolin, *Zeolites* **10**, 61 (1990).
42. D. W. Breck, *Zeolite Molecular Sieves*, p. 98 Wiley-Interscience, New York (1974).
43. W. J. Mortier and H. J. Bosmans, *J. Phys. Chem.* **75**, 3327 (1971).
44. W. J. Mortier and H. J. Bosmans and J. B. Uytterhoeven, *J. Phys. Chem.* **76**, 650 (1972).
45. T. Takaishi, *Zeolites* **17**, 389 (1996).
46. D. H. Olson, *Zeolites* **15**, 439 (1995).
47. C. P. Grey, F. I. Poshni, A. F. Gualtieri, P. Norby, J. C. Hanson and D. R. Corbin, *J. Am. Chem. Soc.* **119**, 1981 (1997).
48. N. J. Turro, *Pure Appl. Chem.* **58**, 1219 (1986).
49. R. G. Weiss, V. Ramamurthy and G. S. Hammond, *Acc. Chem. Res.* **26**, 530 (1993).
50. V. Ramamurthy, R. G. Weiss and G. S. Hammond, *Adv. Photochem.* **18**, 67 (1993).
51. Calculations of polyhedral volumes were performed using a modification of the POLYVOL Program [D. Swanson and R. C. Peterson, *The Canadian Mineralogist* **18**(2), 153 (1980); D. K. Swanson and R. C. Peterson, POLYVOL Program Documentation, Virginia Polytechnic Institute, Blacksburg, VA] assuming the radius of the TO_2 unit to be 2.08 Å (equivalent to that of quartz).
52. V. N. Abramov, A. V. Kiselev and V. I. Lygin, *Russ. J. Phys. Chem. Eng. Trans.* **37**, 613 (1963).
53. K. T. Geodakyan, A. V. Kiselev and V. I. Lygin, *Russ. J. Phys. Chem. Eng. Trans.* **41**, 227 and 476 (1967).
54. C. L. Angell and M. V. Howell, *J. Colloid. Interfac. Sci.* **28**, 279 (1968).
55. B. Coughlan, W. M. Carroll, P. O'Malley and J. Nunan, *J. Chem. Soc., Faraday Trans. I* **77**, 3037 (1981).
56. A. de-Mallman and D. Barthomeuf, In *Proceedings of the 7th International Zeolite Conference*, ed. by Y. Murakami, A. Iijima and J. W. Ward, p. 609. Tokyo (1986).
57. A. de-Mallman and D. Barthomeuf, *Zeolites* **8**, 292 (1988).
58. A. de-Mallman and D. Barthomeuf, *J. Chem. Soc. Chem. Commun.* 129 (1989).
59. A. de-Mallman and D. Barthomeuf, *J. Chem. Soc. Chem. Commun.* 476 (1986).
60. A. de-Mallman, S. Dzwigaj and D. Barthomeuf, In *Zeolites: Facts, Figures, Future*, ed. by P. A. Jacobs and R. A. van-Santen, p. 935. Elsevier, Amsterdam (1989).
61. P. J. O'Malley, *Chem. Phys. Lett.* **166**, 340 (1990).

62. J. J. Freeman and M. L. Unland, *J. Catalysis* **54**, 183 (1978).
63. M. L. Unland and J. J. Freeman, *J. Phys. Chem.* **82**, 1036 (1978).
64. M. Primet, E. Garbowski, M. V. Mathieu and B. Imelik, *J. Chem. Soc., Faraday Trans. I* **76**, 1942 (1980).
65. H. Lechert and K. P. Wittern, *Ber. Bunsenges, Phys, Chem.* **82**, 1054 (1978).
66. V. Y. Borovkov, W. K. Hall and V. B. Kazanski, *J. Catalysis* **51**,437 (1978).
67. R. Ryoo, S. B. Liu, L. C. de-Menorval, K. Takegoshi, B. Chmelka, M. Trecoske and A. Pines, *J. Phys. Chem.* **91**, 6575 (1987).
68. L. C. de-Menorval, D. Raftery, S. B. Liu, K. Takegoshi, R. Ryoo and A. Pines, *J. Phys. Chem.* **94**, 27 (1990).
69. S. B. Hong, H. M. Cho and M. E. Davis, *J. Phys. Chem.* **97**, 1629 (1993).
70. B. F. Chmelka, J. G. Pearson, S. B. Liu, R. Ryoo, L. C. de-Menorval and A. Pines, *J. Phys. Chem.* **95**, 303 (1991).
71. J. G. Pearson, B. F. Chmelka, D. N. Shykind and A. Pines, *J. Phys. Chem.* **96**, 8517 (1992).
72. S. B. Liu, L. J. Ma, M. W. Lin, J.-F. Wu and T.-L. Chen, *J. Phys. Chem.* **96**, 8120 (1992).
73. A. N. Fitch, H. Jobic and A. Renouprez, *J. Chem. Soc., Chem. Commun.* 284 (1985).
74. M. Czjek, T. Vogt and H. Fuess, *Angew. Chem. Int. Ed. Engl.* **28**, 770 (1989).
75. A. Renouprez, H. Jobic and R. C. Oberthur, *Zeolites* **5**, 222 (1985).
76. H. Jobic, A. Renouprez, A. N. Fitch and H. J. Lauter, *J. Chem, Soc., Faraday Trans. I* **83**, 3199 (1987).
77. G. V. Tsitsishvili and T. G. Andronikashvili, In *Molecular Sieve Zeolites II*, ed. by E. M. Flanigen and L. B Sand, p. 217. American Chemical Society, Washington, DC (1971).
78. W. Basler and H. Lechert, *Ber. Bunsenges. Phys. Chem.* **78**, 667 (1974).
79. A. de-Mallmann and D. Barthomeuf, *J. Phys. Chem.* **93**, 5636 (1989).
80. J. Sauer and D. Deininger, *Zeolites* **2**, 114 (1982).
81. P. Demontis, S. Yashonath and M. L. Klein, *J. Phys. Chem.* **93**, 5016 (1989).
82. A. N. Fitch, H. Jobic and A. Renouprez, *J. Phys. Chem.* **90**, 1311 (1986).
83. M. Czjzek, T. Vogt and H. Fuess, *Angew. Chem. Int. Ed. Engl.* **28**, 770 (1989).
84. C. Mellot, M. H. Simonot-Grange, E. Pilverdier, J. P. Bellat and D. Espinat, *Langmuir* **11**, 1726 (1995).
85. M. Czjzek, T. Vogt and H. Fuess, *Zeolites* **12**, 237 (1992).
86. C. Kirschhock and H. Fuess, *Zeolites* **17**, 381 (1996).
87. R. Goyal, A. N. Fitch and H. Jobic, *J. Chem. Soc., Chem. Commun.* 1152 (1990).
88. R. Burmeister, H. Schwarz and B. Boddenberg, *Ber. Bunsenges. Phys. Chem.* **93**, 1309 (1989).
89. M. J. Sefcik, *J. Am. Chem. Soc.* **101**, 2164 (1979).
90. W. D. Hoffman, *Z. Phys. Chem. (Leipzing)* **257**, 315 (1976).
91. H. Lechert, W. Haupt and K. P. Wittern, *J. Catal.* **43**, 356 (1976).
92. H. Lechert and H.-J. Hennig, *Z. Natureforsch.* **29A**, 1065 (1974).
93. M. Nagel, D. Michel and D. Geschke, *J. Colloid Interface Sci.* 36, 254 (1971).
94. J. B. Nagy, E. G. Derouane, H. A. Resing and G. R. Miller, *J. Phys. Chem.* **87**, 833 (1983).
95. B. L. Silbernagel, A. R. Garcia, J. M. Newsam and R. Hulm, *J. Phys. Chem.* **93**, 6505 (1989).
96. R. Eckman and A. J. Vega, *J. Am. Chem. Soc.* **105**, 4841 (1983).
97. R. Eckman and A. J. Vega, *J. Phys. Chem.* **90**, 4679 (1986).
98. B. Boddenberg and R. Burmeister, *Zeolites* **8**, 480 (1988).

99. R. Burmeister, H. Schwartz and B. Boddenberg, *Ber. Bunsenges. Phys. Chem.* **93**, 1309 (1989).
100. B. Boddenberg, R. Burnmeister and G. Spaeth, In *Zeolites as Catalysts, Sorbents, and Detergent Builders*, ed. by H. Karge and J. Weitkamp, p. 533. Elsevier, Amsterdam (1989).
101. I. Kustanovich, D. Fraenkel, Z. Luz, S. Vega and H. Zimmermen, *J. Phys, Chem.* **92**, 4134 (1988).
102. I. Kustanovich, H. Weith, Z. Luza and S. Vega, *J. Phy. Chem.* **93**, 7427 (1989).
103. D. L. Hasha, V. W. Miner, J. M. Garces and S. C. Rock, In *Catalyst Characterization Science: Surface and Solid State Chemistry*, ed. by M. L. Deviney and J. L. Gland, p. 485. American Chemical Society, Washington, DC (1985).
104. M. A. Happ, V, Ramamurthy, D. R. Corbin and C. Dybowski, *J. Phys. Chem.* **96**, 2629 (1996).
105. A. Declemy, C. Rulliere and P. Kottis, *Laser Chem.* **10**, 413 (1990).
106. M. J. Weaver, G. E. McManis, W. Jarzeba and P. F. Barbara, *J. Phys. Chem.* **94**, 1715 (1990).
107. D. -G. Su and J. D. Simon, *J. Phys. Chem.* **93**, 753 (1989).
108. J. Krager and D. M. Ruthven, *Diffusion in Zeolites*. John Wiley, New York (1992).
109. V. Ramamurthy, *Mol. Cryst. Liq. Cryst.* **211**, 211 (1992).
110. V. Ramamurthy, D. R. Sanderson and D. F. Eaton, *J. Phys. Chem.* **97**, 13380 (1993).
111. M. A. Morgan and G. C. Pimmental, *J. Phys. Chem.* **93**, 3056 (1989).
112. F. A. Carrol and F. H. Quina, *J. Am. Chem. Soc.* **98**, 1 (1976).
113. H. A. Morrison, T. Nylund and F. Palensky, *J. Chem. Soc. Chem. Commun.* 4 (1976).
114. R. B. Cundall and J. R. Voss, *J. Chem. Soc, Chem. Commun.* 902 (1968).
115. H.Morrison and A. Miller, *Tetrahedron* **37**, 3405 (1981).
116. F. Scully, T. Nylund, F. Palensky and H. Morrison, *J. Am. Chem. Soc.* **100**, 7352 (1978).
117. A. Nakajima, *Bull. Chem. Soc, Jpn.* **46**, 2602 (1973).
118. K. Hara and W. R. Ware, *Chem. Phys.* **51**, 51 (1980).
119. A. Nakajima, *Bull. Chem. Soc. Jpn.* **44**, 3272 (1971).
120. K. Kalyanasundaram and J. K. Thomas, *J. Am. Chem. Soc.* **99**, 2039 (1977).
121. D. C. Dong and M. A. Winnik, *Photochem. Photobiol.* **35**, 17 (1982).
122. W. R. Ware, In *Photochemistry in Organized and Constrained Media*, ed. by V. Ramamurthy, p. 563. VCH, New York (1991).
123. M. H. Gehlen and F. C. De Schryver, *Chem. Rev.* **93**, 199 (1993).
124. J. A. Rabo, *Cat. Rev. -Sci. Eng.* **23**, 293 (1981).
125. E. Dempsey, *Molecular Sieves*, p. 293. Society of Chemical Industry, London (1968).
126. E. Preuss, G. Linden and M. Peuckert, *J. Phys. Chem.* **89**, 2955 (1989).
127. Y.-Y. Huang, J. E. Benson and M. Boudart, IEC Fundam. **8**, 346 (1969).
128. J. W. Ward and H. W. Habgood, *J. Phys. Chem.* **70**, 1178 (1966).
129. C. L. Angel, *J. Phys. Chem.* **70**, 2420 (1966).
130. C. L. Angell and P. C. Schaffer, *J. Phys. Chem.* **70**, 1413 (1966).
131. G. H. Smudde Jr, T. L. Slager, C. G. Coe, J. E. MacDougall and S. J. Weigel, *Appl. Spectrosc.* **49**, 1747 (1995).
132. Y. Y. Huang, *J. Am. Chem. Soc.* **95**, 6636 (1973).
133. J. W. Ward, *J. Catalysis* **10**, 34 (1968).
134. H. Sugihara, K. Shimokoshi and I. Yasumori, *J. Phys. Chem.* **81**, 669 (1977).
135. J. A. R. Coope, C. L. Gardner, C. A. McDowell and A. I. Pelman, *Mol. Phys.* **21**, 1043 (1971).

136. K. Shimokoshi, H. Sugihara and I. Yasumori, *J. Phys. Chem.* **78**, 1770 (1974).
137. J. W. Ward, In *Zeolite Chemistry and Catalysis*, ed. by J. A. Rabo, p. 118. American Chemical Society, Washington, DC (1976).
138. P. H. Kasai and R. J. Bishop Jr, *J. Am. Chem. Soc.* **94**, 5560 (1972).
139. P. H. Kasai and R. J. Bishop Jr, *J. Phys. Chem.* **77**, 2308 (1973).
140. J. A. Rabo, C. L. Angell, P. H. Kasai and V. Schomaker, *Discuss. Faraday Soc.* **41**, 328 (1966).
141. K. Kalyanasundaram, In *Photochemistry in Organized and Constrained Media*, ed. by V. Ramamurthy p. 39. VCH Publishers, New York, (1991).
142. V. Ramamurthy, D. R. Sanderson and D. F. Eaton, *Photochem. Photobiol.* **56**, 297 (1992).
143. V. Ramamurthy and D. F. Eaton, In *Proceedings of the 9th International Zeolite Conference*, ed. by R. von-Ballmoos, J. B. Hggins and M. M. J. Treacy. p. 587. Butterworth-Heinemann, Boston (1992).
144. C. Reichardt, *Solvents and Solvent Effects in Organic Chemistry.* VCH Publishers, New York (1988).
145. Y. Wang and K. B, Eisenthal, *J. Chem. Phys.* **77**, 6076 (1982).
146. M. V. Auweraer, Z. Grabowski and W. Rettig, *J. Phys. Chem.* **95**, 2083 (1991).
147. B. Baretz and N. J. Turro, *J. Photochem.* **24**, 201 (1984).
148. K. B. Yoon and J. K. Kochi, *J. Phys. Chem.* **95**, 1348 (1991).
149. P. K. Dutta and W. Turbeville, *J. Phys. Chem.* **95**, 4087 (1991).
150. K. K. Iu and J. K. Thomas, *Langmuir* **6**, 471 (1990).
151. A. Corma, *Chem. Rev.* **95**, 559 (1995).
152. D. Atkinson and G. Curthoys, *Chem. Soc. Rev.* **8**, 475 (1979).
153. V. B. Kazansky, *Acc. Chem. Res.* **24**, 379 (1991).
154. J. W. Ward, In *Zeolite Chemistry and Catalysis*, p. 118. American Chemical Society Washington, DC (1976).
155. J. A. Rabo and G. J. Gajda. *Catal. Rev. Sci. Eng.* **31**, 385 (1989).
156. C. N. Salterfield, In *Heterogeneous Catalysis in Industrial Practice*, Ch. 7. McGraw Hill, New York (1991).
157. W. E. Farneth and R. J. Gorte, *Chem. Rev.* **95**, 615 (1995).
158. H. G. Karge, In *Catalysis and Adsorption by Zeolites*, ed. by G. Öhlmann *et al.*, p. 133. Elsevier Science, Amsterdam (1991).
159. R. A. van-Santen and G. J. Kramer, *Chem. Rev.* **95**, 637 (1995).
160. V. J. Rao, D. L. Perlstein, R. J. Robbins, P. H. Lakshminarasimhan, H.-M. Kao, C. P. Grey and V. Ramamurthy, *Chem. Commun.* 269 (1998).
161. R. J. Robbins and V. Ramamurthy, unpublished results.
162. J. W. Ward, *J. Catalysis* **14**, 365 (1969).
163. J. W. Ward, *J. Phys. Chem.* **72**, 4211 (1968).
164. J. W. Ward, *J. Catalysis* **10**, 34 (1968).
165. J. B. Uytterhoeven and R. Schoonheydt, *J. Catalysis* **13**, 425 (1969).
166. P. E. Eberly Jr, *J. Phys. Chem.* **72**, 1042 (1968).
167. H. Hattori and T. Shiba, *J. Catalysis* **12**, 111 (1968).
168. H.-M. Kao and C. P. Grey, *J. Am. Chem. Soc.* **119**, 627 (1977).
169. H.-M. Kao and C. P. Grey, *J. Phys. Chem.* **100**, 5105 (1996).
170. H.-M. Kao and C. P. Grey, *Chem. Phys. Lett.* **259**, 459 (1996).
171. L. M. Kustov, V. Y. Borovkov and V. B. Kazansky, *J. Catalysis* **72**, 149 (1981).
172. N. Katada, H. Igi, J. H. Kim and M. Niwa, *J. Phys. Chem.* **101**, 5969 (1997).
173. U. Lohse, B. Parlitz and V. Patzelová, *J. Phys. Chem.* **93**, 3677 (1989).
174. H.-M. Kao, C. P. Grey, K. Pitchumani, P. H. Lakshminarasimhan and V. Ramamurthy, *J. Phys. Chem.* **102**, 5627 (1998).

175. K. J. Thomas and V. Ramamurthy, *Langmuier*, **14**, 6687 (1998).
176. V. B. Kazansky, In *Catalysis and Adsorption in Zeolites*, ed. by G. Öhlmann *el al.*, p. 117. Elsevier Science, Amsterdam (1991).
177. H. Pfeifer, *Colloids and Surfaces* **36**, 169 (1989).
178. D. Coster, A. L. Blumenfeld and J. J. Fripiat, *J. Phys. Chem.* **98**, 6201 (1994).
179. D. J. Coster, A. Bendada, F. R. Chen and J. J. Fripiat, *J. Catalysis* **140**, 497 (1993).
180. A. Corma, V. Fornés and F. Rey, *Zeolites* **13**, 56 (1993).
181. V. Gruver and J. J. Fripiat, *J. Phys. Chem.* **98**, 8549 (1994).
182. M. Huang, A. Adnot and S. Kaliaguine, *J. Catalysis* **139**, 322 (1992).
183. D. Barthomeuf, *Catalysis Rev.* **52**, 521 (1996).
184. D. Barthomeuf, In *Catalysis and Adsorption by Zeolites*, ed. by G. Öhlmann *et al.*, p. 157. Elsevier Science, Amsterdam (1991).
185. D. Barthomeuf, *J. Phys. Chem.* **88**, 42 (1984).
186. M. Huang and S. Kaliaguine, *J. Chem. Soc. Faraday Trans.* **88**, 751 (1992).
187. S. L. Suib and A. Kostapapas, *J. Am. Chem. Soc.* **106**, 7705 (1984).
188. X. Liu, K.-K. Iu and J. K. Thomas, *J. Phys. Chem.* **93**, 4120 (1989).
189. X. Liu and J. K. Thomas, *Langmuir* **9**, 727 (1993).
190. F. L. Cozens, M. Régimbald, H . García and J. C. Scaiano, *J. Phys. Chem.* **100**, 18165 (1996).
191. A. N. Abramov, A. V. Kiselev and V. I. Lygin, *Russ. J. Phys. Chem. Eng. Trans.* **37**, 613 (1963).
192. C. L. Angell and M. V. Howell, *J. Colloid. Sur. Sci.* **28**, 279 (1968).
193. B. Coughlan, W. M. Carroll, P. O'Malley and J. Nunan, *J. Chem. Soc., Faraday Trans. I* **77**, 3037 (1981).
194. G. I. Denisenko, M. A. Piontkovskaya and I. E. Neimark, *Ukrainskii Khimicheskii Zhurnal* **36**, 260 (1970).
195. G. I. Denisenko and V. A. Lisovenko, *Zh. Prikl. Spektrosk.* **14**, 702 (1971).
196. G. I. Denisenko, *Zh. Prikl. Spektrosk.* **9**, 307 (1968).
197. K. T. Geodakyan, A. V. Kiselev and V. I. Lygin, *Russ, J. Phys. Chem. Eng. Trans.* **41**, 227 and 476 (1967).
198. V. Ramamurthy, D. R. Corbin and L. J. Johnston, *J. Am. Chem. Soc.* **114**, 3870 (1992).
199. E. J. Baum, J. K. S. Wan, J. K. S. and J. N. Pitts, *J. Am. Chem. Soc.* **88**, 2652 (1966).
200. V. Ramamurthy, D. R. Corbin, N. J. Turro and Y. Sato. *Tetrahedron Lett.* **30**, 5829 (1989).
201. V. Ramamurthy, X. G. Lei, N. J. Turro, T. R. Lewis and J. R. Scheffer, *Tetrahedron Lett.* **32**, 7675 (1991).
202. V. Jayathirtha Rao, S. R. Uppili, D. R. Corbin, S. Schwarz, S. R. Lustig and V. Ramamurthy, *J. Am. Chem. Soc.* **120**, 2480 (1998).
203. D. I. Schuster, In CRC *Handbook or Organic Photochemistry and Photobiology*, ed. by W. Horspool and P-S. Song, p. 579. CRC Press, Boca Raton (1995).
204. K. Pitchumani, M. Warrier, J. R. Scheffer and V. Ramamurthy, *Chem. Commun.* 1197 (1998).
205. D. F. Evans, *J. Chem. Soc.* 345 (1953).
206. A. U. Munck and J. F. Scott. *Nature* **177**, 587 (1956).
207. H. Tsubomura and R. S. Mulliken, *J. Am. Chem. Soc.* **82**, 5966 (1960).
208. E. C. Lim and V. L. Kowalski, *J. Chem. Phys.* **36**, 1729 (1962).
209. J. Jortner and U. Sokolov, *J. Phys. Chem.* **65**, 1633 (1961).
210. H. Ishida, H. Takahashi, H. Sato and H. Tsubomura, *J. Am. Chem. Soc.* **92**, 275 (1970).
211. H. Bradley Jr and A. D. King Jr, *J. Chem. Phys.* **47**, 1189 (1967).

212. M. Kristiansen, R. D. Scurlock, K.-K. Iu and P. R. Ogilby, *J. Phys. Chem.* **95**, 5190 (1991).
213. R. D. Scurlock and P. R. Ogilby, *J. Phys. Chem.* **93**, 5493 (1989).
214. E. A. Gooding, K. R. Serak and P. R. Ogilby, *J. Phys. Chem.* **95**, 7868 (1991).
215. Y. Kuriyama, P. R. Ogilby and K. V. Mikkelsen, *J. Phys. Chem.* **98**, 11918 (1994).
216. T. Kanno, M. Hisaoka, H. Sakuragi and K. Tokumaru, *Bull. Chem. Soc. Jpn.* **54**, 2330 (1981).
217. T. Komatsu, M. Tsuchiya, G.-I. Furusawa, Y. Kuriyama, H. Sakuragi, F. Nakanishi and K. Tokumaru, *Bull. Chem. Soc. Jpn.* **68**, 277 (1995).
218. J. C. W. Chien, *J. Phys. Chem.* **69**, 4317 (1965).
219. M. Hori, H. Itoi and H. Tsubomura, *Bull. Chem. Soc. Jpn* **43**, 3765 (1970).
220. H. Tsubomura, T, Yagishita and H. Toi, *Bull. Chem. Soc. Jpn* **46**, 3051 (1973).
221. C. von-Sonntag, K. Neuwald, H.-P. Schuchmann, F. Weeke and E. Janssen, *J. Chem. Soc., Perkin Trans. II* 171 (1975).
222. M. Kojima, H. Sakuragi and K. Tokumaru, *Tetrahedron Lett.* **22**, 2889 (1981).
223. K. Onodera, G.-I. Furusawa, M. Kojima, M. Tsuchiya, S. Aihara, R. Akaba, H. Sakuragi and K. Tokumaru, *Tetrahedron* **41**, 2215 (1985).
224. M. Kojima, H. Sakuragi and K. Tokumaru, *Bull. Chem. Soc. Jpn.* **62**, 3863 (1989).
225. T. Kodaira, K. Hashimoto, Y. Sakanaka, S. Tanihata and K. Ikeda, *Bull. Chem. Soc. Jpn* **51**, 1487 (1977).
226. W. A. Pryor and R. A. Patsiga, *Spectrosc. Lett.* **2**, 61 (1969).
227. J. Betts and J. C. Robb, *J. Chem. Soc., Faraday Trans.* **65**, 2144 (1969).
228. S. Hashimoto and H. Akimoto, *J. Phys. Chem.* **90**, 529 (1986).
229. H. Sun, F. Blatter and H. Frei, *J. Am. Chem. Soc.* **116**, 7951 (1994).
230. F. Blatter, F. Moreau and H. Frei, *J. Phys. Chem.* **98**, 13403 (1994).
231. F. Blatter and H. Frei. *J. Am. Chem. Soc.* **115**, 7501 (1993).
232. F. Blatter and H. Frei. *J. Am. Chem. Soc.* **116**, 1812 (1994).
233. H. Sun, F. Blatter and H. Frei, *J. Am. Chem. Soc.* **118**, 6873 (1996).
234. H. Frei, F. Blatter and H. Sun, *Chemtech.* **26**, 24 (1996).
235. F. Blatter, H. Sun and H. Frei, *Cat. Lett.* **35**, 1 (1995).
236. K. B. Yoon, *Chem. Rev.* **93**, 321 (1993).
237. K. B. Yoon and J. K. Kochi, *J. Am. Chem. Soc.* **111**, 1128 (1989).
238. K. B. Yoon and J. K. Kochi, *J. Phys. Chem.* **95**, 3780 (1991).
239. K. B. Yoon and J. K. Kochi, *J. Phys. Chem.* **95**, 1348 (1991).
240. K. B. Yoon and J. K. Kochi, *J. Am. Chem. Soc.* **110**, 6586 (1988).
241. S. Sankararaman, K. B. Yoon, T. Yabe and J. K. Kochi, *J. Am. Chem. Soc.* **113**, 1419 (1991).
242. K. B. Yoon, S. M. Hubig, J. K. Kochi, *J. Phys. Chem.* **98**, 3865 (1994).
243. K. B. Yoon, T. J. Huh, D. R. Corbin and J. K. Kochi, *J. Phys. Chem.* **97**, 6492 (1993).
244. K. B. Yoon, T. J. Huh and J. K. Kochi, *J. Phys. Chem.* **99**, 7042 (1995).
245. V. Ramamurthy, D. R. Corbin, N. J. Turro and Y. Sato, *Tetrahedron Lett.* **30**, 5829 (1989).
246. D. N. Stamires and J. Turkevich, *J. Am. Chem. Soc.* **86**, 749 (1964).
247. V. Ramamurthy, X. G. Lei, N. J. Turro, T. R. Lewis and J. R. Scheffer, *Tetrahedron Lett.* **32**, 7675 (1991).
248. S. Shih, *J. Catalysis* **79**, 390 (1983).
249. D. R. Corbin, D. F. Eaton and V. Ramamurthy, *J. Am. Chem. Soc.* **110**, 4848 (1988).
250. F. R. Dollish and W. K. Hall, J. Phys. Chem. **71**, 1005 (1967).
251. X. Lin, K.-K. Iu, J. K. Thomas, H. He and J. Klinowski, *J. Am. Chem. Soc.* **116**, 11811 (1994).
252. F. R. Chen and J. J. Fripiat, *J. Phys. Chem.* **96**, 819 (1992).

253. H. Kashiwagi, *Bull. Chem. Soc. Jpn.* **64**, 2196 (1991).
254. A. E. Hirschler, W. C. Neikam, D. S. Barmby and R. L. James, *J. Catalysis* **4**, 628 (1965).
255. K. Pitchumani, D. R. Corbin and V. Ramamurthy, *J. Am. Chem. Soc.* **118**, 8152 (1996).
256. K. Pitchumani, P. H. Lakshminarasimhan, G. Turner, M. G. Bakker and V. Ramamurthy, *Tetrahedron Lett.* **38**, 371 (1997).
257. K. Pitchumani, P. H. Lakshminarasimhan, N. Prevost, D. R. Corbin and V. Ramamurthy, *J. Chem. Soc., Chem. Commun.* 181 (1997).
258. C. J. Rhodes, *Colloids Surf. A: Phys. Chem. Eng. Aspects* **72**, 111 (1993).
259. D. C. Doetschman, D. W. Dwyer, J. D. Fox, C. K. Frederick, S. Scull, G. D. Thomas, S. G. Utterback and J. Wei, *Chem. Phys.* **185**, 343 (1994).
260. J.-P. Lange, A. Gutsze and H. G. Karge, *J. Catal.* **114**, 136 (1988).
261. H. Tokunaga, Y. Ono and T. Keii, *Bull. Chem. Soc. Jpn* **46**, 3569 (1973).
262. F. R. Chen, S. Sheng, W. Zhang and X. Guo, *J. Chem. Soc. Faraday Trans* **88**, 887 (1992).
263. H. Caldararu, A. Caragheorgheopol and R. Russu, *Colloids Surf. A: Phys. Chem. Eng. Aspects* **72**, 37 (1993).
264. P. L. Corio and S. Shih, *J. Catalysis* **18**, 126 (1970).
265. Y. Kurita, T. Sonoda and M. Sato, *J. Catalysis* **19**, 82 (1970).
266. C. J. Rhodes, *J. Chem. Soc., Faraday Trans.* **87**, 3179 (1991).
267. C. J. Rhodes and M. Standing, *J. Chem. Soc., Perkins Trans. II* 1455 (1992).
268. R. Crockett and E. Roduner, *J. Chem. Soc., Perkins Trans. II* 1503 (1993).
269. C. J. Rhodes, *J. Chem. Soc., Chem. Commun.* 900 (1991).
270. F. R. Chen and J. J. Fripiat, *J. Phys. Chem.* **97**, 5796 (1993).
271. E. Roduner and R. Crockett. *J. Phys. Chem.* **97**, 11853 (1993).
272. E. Roduner, L.-M. Wu, R. Crockett and C. J. Rhodes, *Catal. Lett.* **14**, 373 (1992).
273. J. C. Védrine, A. Auroux, V. Bolis, P. Dejaifve, C. Naccache, P. Wierzchowski, E. G. Derouane, J. B. Nagy, J-P. Gilson, J. H. C. van Hooff, J. P. van den Berg and J. Wolthuizen, *J. Catalysis* **59**, 248 (1979).
274. A. E. Hirschler, *J. Catalysis* 196 (1966).
275. A. Corma, V. Fornés, H. García, V. Martí and M. A. Miranda, *Chem. Mat.* **7**, 2136 (1995).
276. M. L. Cano, A. Corma, V. Fornés and H. García, *J. Phys. Chem.* **99**, 4241 (1995).
277. X. Liu, K.-K. Iu, J. K. Thomas, H. He and J. Klinowski, *J. Am. Chem. Soc.* **116**, 11811 (1994).
278. V. L. Zholobenko, L. M. Kustov and V. B. Kazanskii, *Kinet. Catal.* **30**, 901 (1989).
279. A. Slinkin, A. V. Kucherov, D. A. Kondrat'ev, T. N. Bondarenko, A. M. Rubinshtein and Kh. M. Minachev, *Kinet. Katal.* **27**, 364 (1986).
280. V. Ramamurthy, J. V. Caspar and D. R. Corbin, *J. Am. Chem. Soc.* **113**, 594 (1991).
281. V. A. Bolshov and A. M. Volodin, *React Kinet. Catal Lett*, **43**, 87 (1991).
282. S. S. Pollack, R. F. Sprecher and E. A. Frommell, *J. Mol. Catal.* **66**, 195 (1991).
283. A. A. Slinkin, A. V. Kucherov, D. A. Kondrat'ev, T. N. Bondarenko, A. M. Rubinshtein and Kh. M. Minachev, *Kinet. Katal.* **27**, 156 (1986).
284. P. S. Lakkaraju, D. Zhou and H. D. Roth, *Chem. Commun.* 2605 (1996).
285. P. S. Lakkaraju, J. Zhang and H. D. Roth, J. *Chem. Soc., Perkin Trans. II* 2319 (1993).
286. P. S. Lakkaraju, J. Zhang and H. D. Roth, *J. Phys. Chem.* **98**, 2722 (1994).
287. J. V. Caspar, V. Ramamurthy and D. R. Corbin, *J. Am. Chem. Soc.* **113**, 600 (1991).

288. E. Roduner, R. Crockett and L.-M. Wu, *J. Chem. Soc., Faraday Trans.* **89**, 2101 (1993).
289. M. Kojima, H. Takeya, Y. Kuriyama and S. Oishi, *Chem. Lett.* 997 (1997).
290. J. K. Thomas, *Chem. Rev.* **93**, 301 (1993)
291. X. Liu, G. Zhang and J. K. Thomas, *J. Phys. Chem.* **99**, 10024 (1995).
292. K.-K. Iu and J. K. Thomas, *Colloids Surf.* **63**, 39 (1992).
293. X. Liu, Y. Mao, S. A. Ruetten and J. K. Thomas, *Solar Energy Mater. Solar Cells* **38**, 199 (1995).
294. X. Liu, K.-K. Iu and J. K. Thomas, *J. Phys. Chem.* **98**, 7877 (1994).
295. K.-K. Iu and J. K. Thomas, *J. Phys. Chem.* **95**, 506 (1991).
296. X. Liu, K.-K. Iu and J. K. Thomas, *Chem. Phys. Lett.* **204**, 163 (1993).
297. S. Hashimoto, *Chem. Phys. Lett.* **252**, 236 (1996).
298. V. A. Bolshov and A. M. Volodin, *React. Kinet. Catal. Lett.* **46**, 337 (1992).
299. F. Gessner and J. C. Scaiano, *J. Photochem. Photobiol. A: Chem.* **67**, 91 (1992).
300. I. K. Lednev, N, Mathivanan and L. J. Johnston, *J. Phys. Chem.* **98**, 11444 (1994).
301. K.-K. Iu and J. K. Thomas, *J. Photochem. Photobiol. A: Chem.* **79**, 103 (1994).
302. F. L. Cozens, R. Bogdanova, M. Régimbald, H. García, V. Marti and J. C. Scaiano, *J. Phys. Chem. B* **101**, 6921 (1997).
303. S. Hashimoto, *J. Chem. Soc., Faraday Trans.* **93**, 4401 (1997).
304. M. Alvaro, H. García. S. García, F. Márquez and J. C. Scaiano. *J. Phys. Chem. B* **101**, 3043 (1997).
305. H. J. D. McManus, C. Finel and L. Kevan, *Radiat. Phys. Chem.* **45**, 761 (1995).
306. K. Pitchumani and V. Ramamurthy, *J. Chem. Soc., Chem. Commun.* 2763 (1977).
307. V. Jayathirtha Rao, N. Prevost, V. Ramamurthy, M. Kojima and L. Johnston, *J. Chem. Soc., Chem. Commun.* 2209 (1997).
308. H. García, S. García, J. P. Prieto and J. C. Scaiano, *J. Phys. Chem. B* **100**, 18158 (1996).
309. M. L. Cano, F. L. Cozens, V. Fornes, H. García and J. C. Scaiano, *J. Phys. Chem. B* **100**, 18145 (1996).
310. S. Okamoto, H. Nishiguchi and M. Anpo, *Chem. Lett.* 1009 (1992).
311. F. L. Cozens, H. García and J. C. Scaiano, *J. Am. Chem. Soc.* **115**, 11134 (1993).
312. A. Corma, H. García, S. Iborra, V. Marti, M. A. Miranda and J. Primo, *J. Am. Chem. Soc.* **115**, 2177 (1993).
313. J. C. Scaiano, M. Kaila and S. Corrent, *J. Phys. Chem.* **101**, 8564 (1997).
314. V. Ramamurthy. J. V. Caspar, D. F. Eaton, E. W. Kuo and D. R. Corbin, *J. Am. Chem. Soc.* **114**, 3882 (1992).
315. J. V. Casper, V. Ramamurthy and D. R. Corbin, *Coordination Chem. Rev.* **97**, 225 (1990).
316. V. Ramamurthy, J. V. Caspar, D. R. Corbin and D. F. Eaton, *J. Photochem. Photobiol, A: Chem.* **50**, 157 (1989).
317. S. Hashimoto, N. Fukazawa, H. Fukumura and H. Masuhara, *Chem. Phys. Lett.* **223**, 493 (1994).
318. V. Ramamurthy, J. V. Casper, D. R. Corbin, B. D. Schlyer and A. H. Maki, *J. Phys. Chem.* **94**, 3391 (1990).
319. G. Heinrich, G. Holzer, H. Blume and D. Schulte-Frohlinde, *Z. Naturforsch.* **25b**, 496 (1970).
320. V. Ramamurthy and D. R. Sanderson, unpublished results.
321. H. Nishiguchi and M. Anpo, *J. Photochem. Photobiol. A: Chem.* **92**, 85 (1995).
322. D. O. Cowan and R. L. Drisko, *Elements of Organic Photochemistry*, p. 435. Plenum Press, New York (1976).

323. V. Ramamurthy, D. R. Corbin, C. V. Kumar and N. J. Turro, *Tetrahedron Lett.* **31**, 47 (1990).
324. K. Pitchumani, M. Warrier, J. R, Scheffer and V. Ramamurthy, *Chem. Commun.* (submitted).
325. R. Sadeghpoor, M. Gandhi, H. Mahmoudi Najafi and F. Farzneh, *J. Chem. Soc., Chem. Commun.* 329 (1998).
326. M. A. Anderson and C. B. Grissom, *J. Am. Chem. Soc.* **118**, 9552 (1996).
327. T. L. Pettit and M. A. Fox, *J. Phys. Chem.* **90**, 1353 (1986).
328. F. Gessner, A. Olea, J. H. Lobaugh, L. J. Johnston and J. C. Scaiano, *J. Org. Chem.* **54**, 259 (1989).
329. M. V. Baldovi, F. L. Cozens, V. Fornes, H. Garcia and J. C. Scaiano, *Chem. Mater.* **8**, 152 (1996).
330. M. V. Baldovi, A. Corma, H. Garcia and V. Marti, *Tetrahedron Lett.* **35**, 9447 (1994).
331. J. C. Scaiano, N. C. de-Lucas, J. Andraos and H. García, *Chem. Phys. Lett.* **233**, 5 (1995).
332. V. Ramamurthy, D. R. Sanderson and D. F. Eaton, *J. Am. Chem. Soc.* **115**, 10438 (1993).
333. R. J. Robbins and V. Ramamurthy, *Chem. Commun.* 1071 (1997).
334. X. Li, and V. Ramamurthy, *J. Am. Chem. Soc.* **118**, 10666 (1996).
335. K. Pitchumani, J. N. Gamlin, V. Ramamurthy and J. R. Scheffer, *Chem. Commun.* 2049 (1996).
336. K. Pitchumani. M. Warrier, V. Ramamurthy and J. R. Scheffer, *J. Chem. Soc., Chem. Commun.* (in press).
337. H. H. Wasserman and R. W. Murray (eds), *Singlet Oxygen*. Academic Press, New York (1979).
338. P. K. Dutta and J. A. Incavo, *J Phys. Chem.* **91**, 4443 (1987).
339. Y. I. Kim and T. E. Mallouk, *J. Phys. Chem.* **96**, 2879 (1992).
340. P. K. Dutta and W. Turbeville, *J Phys. Chem.* **96**, 9410 (1992).
341. P. K. Dutta and M. Borja, *J. Chem. Soc., Chem. Commun.* 1568 (1993).
342. E. S. Brigham, P. T. Snowden, Y. I. Kim and T. E. Mallouk, *J. Phys. Chem.* **97**, 8650 (1993).
343. P. K. Dutta and M. Borja, *Nature* **387**, 162 (1997).
344. M. Sykora and J. R. Kincaid, *Nature* **387**, 162 (1997).
345. M. Ledney and P. K. Dutta, *J. Am. Chem. Soc.* **117**, 7687 (1995).
346. J. S. Kruegar, C. Lai, Z. Ii, J. E. Mayer and T. E. Mallouk, In *Inclusion Phenomena and Molecular Recognition*, ed. by J. L. Atwood p. 365. Plenum, New York (1990).
347. E. H. Yonemoto, Y. I. Kim. R. H. Schmehl, J. O. Wallin, B. A. Shoulders, B. R. Richardson and T. E. Mallouk, *J. Am. Chem. Soc.* **116**, 10557 (1994)
348. A. Corma, V. Fornes, H. Garcia, M. A. Miranda, J. Primo and M.-J. Sabater, *J. Am. Chem. Soc.* **116**, 2276 (1994).
349. A. Corma, V. Fornes, H. García, M. A. Miranda and M. J. Sabater, *J. Am. Chem. Soc.* **116**, 9767 (1994).
350. M. L. Cano, A. Corma, V. Fornes, H. García, M. A. Miranda, J. Primo, M.-J. Sabater, C. Baerlocher and C. Lengauer, *J. Am. Chem. Soc.* **118**, 11006 (1996).
351. V. Fornes, H. Garcia, M. A. Miranda, F. Mojarrad, M. J. Sabater and N. N. E. Suliman, *Tetrahedron* **52**, 7755 (1996).
352. J. C. Scaiano, S. García and H. García, *Tetrahedron Lett.* **38**, 5929 (1997).
353. X. Li and V. Ramamurthy, *Tetrahedron Lett.* **37**, 5235 (1996).
354. L. Brancaleon, D. Brousmiche, V. J. Rao, L. J. Johnston and V. Ramamurthy, *J. Am. Chem. Soc.* (in press).

355. M. Vaida, R. Popovitz-Biro, L. Leserowitz and M. Lahav, in *Photochemistry in Organized and Constrained Media*, ed. by V. Ramamurthy, p. 247. VCH New York (1991).
356. J. N. Gamlin, R. Jones, M. Leibovitch, B. Patrick, J. R. Scheffer and J. Trotter, *Acc. Chem. Res.* **29**, 203 (1996).
357. F. Toda, *Mol. Cryst. Liq. Cryst.* **187**, 41 (1990).
358. M. Leibovitch, G. Olovsson, G. Sundarababu, V. Ramamurthy, J. R. Scheffer and J. Trotter, *J. Am. Chem. Soc.* **118**, 1219 (1996).
359. G. Sundarababu, M. Leibovitch, D. R. Corbin, J. R. Scheffer and V. Ramamurthy, *J. Chem. Soc., Chem. Commun.* 2159 (1996).
360. A. Joy, J. R. Scheffer, D. R. Corbin and V. Ramamurthy, *J. Chem. Soc., Chem. Commun.* (in press).
361. A. Joy, R. Robbins, K. Pitchumani and V. Ramamurthy, *Tetrahedron Lett.* 8825 (1997).

Chapter 4

Statistically Significant Conclusions from the Cambridge Structural Database

K. WHEELER
Department of Chemistry, Delaware State University, Dover, DE, USA

1. INTRODUCTION

It is well documented that the arrangement and conformation of molecules control not only chemical dynamics, but also the physical and electronic properties of crystalline materials (i.e. non-linear optics, conductivity, and reactivity). Therefore, the ability to predict three-dimensional structure is crucial to the progress of the design of materials. To date, however, this "Holy Grail" remains unclaimed owing to the ill-defined nature of the forces that govern molecular arrangements. The predictability of molecular structures will only be possible when these forces are identified and quantified in a systematic fashion.

Even though the prediction of packing motifs for a given compound is serendipitous, many of the factors that affect molecular arrangements are gradually being elucidated[1,2]. An effective approach towards uncovering these factors is the analysis of intra- and intermolecular geometric criteria of specific arrangements; a wealth of useful chemical information is available from the three-dimensional crystal structure data[3]. One application that applied such an approach investigated the nature of intramolecular C=O···N interactions

Organised Molecular Assemblies in the Solid State Edited by James K. Whitesell
© 1999 John Wiley & Sons Ltd.

towards understanding the reaction coordinate for the addition of an amino group to a carbonyl[3–10]. Bürgi and Dunitz analyzed crystal structures comprised of structural types **1** and **2**. From analysis of these structures, the C═O···N distances differ markedly (i.e. 1.49 to 2.91 Å) as a result of molecular conformations and the constitution of the tether connecting the amine and carbonyl groups. Since the motion of molecules in the crystalline state tends to be static, it is reasonable to view each of these structures as a specific representation or "snap-shot" along the reaction coordinate. Therefore, investigation of structural parameters for each example should provide insight into the mechanistic details of this addition reaction.

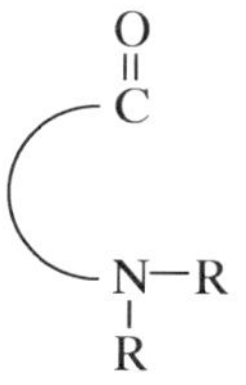

1

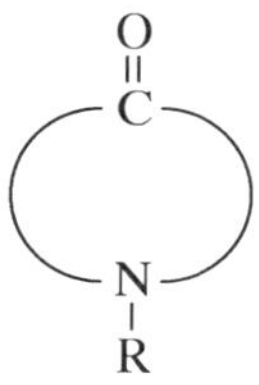

2

Analysis of fourteen structures of types **1** and **2** resulted in the acquisition of relevant geometrical parameters. In all cases, the N, C, and O atoms lie in an approximate local mirror plane for the RR′N···C═O interaction; however, the C atom differs significantly from the plane defined by R, R′ and O. Also, from analysis of the relative positions of the N, C, and O atoms (Figure 4.1) it is evident that a close approach of the nucleophile to the carbonyl C atom results in a distortion of the two alkyl substituents R and R′ from the defined plane and a longer C—O bond distance. Thus, the closer in proximity of the nucleophile to the substrate, the larger the nonplanarity of the RR′C═O system.

From analysis of the data, the authors concluded that the nucleophile approaches the substrate in a specific orientation and direction. If the lone-pair orbital is defined by the local three-fold axis of the tertiary amine group, the direction of the orbital coincides with the direction of the approach of the amine to the carbonyl carbon atom. This approach is best described by a straight line not perpendicular to the C—O, but rather at an angle of ~107°. This study concluded that a strong preference exists for the orientation of nucleophilic attack in these systems.

As evident from this example, analyses based on crystal structures can provide vital information about specific aspects of molecular structure and processes. Even so, the significance of such a study is limited by the quality and relative number of examples analyzed; statistically significant studies are only meaningful from analysis of a large number of structurally-related compounds. Prior to the availability of refined computer databases, the information for structural analyses were generally retrieved from the literature manually, making scientific information intractable. With the advances in computer

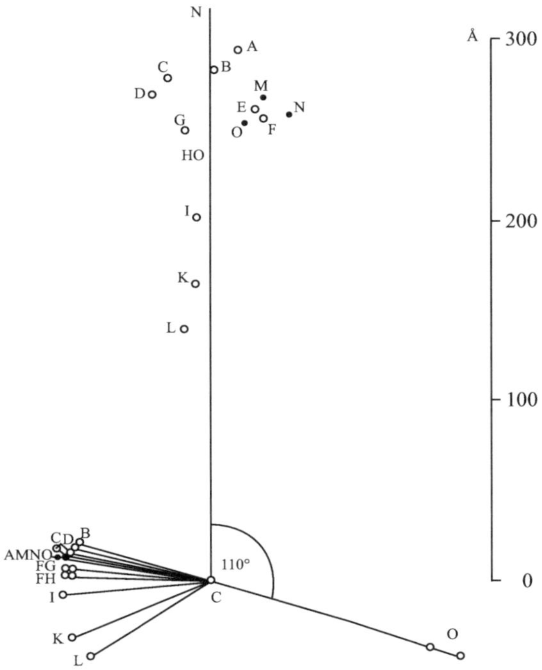

Figure 4.1 Relative positions of N, C, and O atoms and of the RCR' plane in 14 molecules showing N···C=O interactions. (Reproduced by permission from Bürgi *et al.*[5].)

hardware and software, the notion of accessing a large body of evaluated structural data for crystalline solids is now a reality. At first glance, the compilation of structural data in a computer database appears to be a minor event; however, because of the ease of data accessibility and manipulation, this improvement is viewed as a revolutionary development by structural scientists.

Since a large collection of high quality crystallographic data exists, several computer databases provide the necessary storage space for information. The five main structural databases include; NBS Crystal Data (basic cell parameters)[11], NRC Metals Crystallographic Database[12], Brookhaven Protein Data Bank[13], Inorganic Crystal Structure Database[14] and the Cambridge Structural Database[15,16] (organic and organometallic structures). Each database is specific for data type or material classification. Because the scope of this chapter focuses on understanding organic based organized molecular assemblies, this work explores exclusively the Cambridge Structural Database (hereafter, the CSD).

The CSD provides a medium not previously available to the scientific community. In the early stages of understanding molecular arrangements,

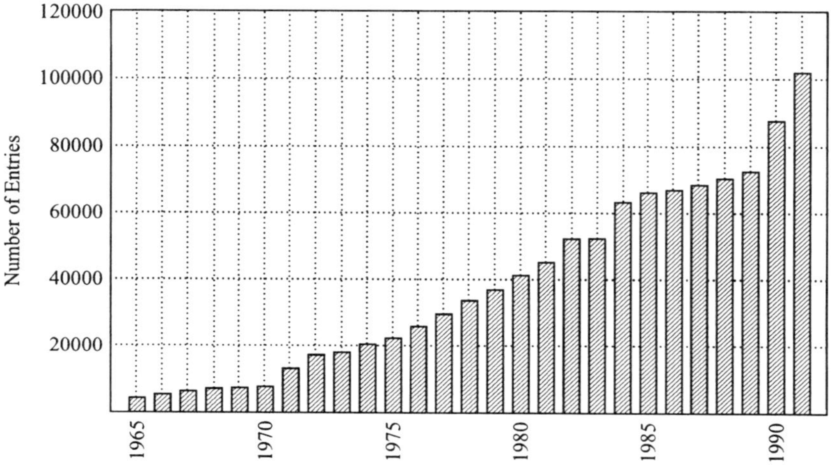

Figure 4.2 Growth of structures. (Reproduced by permission from Orpen *et al.*[17].)

extrapolation of "structure-property" relationships from crystallographic data was attempted from a very small number of extant structures. In many of these cases, the statistical significance of the results is unclear owing to the quality and small sampling of data. Today, due to the large body of precise structural data now available, statistically significant conclusions, regarding structural features, are possible. Currently, over 100 000 entries comprise the CSD (Figure 4.2), with about 10 000 structures added yearly. The systematic use of this structural pool is largely responsible for many developments in recent years (e.g. crystal engineering, host–guest inclusion compounds, Langmuir–Blodgett films, crystal engineering, non-linear optics, ceramics, protein conformation, and reactivity).

The number of users determine the success of any computer database. Even though in theory the number of users is a measurable quantity, in practice it is a very difficult number to quantify. Tabulation of the number of citations from the CSD provides a method for monitoring the CSD interest[18]. From analysis of Figure 4.3, a dramatic increase can be seen in the number of CSD citations since its debut in 1983[15,16]. These results do not necessarily correlate to an increase in usage, because every citation does not utilize the CSD systematically. In fact, many of the citations only reference the CSD in a precursory manner. More correctly, the results merely serve to indicate an increased awareness in the utility of the CSD computer database.

Several recent reviews provide insight into the capability and utility of the CSD[1]. It is impossible to present the CSD without some overlap of the past review material. Therefore, the intent of this chapter is to extend and update the previous reviews as well as present the CSD in a manner comprehendible to the

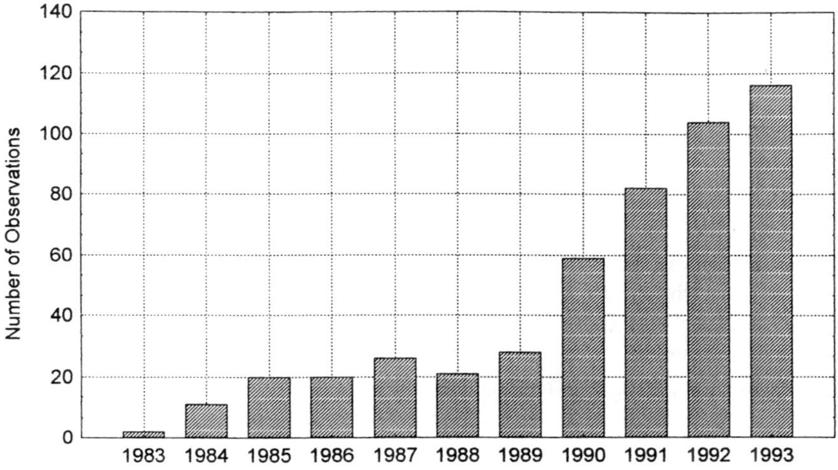

Figure 4.3 Interest in the CSD.

entire scientific community. The first section outlines the general components of the CSD software package. An exhaustive approach towards understanding each platform will not be discussed in this section, since it is impractical to provide the necessary detail; several satisfactory sources already exist[19,20]. It is, however, warranted that the information provided will help potential users understand the methods used to store, access and manipulate data. The second part of this chapter investigates the utility of the CSD by selecting representative examples in several categories. These categories include:

- Mean molecular dimensions and molecular geometries
- Hydrogen bonding
- Crystallographic applications
- Crystal packing forces: crystal engineering

The investigation of the utility of the CSD is best discussed by analysis of the literature examples. This study is not conclusive but rather meant to provide an avenue into the practical and statistical uses of the CSD based on literature examples to the end of 1994.

2. CAMBRIDGE STRUCTURAL DATABASE: AN OVERVIEW

2.1 The CSD Background

The Cambridge Crystallographic Data Centre (CCDC)[21] initially produced a compendium of all organic, organometallic and metal complex crystal structures published in the period 1960–1965[22]. This publication eventually led to a computerized database version in 1983—the Cambridge Structural Database[15,16]. Since its inception, the CSD has evolved over the last two decades into a powerful package for crystal structure analysis. The CSD not only provides the scientific community with a depository for crystallographic data, but also the ability to manipulate and graphically visualize desired structures and X-ray structural information. The CSD records include bibliographic, two-dimensional chemical and three-dimensional structural results from crystallographic analysis of organic, organometallics and metal complexes. Both X-ray and neutron studies are included for compounds containing up to approximately 500 atoms. The CSD is a fully retrospective over some 70 years and is updated on a current and continuing basis. Over 100 000 entries comprise the CSD system, with some 10 000 entries added yearly. The following overview covers the basic principles of the CSD (i.e. content, search techniques, structure visualization and methods for statistical analysis) with the intent to provide the reader with a fundamental understanding of the CSD system for further discussion on statistical analyses in the third section.

2.2 The CSD System Structure

The most important component of the CSD system is the database itself. The database consists of a single binary file (ASER), that contains all the information for each entry in the database. Each entry contains three basic types of information: one-dimensional textual and simple numeric information; two-

Table 4.1 Data types deposited in the CSD.

One-dimensional data	Two-dimensional data	Three-dimensional data
Bibliographic items • reference code • compound name • journal reference • authors *Chemical text strings* • molecular formula *Numeric items* • unit cell parameters • crystal density • *R* factor • space group number • *Z* value	*Atom properties* • atom sequence number • element symbol • number of connected non-H atoms atoms • net charge • two-dimensional coordinates *Bond properties* • atom connectivity • bond type	*Symmetry positions* • space group symmetry operators *Atomic coordinates* *Crystal atom connectivity*

dimensional chemical properties; and three-dimensional crystallographic data. A few examples of each data type are illustrated in Table 4.1.

Since the ASER file contains a large amount of information for each entry, extraction is possible for entries with specific properties (e.g. atom type, authors, bond lengths, and three-dimensional structures). This extraction process has been made user-friendly by the recent advances of a menu-driven graphical user interface to the CSD; a dramatic change over the earlier batch-mode versions of the CSD. The three main facilities within the CSD provide search techniques (QUEST3D), structure visualization (PLUTO) and methods for statistical analysis (GSTAT and VISTA). As illustrated in Figure 4.4, the search platform (QUEST3D) provides the necessary output files as input for PLUTO, GSTAT and VISTA. These facilities will be discussed in the following sections.

As stated previously, early versions of the CSD operated almost exclusively through batch-mode processes. As more facilities were added, the CSD system

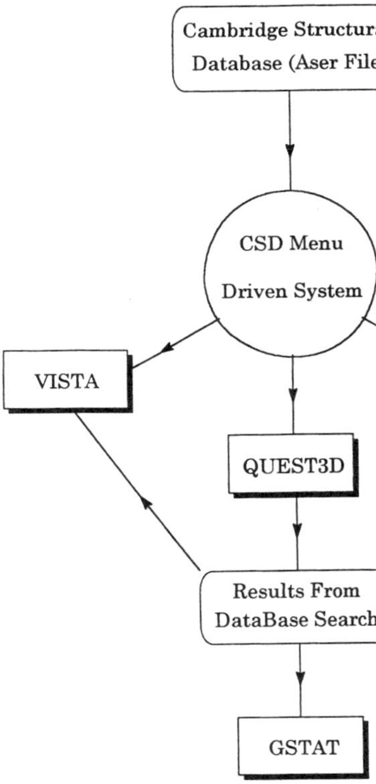

Figure 4.4 Schematic representation of the CSD platforms.

became more complex and difficult to operate, confusing the user with long lists of field codes and commands. Presently, the complete CSD package provides a friendly interactive menu, improved information content and rapid data manipulation capabilities. These developments will enable established CSD devotees and novice users to glean the desired information from the structural database.

2.2.1 QUEST3D

QUEST3D is a fully interactive search program that uses menu driven graphics for both query construction and the display of results. User defined tests utilize the information contained in the ASER file; therefore, queries must include either simple one-dimensional numeric or text, two-dimensional chemical connectivity or three-dimensional geometrical calculations. The preceding section provided several examples of each test category. Interactive construction of two-dimensional skeletal structures provide a basis for formulating chemical connectivity tests. Once a fragment or molecular structure is constructed, additional constraints (e.g. coordination number, charge and number of attached hydrogen atoms) provide a more focused structural question. In a similar manner, entries can also be recovered by three-dimensional tests. This search type permits the selection of structures in the CSD on the basis of limiting values such as interatomic distances (bonded and non-bonded), ring-distortion parameters and torsional angles. Search questions are not restricted to a specific category (i.e. one-, two- and three-dimensional); search questions may and usually do include a combination of the preceding search categories.

Processing the defined search question over the entire database results in one or more hits (entries form the ASER file). Since QUEST3D provides an interactive menu system, searches result in immediate graphical visualization of these query hits. Each hit may then be displayed as two- and/or three-dimensional diagrams. Furthermore, manipulation of these structural illustrations (e.g. rotation, translation, and variable scaling) result in closer inspection of the queries and allow hits to either be saved, rejected or used as the query for further searches. Novices quickly realize that incorrect and ill-defined queries produce either enormous or nonexistent output, whereas correct search strategies use well constructed test questions—a basic tenet of the CSD usage.

QUEST3D also allows for several types of output files. The categories of savable information include a subset of the original CSD binary search file (ASER), bibliographic information (FBIB), chemical connectivity (FCON), input for external molecular modelling packages (MODEL), two-dimensional chemical diagrams and displayed reference information in POSTSCRIPT format (POSTSCRIPT), reference codes for all hits (REFC) and a data file (FDAT). The latter file (FDAT) receives the most use since it is used as input to the programs GSTAT, VISTA and PLUTO. Each entry in the FDAT file contains

```
#AABHTZ  13770506        18  9  0  0  0  1  5 35  0  0 3713210000000000000000076
 11372 10272  7359 10875  7107  9616333222 9 5 9 6 4 8  0  0  2P-1        240
R=0.0410
211 0121 0112 0
C  68H  230  68N  68CL  99
CL1    -33550    9980  10610 CL2    -64070 -30840  32700 C1     -47750   3880  23070
C2     -57270   13370  34240 C3     -68450   9160  44830 C4     -70210  -4450  44470
C5     -60690  -13870  32950 C6     -49040 -10070  21810 C7     -38440 -19380   8860
N8     -36680  -29770  13510 N9     -26290 -38009    990 C10    -18570 -36530 -17520
N11    -21960  -38590 -33470 N12    -11480 -36040 -48430 C13     -2540 -32560 -40600
N14     -6420  -32680 -21040 N15       230 -28370  -7410 C16     -3050 -16060   7440
C17      4050  -12440  22150 O18    -11020  -8830   7930 C19    -23940 -48600   7180
C20    -33070  -50930  25480 O21    -14540 -55420  -2960 H2     -55800  23200  34700
H3     -75200   15700  53100 H4     -78400  -7800  53000 H7     -32600 -17500  -3200
H13      5700  -29600 -46900 H15      4600 -34000  -5700 H171     8100  -3600  21700
H172    10500  -19800  18900 H173     -600 -10700  33300 H201   -31300 -59800  27500
H202   -32900  -45100  37400 H203   -41300 -52500  24300
  3 7 4 5 6 7 8 3 8 910111213141216171818112121 4 5 6 915171919192222221516
```

Directory Information
Unit Cell Parameters
Symmetry Position
Radius Values

Atomic Coordinates

Connectivity Table

Figure 4.5 An exemplary FDAT file showing components.

directory information, unit cell parameters, symmetry positions, radius values, atomic coordinates and chemical connectivity (Figure 4.5).

2.2.2 PLUTO

The PLUTO program is a general plotting package, first written in the 1970s, for the display of crystal and molecular structures[23]. The program operates on FDAT files generated from the QUEST3D search program. Even though QUEST3D contains graphical output features for display of three-dimensional coordinates, PLUTO provides extra options for publication and teaching purposes. Along with many of the other changes in the CSD system, PLUTO has been given a new look with a friendly interactive menu. This allows users easy access to a wide range of drawing styles for molecules and packing diagrams. In the near future, the CCDC anticipates to integrate the PLUTO functions in the QUEST3D display menu. PLUTO's interactive commands enable the user to move to any given entry in the data set, and then examine this entry in such detail as required. These simple commands enable the user to view rotation about the plot axes and explore various molecular motifs with minimum effort.

The main function of PLUTO continues to be for elucidation of structural features of entries selected from the database. The program is particularly useful for exploring non-bonded interactions between molecules in the crystal environment. This is an area that is often difficult to visualize. Facilities are

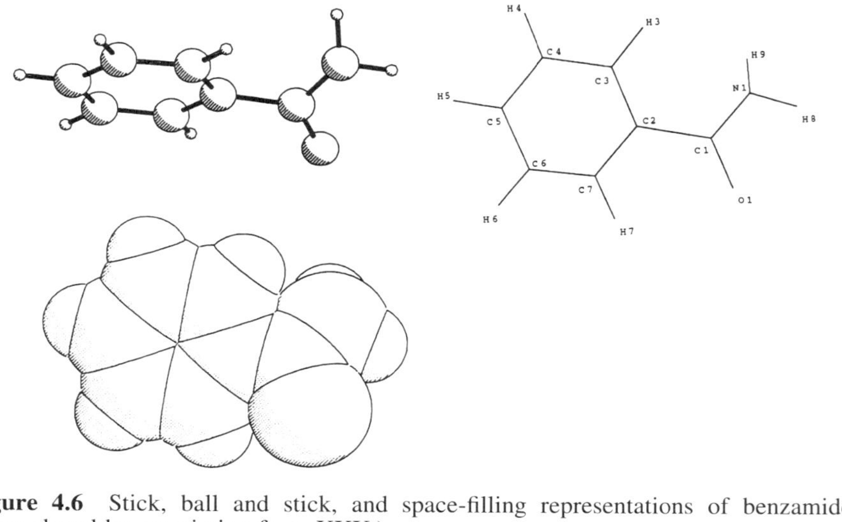

Figure 4.6 Stick, ball and stick, and space-filling representations of benzamide. (Reproduced by permission from XXX.)

provided to select the basic molecular residues and to obtain a unique set of interactions with other neighboring molecules. One can then explore outward from the base residue of interest. This leads to easy formation of networks of molecules connected by hydrogen-bonds, or ionic interactions to metals.

PLUTO provides three basic styles of illustrations (Figure 4.6):

- stick diagrams
- ball and stick diagrams
- space-filling models

These styles may either be plotted for a single molecule or for an assembly of molecules. Smooth, real-time rotation of stick molecules is achieved on several types of workstations. This allows for an effective analysis of molecules and packing motifs. The presence of PLUTO within the CSD system, provides users with stand-alone plotting capabilities that enhance the capabilities now available within QUEST3D.

2.2.3 GSTAT and VISTA

The GSTAT and VISTA facilities are multi-functional geometry programs within the CSD system. These facilities enable users to construct statistically significant conclusions by analysis of potentially large data samples of crystallographic data. In addition to visual and statistical analyses, these programs perform the following tasks:

- geometrical calculations
- location of three-dimensional substructural fragments
- geometrical calculations for molecular fragments
- output of atomic coordinates

VISTA (VIsual STAtistics), identical to GSTAT in principle, was released in April 1994. Compared to the batch-mode program GSTAT, VISTA complies with the new graphical standards set initially by QUEST3D and later PLUTO. VISTA provides the user with a menu-driven, graphical interface to the CSD system. VISTA's design allows the user to manipulate data with great ease and with the added advantage of immediate graphic visualization. In this first release, VISTA concentrates on the basics of geometrical and statistical calculations, and will eventually be capable of all GSTAT functions. At that time, GSTAT will most likely become unnecessary and removed from the system.

Since this first version of VISTA provides only the basic functions, the majority of statistical functions are still performed with GSTAT. Therefore, the following overview refers only to the GSTAT program. This discussion includes the basic operations, geometry calculations for complete entries and substructural fragments and finally visual, statistical and numerical analysis of fragment geometry.

GSTAT requires two files for the analyses of the CSD data; an FDAT file and an instruction file (*.DAT). The latter file contains alphanumeric information to control the operations of GSTAT, whereas, the FDAT file is prepared from a QUEST3D search. The output files generated by GSTAT operations may contain atomic coordinates (fractional or cartesian), fragment geometry tables, reference codes, principle component axis data, averaged coordinates and results from matching molecular fragment data.

The simplest type of GSTAT operations include calculation of basic geometry parameters for entire FDAT entries:

- calculation of intramolecular geometry—bond lengths and angles, and torsional angles
- calculation of coordination spheres—generates distances and angles within coordination spheres of specific radii
- calculation of intermolecular distances between user specified elements (non-bonded contacts)

Separate commands exist to control the inclusion or omission of specific CSD entries or elements in the calculations. Furthermore, user defined qualifiers placed in the instruction cards provide additional restrictions on output files.

The next level of structural calculations provide information on substructure fragment geometry. From a series of command lines in the instruction file, a description of a molecular fragment may be specified. A batch-mode process utilizing this instruction file "flags" all entries that possess the defined molecular fragment. Calculations that involve this data set may be manipulated, as described previously.

A typical search might look like the analysis of the non-bonded contacts to nitro moieties—more specifically, only three-center hydrogen bonds to nitro groups of type **3**. The analysis begins by structurally defining the nitro and hydrogen interaction by a series of line commands in the instruction file. Utilizing this criterion for the CSD search results not only in hits with the expected hydrogen-bond motif, but also *all* permutations of hydrogen bonds to the nitro functional group. The addition of further search criteria (i.e. $\angle N—O_1 \cdots H \approx \angle N—O_2 \cdots H$; $O_1 \cdots Z \approx O_2 \cdots Z$; and $O_1 \cdots H \cdots O_2$) to the instruction file provide the necessary constraints to uniquely describe the desired three-center structural pattern.

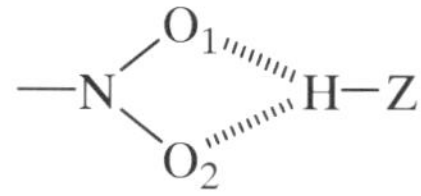

3

Initiation of the GSTAT facility based on the above criteria, results in a search of all entries in the FDAT file. Each entry is analyzed by comparison with the defined fragment composition and geometry constraints. Entries that fulfil the defined parameters are then stored in an output file.

In most instances, the definition and structural search based on fragment parameters is only the first part of a GSTAT analysis. Statistical analyses of these data enables the user to understand correlations between structures. These include examination of mean values of parameters, deviation from the mean, generation of correlations between parameters and multivariate statistical analysis. The analyses may be carried out using one or more facilities within GSTAT:

- histograms and scattergrams
- detection and rejection of outliers
- correlation and covariance
- linear regression
- principal component analysis
- cluster analysis

Such statistical analysis are performed by the addition of several command lines in the instruction file. A variety of output forms exist, as defined in the instruction file. The histograms and scattergrams allow for visual analysis of data. VISTA provides current style plots, whereas GSTAT provides antiquated graphs.

GSTAT provides an adequate method for analyzing crystallographic data. It is hoped that this overview provides the user with a basic understanding of the major components of the CSD system; namely, structure retrieval (QUEST3D), structure plotting (PLUTO) and statistical analysis of crystallographic information (GSTAT and VISTA).

3. THE UTILITY AND STATISTICALLY SIGNIFICANT CONCLUSIONS FROM THE CSD

3.1 Statistical Significance of Crystallographic Data

Before proceeding to specific studies that utilize the CSD, it is necessary to discuss the role statistics play on structural analyses. Generally speaking, the ultimate goal of every analysis is to find a relationship between variables (e.g. bond distances and angles). A basic tenet of scientific analyses states that "meaning" can only be represented in terms of relations between specific variables. In simplistic terms, statistics does nothing else but help one evaluate relations between variables.

In statistical studies, the two most elementary properties of every relationship between variables are the sample size and reliability of data. The magnitude is much easier to understand and measure than reliability. For example, if every C—C bond in your sample was found to be longer than every C═C bond, then one could say that the magnitude of the relation between the two variables is very high in your sample. The reliability of a relation is a much less intuitive concept, but still extremely important. It pertains to the "representiveness" of the result found in your specific sample for the entire population. In other words, it says how probable it is that a similar relation would be found if the experiment was replicated with other samples drawn from the same population. Typically, structural studies use all available data rather than a subset of an entire population. Even so, results obtained from statistically meaningful studies performed decades ago should strongly correlate to structures determined after the original study, within the defined criteria. Any agreement between the specified criteria represents a relationship between the variables. From a strict statistical perspective, the reliability of this relationship can then be quantitatively estimated and represented using a standard measure—technically called p-level or statistical *significance* level.

The statistical significance of a result is an estimated measure of the degree of which it is *true*. More technically, the value of the p-level represents a decreasing index of the reliability of a result. The higher the p-level, the less one can rely that the observed relation between variables in the sample is a reliable indicator of the relation between the respective variables in the population. Specifically, the p-level represents the probability of error that is involved in accepting an observed result as valid, that is, as "representative of the population". For example, a p-level of 0.05 indicates that there is a 5% probability that the relation between the variables found in a sample is a fluke.

There is no way to avoid arbitrariness in the final decision as to what level of significance will be treated as really "significant". That is, the selection of some level of significance, up to which the results will be rejected as invalid, is arbitrary. In practice, the final decision usually depends on whether the outcome was predicted a priori or only found post hoc in the course of many analyses and comparisons performed on the data set, on the total amount of consistent supportive evidence in the entire data set, and on "traditions" existing in the particular area of research.

Since crystallographic measurements are fundamental to any structural study, it is essential that these measurements be *correct*. As a result of current state-of-the-art automated diffractometers and fast, powerful methods for interpreting diffraction measurements, crystal structures reported in the literature tend to be statistically reliable—but not always. A few basic criteria exist for assessing the likely correctness of a structure. One measure of the correctness is the so-called discrepancy index (R factor), which denotes how well the calculated structural model fits the observed data. At present, R values in the range of 0.02

to 0.06 are being quoted for the most reliably determined structures (an *R* factor of 0.00 represents a perfect correlation between the calculated model and observed data). Since many resolved structures, even published structures[24], are deemed incorrect, the *R* factor is by no means a definitive index of the reliability of structure analyses.

Structures with acceptable *R* values may then be checked by analysis of residual electron density. Electron density patterns arise from manipulation of diffraction measurements. In simple terms, diffraction occurs from X-ray or neutron scattering as a result of atomic electron clouds or nuclei. A well determined structure accounts for a majority of the observed electron density, whereas, inaccurate structures typically show residual density. In addition to electron density, any anomalies in the molecular geometry or arrangements provide insight to possible associated with the general assignment of the structure.

Errors associated with crystal structures fall into one of two categories, random or systematic. These errors affect both the precision and accuracy of a structure. The precision of an experimental result, usually in terms of an estimated standard deviation, is a measure of the reproducibility of the observed value if the experiment were to be repeated. The standard deviations of the observed results (e.g. distances, angles, etc.) can be estimated by statistical methods. The basic assumption involved in the estimation of standard deviations is that fluctuations in observed quantities are due solely to random errors.

If the distribution of errors is normal, statistical correlations can be used to assess the probability that one observation is "significantly" different from another. For example, consider a hypothetical study of two crystal structures that contain β-ketoester fragments. The ketonic bond distance of each structure was determined to be 1.231 Å and 1.224 Å. In both structures, the bond distances were assigned an estimated standard deviation (e.s.d.) of ± 0.005 Å (hereafter represented as 1.231(5) Å and 1.224(5) Å. Since the two measurements do not differ from one another by more than twice the e.s.d., there exists a 5% probability that the difference between them represents a random fluctuation. Measurements that differ by 2.7 times the e.s.d. represent a 1% probability that the difference is an incidental occurrence. As discussed previously, the level at which two or more measurements are deemed "statistically significant" is somewhat arbitrary. No specific limit has been established for statistically significant correlations; however, literature examples provide precedence for the treatment of crystallographic data. At the lower level, differences between measurements with 2 e.s.d. differences are generally acceptable, however, more conservative approaches implement a 2.7 e.s.d. confidence level.

The examples discussed in the following sections represent a broad range of applications that utilize the CSD. Not every case attempts to find meaning in the results by statistical analysis, at least in the strictest sense of the phrase. Nonetheless, these results provide significant contributions to the understanding of molecular crystals, even though the results are not necessarily statistically significant. For this reason, these notable applications of the CSD are included in this discussion.

3.2 Mean Molecular Dimensions and Molecular Geometries

By far, the purpose of most CSD searches concentrate on the correlation of specific molecular geometries (i.e. bond lengths and angles) for structurally-related compounds. These studies allow identification and assignment of statistically significant uniformities and irregularities in molecular structures. Even though the investigation of bond distances and angles seems simplistic and uneventful, studies in this area often provide deeper insight into chemical dynamics and structural relationships. In the first section of this chapter, the work discussed by Bürgi and Dunitz[4] uncovered important mechanistic details for addition reactions based primarily on the analysis of interatomic structural parameters. This seminal work solidified the significance of geometrical studies, even at a time prior to well-developed structural databases. The present databases, specifically the CSD, provide superior tools to perform such studies.

It has long been realized that most structural parameters (i.e. bond lengths) for a series of related compounds tend to be constant, particularly in saturated compounds. For example, in Sutton's list of selected bond lengths, a value of 1.426(5) Å is given as the length of the C—O bond[25]. The observed e.s.d. of ± 0.005Å represents a consistent C—O bond length over the entire data set studied; however, since the publication of Sutton's tables, many structures reveal statistically significant variations from the average C—O bond length reported[26–28]. These variations, if systematic, can provide valuable mechanistic information from a detailed knowledge of the ground state of each reacting system, particularly the length of the bond being broken.

An extension of the approach pioneered by Bürgi and Dunitz[34] explored the early stages of bond breaking in R_1—OR_2 organic structures[27]. This three-part work by Kirby *et al.* analyzed the CSD for C—O bond length and determined correlations between bond lengths and reactivity.

A study of 22 X-ray crystal structure determinations of tetrahydropyranyl acetals (**4** and **5**) revealed that the C—O bond lengths are not even approximately constant but vary over a range of more than 0.1 Å[28]. This variation is systematic and depends on the effective electronegativity of the oxygen atom of the exocyclic OR group: the more electronegative the group, the longer is the C—OR bond and the shorter the endocyclic bond. This is most evident in the axial form (**4**), since bond length changes can be interpreted in terms of the stabilization of $n-\sigma^*$ interactions between the axial lone pair on the ring oxygen and σ^*_{C-OR}. The equatorial form (**6**) also possesses similar effects on lengths, even though unfavorable n–σ^* stereoelectronic overlap exists. This observation was further explained by invoking σ–σ^* interactions between σ^*_{C-OR} and the antiperiplanar σ-bonding orbital of the remote ring C—O bond.

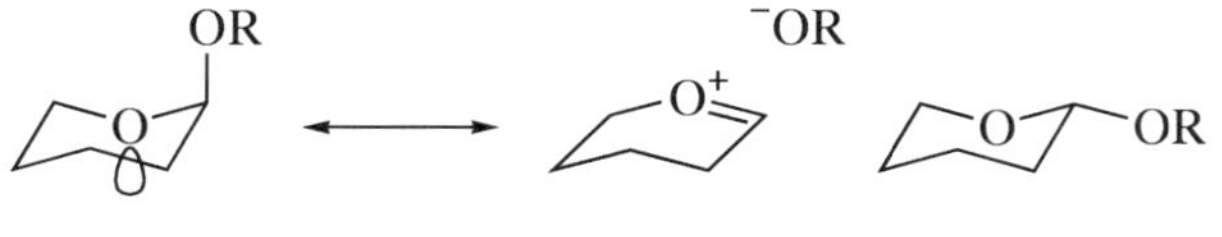

4 **5** **6**

In order to explain the general effect of electronegativity on the C—O bond length, the authors probed 2367 ether and ester fragments extracted from the CSD[27]. The C—O bonds represented in the data were observed to be well-defined with systematic variations in bond lengths. The plots shown in Figure 4.7 suggest a relationship exists between bond length and the effective electronegativity of the oxygen atom. Electronegativities of each ether or ester fragment was determined by the approximate pK_a value of R_2OH. As revealed in figure 4.7, the length of the bond in the R_1—OR_2 system increases with increasing electron withdrawal character in the group OR_2 and also as the group R_1 changes from methyl through primary, secondary, and tertiary alkyl. At first glance, this information suggests the electronegativity of OR_2 as well as the size of R_1 controls the R_1—OR_2 bond length.

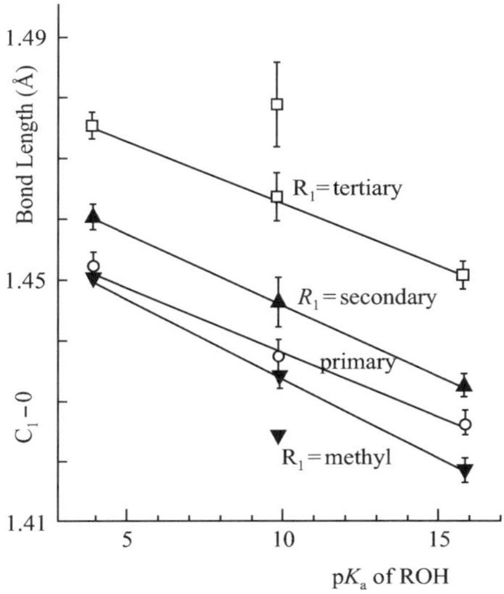

Figure 4.7 Plot of C—O bond lengths versus the approximate pK_a of R_2OH (approximate K_as of 16, 10, and 4 for alcohols, phenols, and carboxylic acids, respectively). Lines are drawn through data sets for R_1 = methyl (▽), primary (○), secondary (△) and tertiary (□). (Reproduced by permission from Kirby *et al.*[27])

From investigation of a series of tertiary esters, Kirby *et al.* determined that electronic effects best account for the lengthening of the C—O bond and not steric effects. If nonbonded (steric) interactions governed the bond length, similar C—O bond lengths should exist. However, from the authors investigation, the variation in the C—O bond ($\Delta = 0.12(1)$ Å) was significant. As a result, the effects on bond length was explained in terms of a variable contribution from the ion pair form (**7**). The ground-state structure depends on the stabilities of both R_1^+ and R_2O^-.

$$R_1\text{—}OR_2 \leftrightarrow R_1^{+}\,{}^{-}OR_2$$

7

Analysis of the structural as well as the kinetic data for the tetrahydropyranyl acetal axial systems (**4**) allowed exploration of the reaction-coordinate for the acetal cleavage reaction **4** → **5**[29]. Since a linear correlation between the length

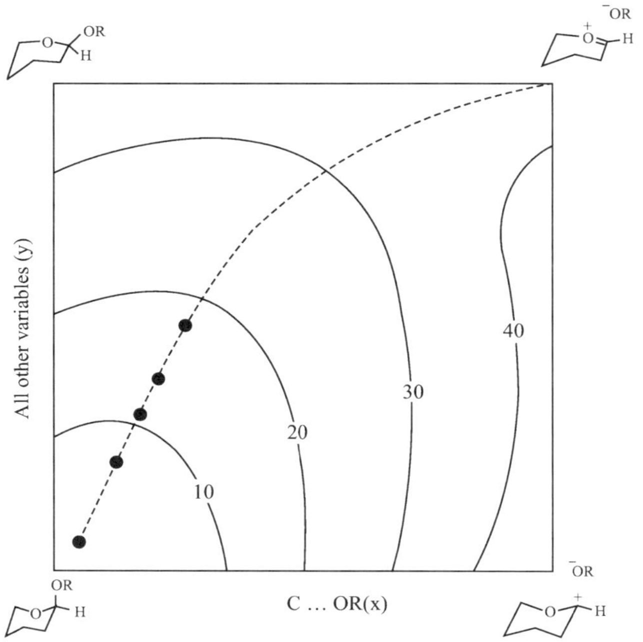

Figure 4.8 Speculative reaction coordinate–energy contour diagram for the conversion of 1 to 2. The energy contour scale (in kcal mol^{-1}) is intended to give only a general indication of the shape of the surface. The reaction coordinate (dashed line) traces the minimum energy path from 1 to 2 and the circles give an approximate indication of the location of the experimental points. (Reproduced by permission from Kirby *et al.*[29].)

of the C—O bond and the free energy of activation for its cleavage exists, the definition of the reactivity based on bond length for this series of compounds is possible. This relationship also implies that there is a substantial linear region in the first half of the reaction coordinate for cleavage of this bond (Figure 4.8). By implementing structural as well as kinetic data, a "true" structure-reactivity relationship could be established for the hydrolysis of **4**.

The utility of the ether functional group has long been realized[30], however, the correlation between reactivity and structure is not clearly understood. Ethers display a significant variance about the oxygen atom as evident from the dimethyl ether ($\angle$C—O—C = 112°)[31], methoxysilane ($\angle$Si—O—C = 121°)[32,33], and disiloxane ($\angle$Si—O—Si = 144°)[34] examples. This observation has been theorized in terms of mixing of oxygen lone-pair (n) and silicon 3d orbitals[34–36], steric effects/electrostatic repulsions[37–41], and delocalization of nonbonding orbitals on oxygen into empty orbitals on silicon[42–47].

In an effort to further explain this structural variance, Shambayati and coworkers[48] combined analysis of the CSD data and results of *ab initio* calculations to provide a clearer explanation for the geometrical preferences and chemical characteristics of silyl ethers. Specifically, structural studies were implemented in order to challenge the accepted notion that angle widening is accompanied by steric congestion[37–41]. A search of the CSD for acyclic Si—O—C_{sp^3} fragments with four-coordinated silicon revealed 127 substructures with an average Si—O—C bond angle of 134.2° and Si—O bond length of 1.62 Å. From inspection of Figure 4.9, the authors report an

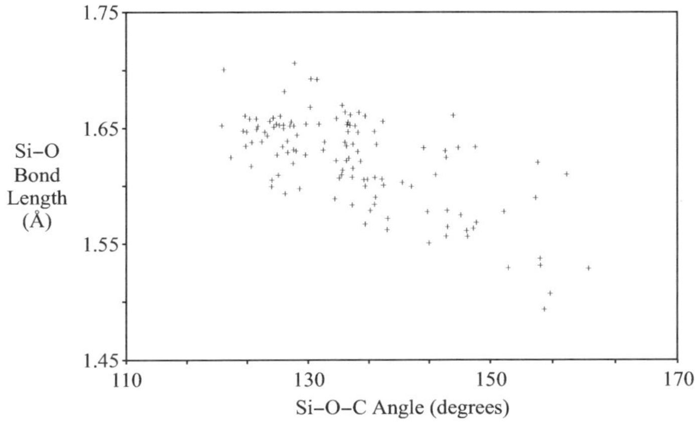

Figure 4.9 Correlation of Si—O bond lengths and Si—O—C bond angles for acyclic Si—O—C fragments with four-coordinated silicon. (Reproduced by permission from Shambayati *et al.*[48].)

inverse relationship between Si—O bond lengths and Si—O—C bond angles. This observation clearly opposes predictions based solely on steric arguments. Thus, it appears that the proposed steric effect is inconsistent with the structural data.

Selected structures of trimethyl silyl ethers further illustrate this inverse bond length/bond angle relationship (see **8**, **9**, **10** and **11** in Figure 4.10). These representative examples indicate that an increase in steric congestion around the trimethyl silyl ether group is accompanied by a widening of the Si—O—C angle and a shortening of the Si—O bonds. This is most apparent from comparison of silyl ethers derived from primary (**8**) and tertiary (**11**) alcohols. The difference in bond angles ($\Delta = 14.87(16)^\circ$) and distances($\Delta = 0.019(19)$Å) indicate a possible structural trend; however, as indicative of the correlation

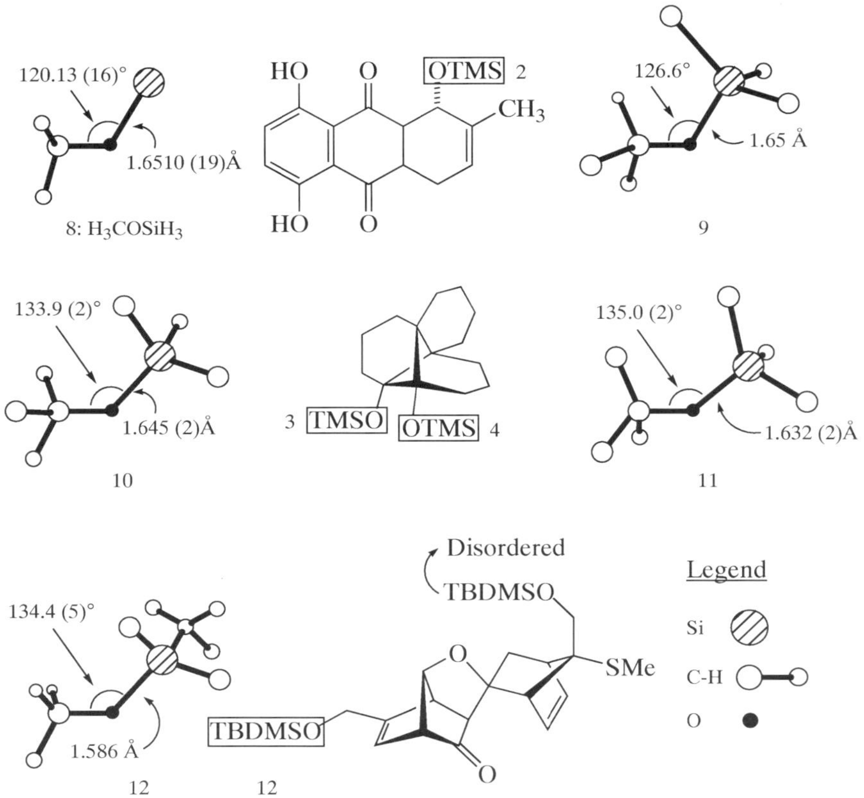

Figure 4.10 Comparison of Si—O—C bond angles and Si—O bond lengths from various crystal structures. (Reproduced by permission from Shambayati *et al.*[48].)

coefficient for Figure 4.9, not all entries exhibit this pattern (see **12** in Figure 4.10). Even so, inspection of the data, as a whole, strengthens the bond angle/bond length relationship based on crystallographic data.

In conjunction with the analyses of crystal structures, *ab initio* MO calculations suggested possible factors for the wide Si—O—C bond angles and the low oxygen basicity of silyl ethers. A frontier orbital analysis of the problem revealed that this structural trend results impart from the higher energy $\pi(SiR_3)$ than $\pi(CR_3)$ group orbitals and poorer mixing of $\pi(SiR_3)$ orbitals with the oxygen lone pair. This results from the required 2p–3p overlap. Additionally, this structural trend occurs from some mixing of the lone-pair orbitals with the relatively low-lying $\pi^*(SiR_3)$ group orbitals. Both effects lead to lower energies and smaller oxygen components for the HOMOs of silyl ethers compared to alkyl ethers and to wider bond angles at oxygen.

3.3 Hydrogen Bonding

Although the hydrogen bond is not a strong bond (its bond energy, that is, the energy of the reaction $XH + Y \rightarrow XHY$, being only about 5 kcal/mole), it has great significance in determining the properties of substances. Because of its small bond energy and the small activation energy involved in its formation and rupture, the hydrogen bond is especially suited to play a part in reactions occurring at normal temperatures. It has been recognized that hydrogen bonds restrain protein molecules to their native configurations, and I believe that as the methods of structural chemistry are further applied to physiological problems it will be found that the significance of the hydrogen bond for physiological is greater than that of any other single structural feature.

L. Pauling[49]

Even from the first edition (1939) of Pauling's *The Nature of the Chemical Bond*[49], it is not difficult to understand the long-standing interest in the subject of hydrogen bonds. His insightful account of the hydrogen bond proved to be well beyond its years. For example, Pauling's estimate of the energy of hydrogen bonds is virtually identical to that calculated decades later[50]. It is quite remarkable that subsequent editions of the now famous work merely serve to incorporate the scientific developments of the time and many of the basic principles covered in that first edition hold true even today.

Currently, a lot of interest in the area of hydrogen bonds centers on crystal engineering, which deals with the prediction, control and exploitation of crystalline patterns. Efforts that focus on this formidable task tend to seek an understanding of hydrogen bonds at the molecular level (i.e. crystal packing patterns)[51]. Since molecular aggregates formed by hydrogen bonds in the solid state are virtually static and well defined, crystallographic techniques provide the necessary tools to elucidate the influence of hydrogen bonding on crystal packing. To achieve a broad understanding of hydrogen bonding in solids, the most successful investigations tend to incorporate a large number of related structures. Until recently this proved to be almost an overwhelming task. However, with the improvements in crystallography techniques and structural databases, access to large amounts of *quality* structural data is now a reality.

Jeffery and Maluszynska investigated 832 hydrogen bonds in 45 crystal structures of barbiturates and 214 crystal structures of purines and pyrimidines collected from the CSD[52]. From this structural data, they determined that certain functional groups tend to be mutually oriented in the crystal in specific patterns. The retrieval criteria implemented for this study included structures from the CSD class groups (i.e. barbiturates, purines and pyrimidines) and excluded structures with metal atoms, unreported hydrogen atom coordinates and agreement factors (R) greater than 0.10.

The criteria established for two-center hydrogen bonds included only those arrangements that consisted of one nearest neighbor (<3 Å) and an $X—H\cdots A > 90°$ interaction. In a similar fashion, the search for three-center bonds included only those arrangements that consisted of two acceptor atoms; with the hydrogen atom within 0.3 Å of the plane defined by the two acceptor atoms and the atom covalently bonded to the hydrogen atom.

After inspection of the CSD data, a comparative study of the barbiturate and purine/pyrimidine structures resulted in a large population of $Nsp^2—H\cdots O{=}C$ arrangements. The barbiturate and purine/pyrimidine mean values for 96 such interactions were 1.856(58) Å and 1.848(103) Å, respectively. Because no significance could be assigned to the difference in these values, the authors combined the results of the two classes for the remainder of the hydrogen-bond study.

Conclusions from this study were drawn from the hydrogen-bond combinations that contained greater than 10 entries; analyses based on small samplings tend to be statistically unfounded. This study revealed a sequence of hydrogen-bond lengths of: $Nsp^2—H < Hsp^3— < O_WH < N(H)H$, with an acceptor sequence of $O_W < O{=}C < Nsp^2 \approx O$.

Investigation of the $R_2—H\cdots O{=}C$, $R_2N—H\cdots Nsp^2$, $RN(H)H\cdots Nsp^2$, and $RN(H)H\cdots O{=}C$ two-center bonds resulted in the histograms shown in Figure 4.11. The peak values* for the bond lengths of the first three interactions (1.84, 1.87 and 1.99 Å) correspond closely to their mean values (1.85, 1.88 and 2.06). Analysis of the last interaction type, $RN(H)H\cdots O{=}C$, revealed a peak value of 1.88 Å; which is 0.1 Å shorter than the mean value, due to the large number of distances between 2.0 and 2.4 Å.

The hydrogen-bond angles for these structures reveal a similar pattern. From analysis of the bond angle distributions in Figure 4.12, the expected trend that

* The peaks in the histograms correspond to a minimum in an effective potential energy for a solid phase system and are useful for crystal structure model building.

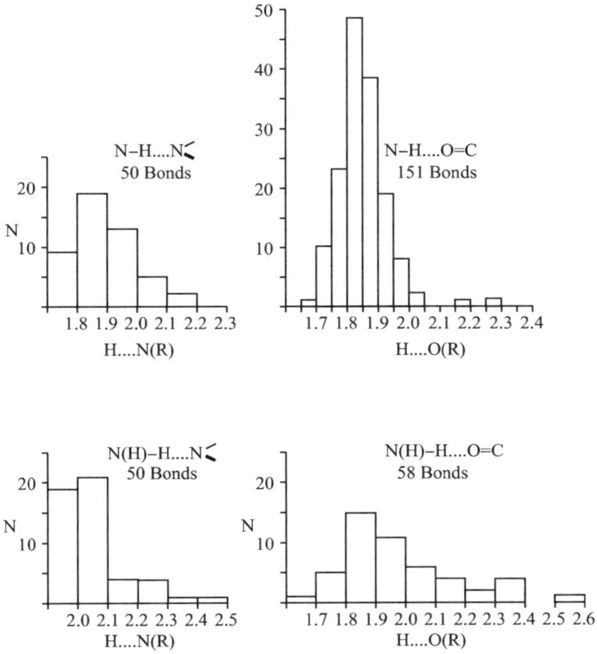

Figure 4.11 Distribution of two-center hydrogen bond lengths for $R_2N{-}H\cdots O{=}C$, $R_2N{-}H\cdots Nsp^2$, $RN(H)H\cdots Nsp^2$, and $RN(H)H\cdots O{=}C$. (Reproduced by permission from Jeffrey *et al.*[52].)

stronger bonds tend to be more linear in nature holds true. Of the entire data set studied 27% consists of three center bonds, and exist in a significant amount of the hydrogen-bond combinations. Types **I** and **II** in Figure 4.13 differ by the attachment of the acceptor groups. The authors refer to these relationships as either unchelated or chelated three-center bonds. No examples of **III** were detected in the study. Type **IV** was observed only as a subset of **V**.

Often, comparison of specific structural features of a compound to the CSD data provides evidence for structural irregularities/normalities as in the study of 1,2,4-triazolo[3,4-*b*][1,3,4]thiadiazoles (**15**). Current interests in the field of cancer chemotherapy include the study of bicyclic [5,5] ring systems containing a 1,2,4-triazole moiety. As a continuing effort to prepare bridgehead nitrogen heterocycles from N-amino heterocycles, Molina and coworkers investigated [1,2,4]triazolo[1,5-*c*]quinazolin-4-ium compounds (**13**). On treatment of **13** with hydrochloric acid, ring opening to afford the 1,2,4-triazoles **14** occurred. Further treatment with phosphorus pentasulfide converted derivatives **14** into the corresponding thiadiazoles **15**, rather than the expected 1,2,4-triazoles **16**. Since an unambiguous structural assignment of **15** could not be achieved from

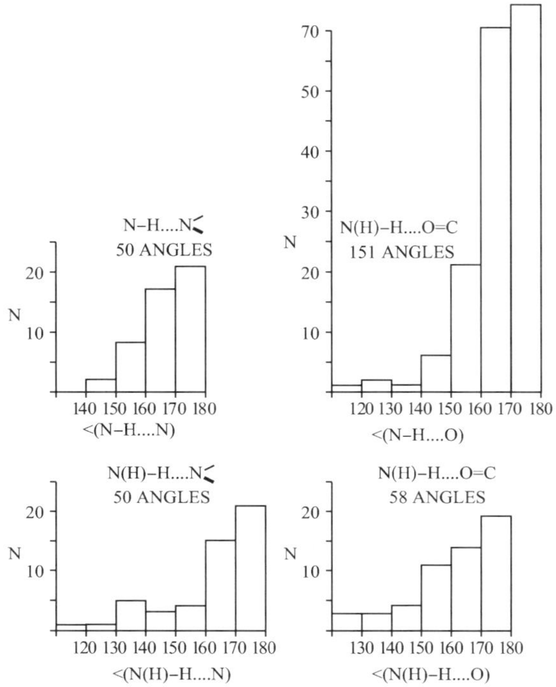

Figure 4.12 Distribution of two-center hydrogen bond angles for $R_2N—H\cdots O{=}C$, $R_2N—H\cdots Nsp^2$, $RN(H)H\cdots Nsp^2$, and $RN(H)H\cdots O{=}C$. (Reproduced by permission from Jeffrey *et al.*[52].)

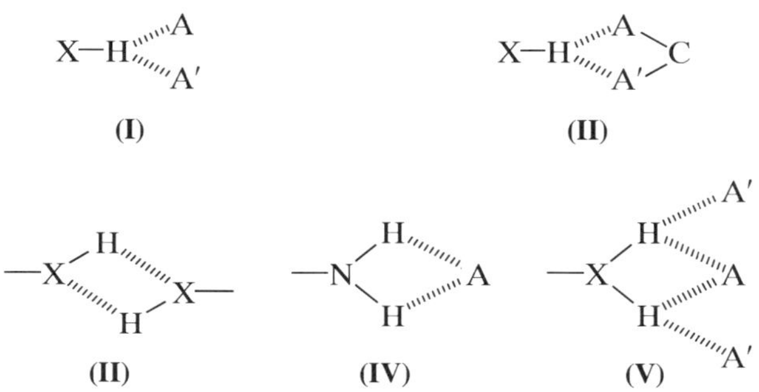

Figure 4.13 Possible hydrogen-bond patterns for barbiturate and purine/pyrimidine crystal structures.

analysis of spectral data alone, an X-ray crystallographic analysis of the toluene derivative of **15** ($R = 4\text{-}CH_3{-}C_6H_4$; hereafter **15a**) was, therefore, performed.

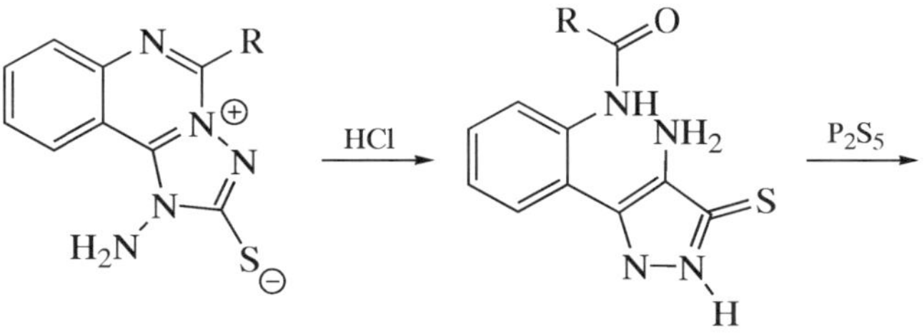

13 **14**

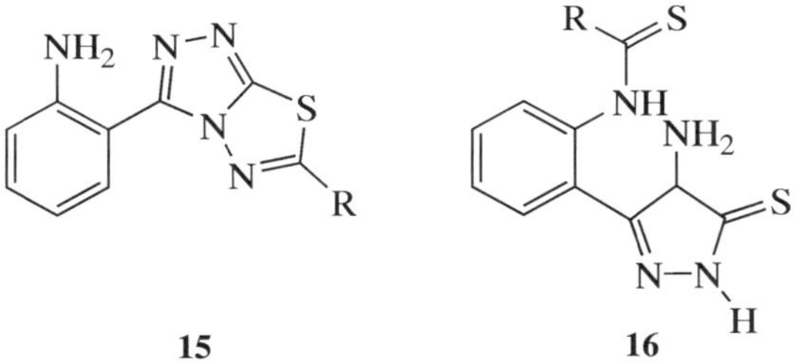

15 **16**

In order to compare the geometrical characteristics and $NH_2 \cdots Nsp^2$ interactions of **15a** with the literature, several CSD searches were performed. The hydrogen-bond study included only those structures that contained: error free data, $NH_2{-}W{-}X{-}Y{-}Nsp^2{-}Z$ fragments (where W, X, Y and Z represent any atom type), $NH_2 \cdots Nsp^2$ contacts less than 3.2 Å, $N{-}H \cdots Nsp^2$ angles greater than 140° (intermolecular) and 120° (intramolecular), organic structures, no disorder and *R* factors less than 0.10. Using these criteria, the authors CSD search resulted in 307 and 26 inter- and intramolecular interactions, respectively. The results from this investigation are presented in Table 4.2.

In compound **15a**, intramolecular N···N (2.706(6) Å) and NH···N (1.86(7) Å and 134(5)°) contacts exist. The difference between these results and the mean CSD N···N and NH···N interactions is not statistically significant; however, the authors suggest that this intramolecular contact may contribute to the planarity of the triazole-analine moiety of **15a**. From analysis of the crystal structure of **15a**, the authors discovered molecules of **15a** packed in strips along the *a* axis. This motif is connected by $N{-}H \cdots Nsp^2$ intermolecular contacts (N···N (3.045(7) Å) and NH···N (2.05(7) Å and 174(6)°)). Comparison of the intermolecular geometry criteria with that from the CSD mean search results also revealed no statistically significant differences.

Table 4.2 CSD inter- and intramolecular results for $NH_2 \cdots Nsp^2$ interactions (Å, °).

	Intermolecular contacts			
	Mean[a]	s.d. sample	s.d. mean	Range
N···N	2.788	0.132	0.026	2.645–3.076
NH···N	2.06	0.15	0.03	1.81–2.40
N—H···N	132	5	1	124–149
	Intramolecular contacts			
	Mean[a]	s.d. sample	s.d. mean	Range
N···N	3.037	0.078	0.004	2.805–3.199
NH···N	2.16	0.12	0.01	1.86–2.43
N—H···N	164	9	1	140–179

[a]Unweighted estimators.

Even though Molina and coworkers discovered that the hydrogen-bond system of **15a** is not statistically different compared to similar structures reported in the literature, this does not suggest that the information gleaned is meaningless. Often, statistically insignificant relationships provide supportive evidence to account for structural features.

3.4 Crystallographic Applications

The orientation of molecules in a typical crystalline lattice is best described by one or more symmetry operators. These operators consist of centers of inversion, rotation axes, mirror planes, translation displacements, screw rotations (a combination of rotation and translation operators) and glide reflections (a combination of mirror and translation operators). If these operators contribute equally to the arrangement of molecules in crystals, the distribution of each operator in all known structures should be uniform. However, the disproportionate occurrence of symmetry operators for organic materials was recognized by Kitaigorodskii[53] in the 1960s. In his analysis of crystal packing, Kitaigorodskii pointed out that organic molecules crystallize preferentially in space groups that allow close packing; most evident by the >50% occurrence of organic structures in the space groups $P2_1/c$ and $P\bar{1}$. The center of inversion represents the mutual feature between these space groups. This suggests a preference for organic molecules to pack in centrosymmetric patterns.

Since Kitaigorodskii's initial study of crystal packing, a 20-fold increase in the number of reported structures has occurred. Even with the increase in structural data, recent studies on crystal packing have only served to substantiate this initial study[54–61]. These studies clearly illustrate the uneven distribution of symmetry elements in organic structures tabulated over all

Table 4.3 Frequency with which each operator is top ranking in each space group. (Reproduced by permission from Filippini *et al.*[56].)

Space group	Z	No. of structures	Frequency (%) T	I	S	G
$P\bar{1}$	2	121	13	87	–	–
$P\bar{1}$	1	37	100			
$P2_1$	2	93	19	–	81	–
$P2_12_12_1$	4	206	26	–	74	–
$P2_1/c$	4	410	15	37	27	21
$P2_1/c$	2	118	45		55	
$Pbca$	8	61	2	16	15	67
$Pbca$	4	22	0		100	

organic centrosymmetric and noncentrosymmetric structures[54,55], by revealing a statistical preference for centrosymmetric molecular arrangements. Of the entries studied in the CSD, 67% exist in centrosymmetric space groups. This value represents a lower limit since the studies include enantiomers with molecular orientations strictly constrained to noncentrosymmetric packing patterns.

Another approach towards understanding the relative importance of symmetry operators quantitatively analyzed operator contributions to the lattice energy[56]. Filippini *et al.* assembled a database, extracted from the CSD, consisting of 1356 hydrocarbon, oxahydrocarbon and azahydrocarbon crystal structures. For each entry, calculation of the packing potential energy[62] provided evidence for the relative importance of the symmetry operators. Analysis of Table 4.3 shows the frequency with which the inversion center (I), translation (T), two-fold screw (S) and glide (G) operators are top ranking in each space group. Filippini *et al.* concluded that those calculations suggest that each symmetry operator contributes unequally to the total lattice energy. Certain operators appear to be silent, that is, they do not contribute significantly to the packing, but appear as unavoidable products of the combination of other operators. A global analysis of the distribution of organic molecules among space groups, together with these results, suggest that the order of importance of operators is: inversion centers > screw axes > glide planes > translation displacements.

One implication of centrosymmetric arrangements was observed by Wallach[63] in 1895. His comparative study of eight pairs of related structures uncovered that racemic crystals possess greater densities than their chiral counterparts—hence Wallach's rule. This preference for tighter molecular packing in racemic crystals has been the topic of much debate and only in recent years has it been adequately addressed[64,65].

The arrangement of molecules in enantiomeric structures *must* differ markedly from racemic structures since the former lack inversion, mirror, and/or glide operators. Since organic molecules do not preferentially orient in mirror relationships, observed by the lack of structures in space group *Pm*, centrosymmetric and/or glide plane relationships must provide an explanation for the difference in densities. In a continuing effort to study Wallach's rule, a comparison of densities of 129 chiral/racemic pairs acquired from the CSD only proved to substantiate the rule[64]. This study revealed that the difference in densities of achiral molecules in centrosymmetric and noncentrosymmetric space groups (i.e. polymorphic forms) were not statistically different. However, racemic crystals composed of enantiomers, pack on the average 1% more tightly than their chiral counterparts. Thus for enantiomeric/racemic pairs, centrosymmetric relationships represent a more favorable packing arrangement of molecules.

Since the proceeding studies suggest a high propensity for organic molecules to arrange in centrosymmetric relationships, what direct evidence exists to support this structural theorem? Whitesell *et al.* demonstrated that the bias for centrosymmetry can be turned to advantage to provide a desirable alignment of molecule[66]. In the absence of hydrogen bonds, molecular shape represents a dominant factor in determining the high preference for centrosymmetric alignments, since molecular dipoles do not correlate significantly to centrosymmetry[54]. Enantiomerically pure sulfoxide **17** was designed with two segments of nearly identical shape (i.e. *N*,*N*-dimethyl and isopropenyl). Even though it is impossible to predict the structure of **17**, or any material, from knowledge of its chemical composition, several truths holds fast: one is that it is physically impossible for chiral molecules to arrange in "true" centrosymmetric relationships. Even so, packing of **17** was predicted to occur in an

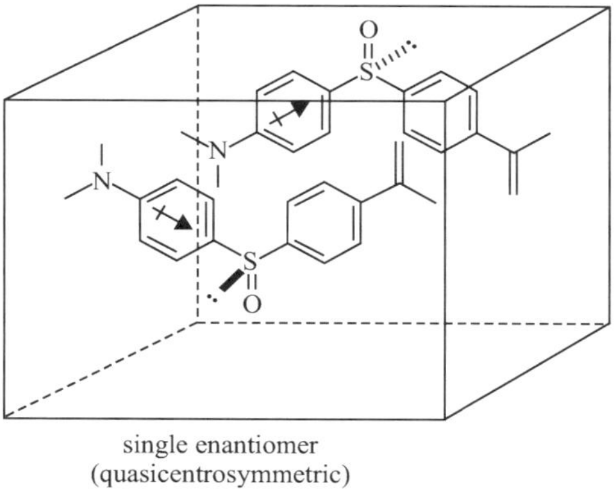

single enantiomer
(quasicentrosymmetric)

Figure 4.14 Possible quasicentrosymmetric packing for a single enantiomer of the nearly symmetric sulfoxide **17**. (Reproduced by permission from Whitesell *et al.*[66].)

approximate centrosymmetric fashion, with pairs of **17** arranged as would be expected for a true racemic crystal (Figure 4.14). The crystal structure of the *S*-**17** enantiomer showed, that indeed, approximate centrosymmetrically related pairs of molecules existed in the crystal in approximate space group $P2_1/c$. This example not only illustrates the dominant factor of centrosymmetric packing, but also the utility of symmetry, in conjunction with molecular shape, as a controllable determinant of orientation in molecular crystals.

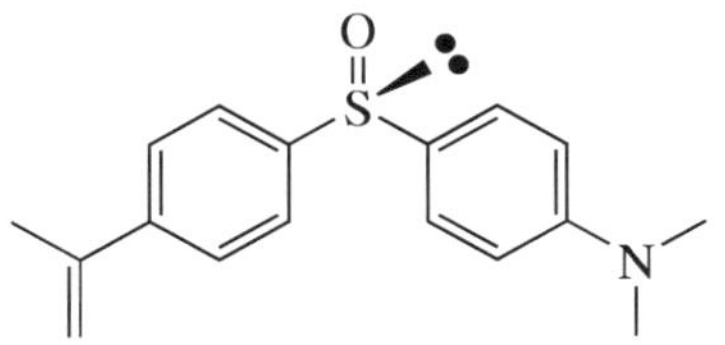

17

In an attempt to elucidate the importance of symmetry operators in crystal packing, Desiraju *et al.* studied 1166 crystal structures in $P\bar{1}$ with multiple molecules per asymmetric unit[67]. Identification of approximate inversion symmetry in those structures was accomplished with an algorithm that calculated interplanar angles between mean planes of all chemical residues. Using this procedure in conjunction with manual inspection of the structures, about 20 entries (2%) revealed symmetry-independent molecules related by an approximate inversion symmetry. The idea that approximate symmetry exists between molecules is not new; studies of macromolecular structures (i.e. proteins, peptides, polymers) frequently reveal near-symmetry relationships[68,69].

As an extension of Desiraju's initial effort to understand the importance of approximate symmetry on molecular arrangements, Davis *et al.* focused on the identification of all approximate symmetry operators in crystal structures[70,71]. This study included structures that contain two or more molecules per asymmetric unit, one distinct molecular residue and fewer than 60 non-hydrogen atoms. A systematic search in the CSD revealed 2,503 such entries in space groups $P1$, $P\bar{1}$, Pc, $P2_1$, $P2_1/c$, $P2_12_12_1$, $Pbca$. These entries constitute 9.4% of the total organic structures for those space groups (26 499 entries).

The methodology used to search for approximate symmetry in these structures employed a series of computational algorithms with the existing CSD structural information. The criteria in each algorithm allowed a description of the relationship between symmetry-independent molecules in terms of a symmetry operator, i.e. inversion, translation, rotation, screw, and mirror.

Shown in Figure 4.15 is the computational approach employed for each symmetry operator applied to a hypothetical molecule. The investigation of inversion, rotation, screw and mirror symmetry relationships entailed calcula-

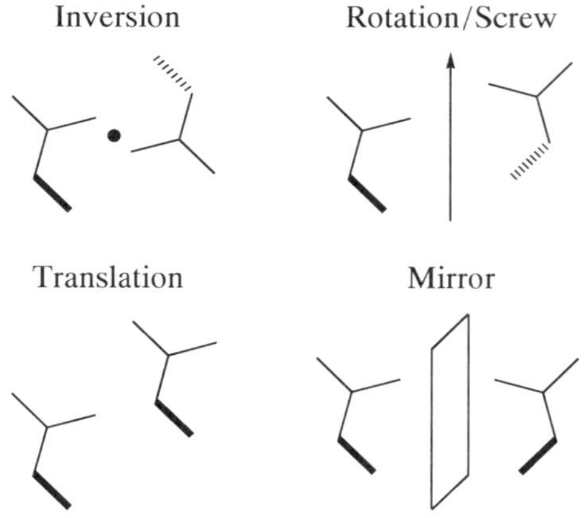

Figure 4.15 Scheme for computational identification of approximate symmetry relationships.

tion of all interatomic midpoints for the symmetry-independent molecular pair. An approximate symmetry relationship exists between molecules (n atoms per molecular fragment) only when n of these midpoints lie on or near the approximate symmetry element (i.e. point; inversion, line; rotation and screw; plane, mirror). In a similar manner for translation symmetry, an approximate symmetry relationship between molecules exists when n interatomic vectors are comparable in length and nearly parallel. The degree of symmetry for each structure was defined by the parametrer ϕ_N, where N = inversion (I), rotation/screw (R), translation (T), and reflection (M)*. As ϕ_N approaches unity, the probability that the relationship between molecules is described by the symmetry operator increases.

As a result of the tolerances imposed on the calculations, the possibility exists that a structure with ϕ_N near unity will not possess an approximate symmetry relationship. In an attempt to identify structures with "true" approximate

---

*

$$\Phi_I = \frac{\text{\# of vector midpoints} \leq 0.35\ \text{Å from the centroid}}{\text{\# of atoms per fragment}}$$

(centroid = average of all atomic positions)

$$\Phi_M = \frac{\text{\# of vector midpoints near a plane}}{\text{\# of atoms per fragment}}$$

$$\Phi_R = \frac{\text{\# of vector midpoints near a line } (\leq 0.5\ \text{Å})}{\text{\# of atoms per fragment}}$$

$$\Phi_T = \frac{\text{\# of atomic vectors nearly parallel } (\leq 3.0^\circ) \text{ and similar in length } (\leq 0.5\ \text{Å})}{\text{\# of atoms per fragment}}$$

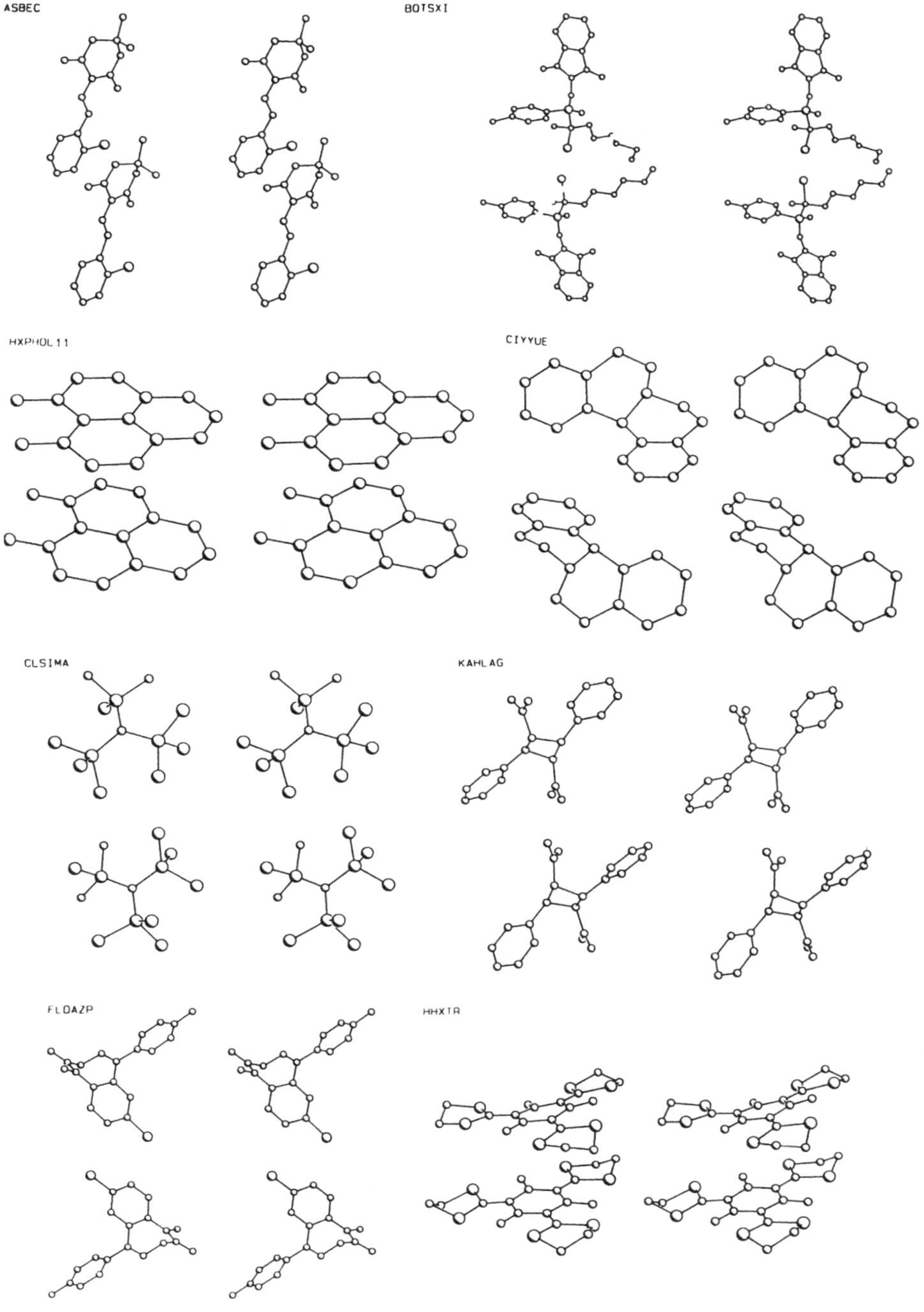

Figure 4.16 Stereoview of approximate symmetry related molecules.

symmetry, molecular arrangements for each structure were analyzed manually. This approach revealed three dominant types of symmetry: inversion, translation and rotation/screw. Not surprisingly, the unfavorable mirror symmetry relationships were not detected from analysis of the entries. The stereoviews of several examples are shown in Figure 4.16. Analysis of Figure 4.16 reveals that in the structures BASBEC ($P2_12_12_1$; $\phi_T = 0.89$), BOTSXI ($P1$; $\phi_R = 0.97$), and HXPHOL11($P2_1$;$\phi_R = 1.13$) a nearly perfect approximate symmetry exists. In CIYYUE ($P2_1$; $\phi_I = 0.83$), CLSIMA (Pc; $\phi_I = 0.100$), and KAHLAG ($P2_1$;$\phi_I = 1.00$) there exists a similar relationship; however, not all atoms pairs related by symmetry are of the same atom type. More commonly, the symmetry operator relates only a portion of the symmetry independent molecules. These variations are often conformational in nature as in FLDAZP ($P2_12_12_1$; $\phi_I = 1.00$) and THHXTR (Pc; $\phi_T = 0.67$). Table 4.4 list the frequency of approximate symmetry operators for the entire data set studied.

The position of approximate symmetry operators can provide information about additional symmetry in molecular arrangements. Structures with symmetry operators that only relate pairs of symmetry-independent molecules contain extra or "local" symmetry. Whereas, operators located at special positions ("global symmetry") create higher crystallographic symmetry, approximate supergroups or an alteration of unit cell size. This class of approximate symmetry in this study was identified by investigation of the position of the approximate inversion centers*. A study of the centers in the structures revealed 137 (20%) cases with approximate inversion symmetry located near special positions (Table 4.5). These centers have x, y, and z values near 0, 1/4, 1/2, 3/4 for groups $P1$, $P\bar{1}$, $P2_1/c$, $P2_12_12_1$ and $Pbca$. For $P2_1$ these

Table 4.4 Frequency of approximate symmetry operators in database study.

Space group	Entries from CSD	Entries with approx.symmetry	Frequency of approx. symmetry operators: *I*	*R*	*T*	Total
$P1$	148	34	15	29	2	46
$P\bar{1}$	671	121	68	134	86	288
$P2_1$	484	162	74	109	56	239
Pc	28	13	12	14	8	34
$P2_1/c$	800	219	202	334	214	750
$P2_12_12_1$	311	88	26	62	17	105
$Pbca$	61	35	20	23	24	67

* Computational methods to locate rotation axes and mirror/glide planes are under investigation.

Table 4.5 Alterations in space groups due to approximate inversion symmetry.

Space group	Higher/lower approximate space group symmetry	Number of of entries
$P1$	$P\bar{1}$	3
$P\bar{1}$	$P\bar{1}$ (reduced)	20
$P2_1$	$P2_1/c$	55
	$P2_1/m$	1
Pc	$P2_1/c$	6
$P2_1/c$	$P2/c$	4
	$P2_1/m$	11
	$P2_1/c$ (reduced)	26
	$C2/m$	2
	$C2/c$	7
$P2_12_12_1$	$Pbca$	2
	$Pnma$	2
$Pbca$	–	–

special values are located at x and z and for Pc at y. The most prominent example is $P2_1$, where 55 entries closely mimic $P2_1/c$ packing.

3.5 Crystal Packing Forces: Crystal Engineering

The ultimate goal behind any crystal packing investigation is the ability to predict packing motifs based entirely on molecular structure. As early as 1965, Kitaigorodskii[72] suggested several factors that influence molecular packing. Since that time, many other factors have been gradually uncovered as evident from various discussions in this book. Even so, the jump to molecular arrangements in crystals as a predictive science is still a monumental task. The following section attempts to illustrate studies that employ the CSD towards understanding crystal packing forces.

Studies that attribute molecular conformation to crystal packing forces tend to view the crystal environment as anisotropic. This assumes that molecular deformations that result from packing forces are relatively small in terms of energy and that they are distributed randomly over the entire crystalline sample. Consequently, the conformation of molecules in the crystalline state should approximate a global minimum of its potential energy surface. In order to investigate the likelihood that molecular conformations correspond to calculated global minima, Brock and Minton studied 101 *ortho* unsubstituted biphenyl fragments extracted from the CSD[73].

This class of compound was selected since only one important intramolecular degree of freedom exists—the torsional angle (ψ) between aryl groups (Figure 4.17). For biphenyl itself, gas phase studies[74] revealed a ψ value of 44(1)°, while in the solid state the ψ value is close to 0° for temperatures above 80K.

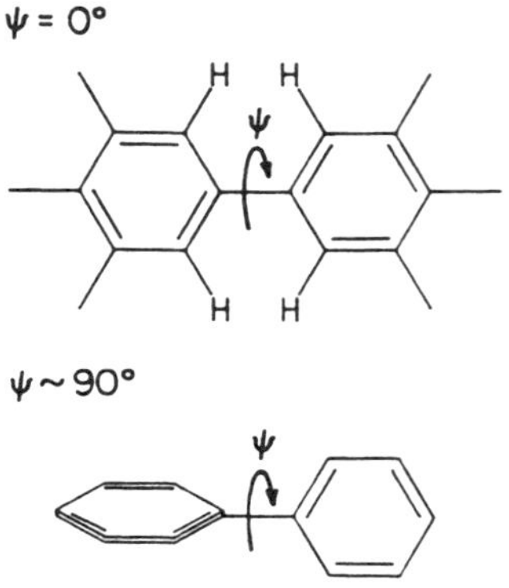

Figure 4.17 Angle ψ in an idealized biphenyl molecule. (Reproduced by permission from Brock and Minton[73].)

The planar conformation of biphenyl is located above the minimum corresponding to the skewed molecule; approximately 1.5 kcal mol^{-1} higher[75]. This structural deviation of biphenyl and other derivatives led the authors to wonder whether biphenyl fragments might represent an important exception to the general rule that sample points cluster in low-energy regions of conformational space.

Examination of biphenyl fragments allowed a fairly direct comparison of the observed and calculated distribution functions for ψ. The *ortho* unsubstituted biphenyls represent a unique class of compounds since these fragments are likely to be affected by systematic packing effects (Figure 4.18). The energy cost of rotation about the central C—C bond is small and a small energy increment can correspond to a large change in the shape of the fragment. There is some evidence that the average value of ψ is at least 10° lower in solution than in the gas phase[76]. That result suggests that the torsional angle ψ may be especially sensitive to intermolecular interactions.

Brock and Minton tabulated 101 ψ values corresponding to 68 structures of 65 compounds. The observed data (Figure 4.19) varies dramatically from the calculated Boltzmann (Figure 4.18) distributions. The former appears to be trimodal with maxima at about $\psi = 0$, 18, 37°, while calculated Boltzmann ψ plots possess single peaks. The largest maximum ($\psi = 37°$) in Figure 4.19 corresponds to the average conformation observed in the gaseous state ($\psi = 44°$)[75]. The peak in the histograms at $\psi = 0°$ is sharper than the maximum at 37° and represents 18 ψ values in the range of 0–2.6°. This maximum corresponds to fewer biphenyl fragments. The data collected by the authors indicate that crystal-packing effects systematically stabilize biphenyl fragments with small ψ values. Fragments with $\psi < 20°$ are very much more common than predicted from Boltzmann calculations. The basic assumption that molecules

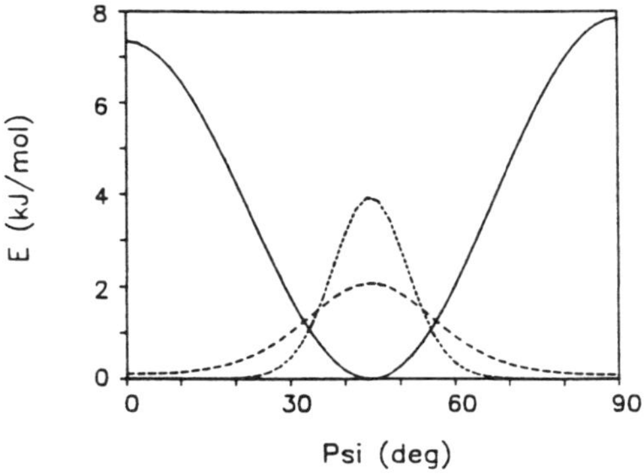

Figure 4.18 Average potential energy curve (—) for the torsional mode of *ortho* unsubstitute biphenyls determined by Bastiansen and Samdal[75]. The associated Boltzmann distributions calculated at 100 K (···) and 300 K (--) are also shown. (Reproduced by permission from Brock and Minton[73].)

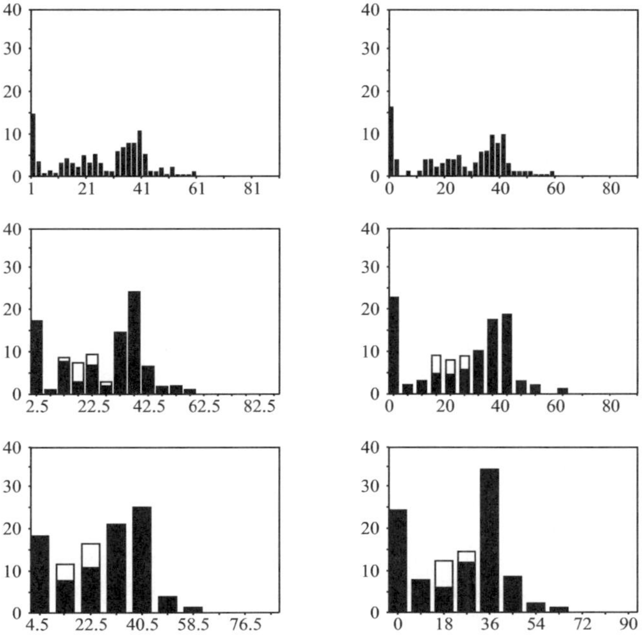

Figure 4.19 Distributions of sample points for bin sizes of 2, 5, 9°. Two choices for the bin boundaries are shown in each case: (a) first bin starts at 0° (left column); (b) first bin is centered at 0° (right column). (Reproduced by permission from Brock and Minton[73].)

crystallize in a conformation near the global minimum of its potential energy surface appears to be invalid for these molecules.

The authors suggest three explanations for the multi minima energy curve: (i) the structures extracted from the CSD do not accurately represent the distribution that would be obtained if more structures were available; (ii) the *meta* and *para* substituents control the molecular conformation to a greater extent than expected and (iii) the solid state systematically favors nearly planar biphenyl fragments because they pack better. The authors invoked the latter to explain the observed distribution.

It is well known that the arrangement of planar molecules into herringbone patterns is very favorable (refer to the discussion on polynuclear aromatic hydrocarbons). Even though biphenyls studied possess a wide variety of substituents, this arrangement accounts for many of the structures. In the cases where the herringbone pattern exists, no entries had ψ values greater than 25°. This observation by Brock and Minton clearly illustrates the importance of crystal packing forces on molecular structure[73].

Another approach towards understanding packing motifs focused on the study of polynuclear aromatic hydrocarbons. Desiraju and Gavezzotti showed that this structural class permits a viable investigation of molecular arrangements, since many of the ill-defined factors that govern molecular arrangements do not exist in these system[77]. For instance, these systems lack strong hydrogen bonds, complex functional groups and molecular flexibility. As a result, a simple approach based on shape descriptors allowed assignment of relationships between molecular structure and crystal packing.

Planar aromatic compounds were classified in terms of four basic structural types: herringbone, γ-type, sandwich, and β-type (Figure 4.20). Each structural pattern involves, to a different extent, C···C and C···H intermolecular interactions. The geometrical requirements for these interactions are quite different. Effective C···H interactions occur between inclined molecules with C—H bonds aligned toward the center of the aromatic rings, while favorable C···C interactions occur between parallel aromatic rings.

The herringbone motif is comprised of C···C interactions between nonparallel nearest neighbor molecules. In a similar fashion, γ-type arrangements also form herringbone patterns; however, the closest C···C interactions exist between parallel molecules. In the third class, the sandwich type arrangement, two molecules form a stack. These stacks are further arranged in a herringbone motif. In addition to C···C interactions in each of these three patterns, favorable intermolecular C···H interactions play an important role in the molecular arrangements. The last structural pattern, β, involves strong C···C interactions with diminutive contributions from C···H contacts; resulting in a layered molecular arrangement.

Since the degree of C···C and C···H interactions constitute the main difference between the four packing motifs, the authors invoked a systematic

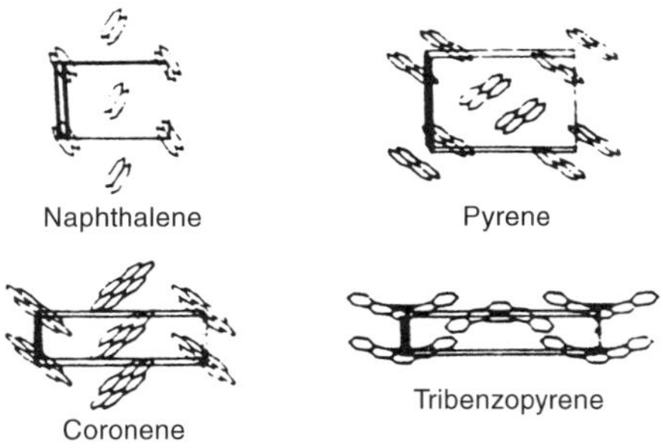

Figure 4.20 The four basic aromatic crystal packings as exemplified by naphthalene (herringbone), coronene (γ), pyrene (sandwich) and tribenzopyrene (β). (Reproduced by permission from Desiraju and Gavezzotti[77].)

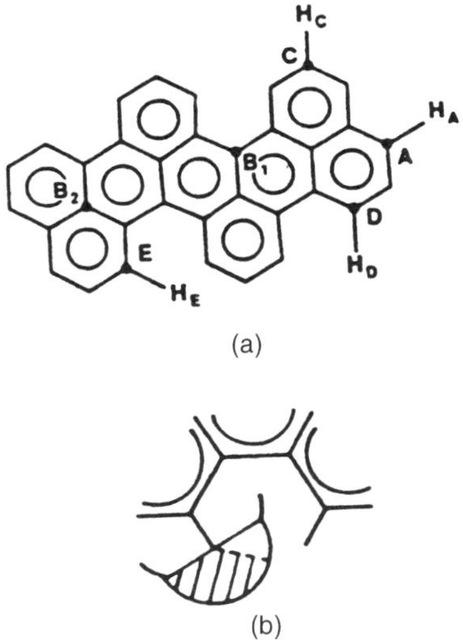

Figure 4.21 (a) Labeling of C and H atoms for the assignment of area coefficients (Table 4.6). (b) Explanation of the partial availability of edge atoms. Only the shaded area is available for intermolecular contacts. (Reproduced by permission from Desiraju and Gavezzotti[77].)

methodology based on the relative positioning of C and H atoms in molecules. It has been shown that atoms may be defined in terms of free atomic van der Waals surfaces[78]. Since these atomic surfaces vary within a given molecule, the assignment of nuclei as stack- or glide-promoting (herringbone) is possible. For example, in Figure 4.21 the hypothetical structure contains several chemically

Table 4.6 Average atomic surface increments, $S_{a,i}$(from the available molecular structures) and empirical availability factors. (Reproduced by permission from Desiraju and Gavezzotti[77].)

Atom type	$S_{a,i}$	% Stack[a]	% Glide[b]	Availability factor[c]
C_A	10.8	50	50	1.00
C_{B1}	5.8	50	50	1.00
C_{B2}	5.8	100	0	1.00
C_C	11.2	50	50	1.00
C_D	10.4	50	50	0.67
C_E	9.2	50	50	0.50
H_A	6.8	0	100	1.00
H_C	6.8	0	100	1.00
H_D	5.6	0	100	0.67
H_E	5.1	0	100	0.50

[a] Total stack are per molecule $= 0.5(10.8NA + NB_1 + 11.2NC + 0.67 \times 10.4ND + 0.50 \times 9.2NE) + 5.8NB_2$, where NA, NB_1, etc. are the number of atoms of the type A, B_1, etc. Note that $NA =$ number of CA atoms $=$ number of HA atoms.
[b] Total glide area per molecule $= 0.5(10.8NA + 5.8NB_1 + 11.2NC + 0.68NA + 6.8NC + 0.67 \times 5.6ND + 0.50 \times 5.1NE)$.
[c] The availability factors are less than unity for intramolecular sterically hindered atoms.

and sterically different C and H atoms (i.e. A through E). In the extreme case, such as B_2, this atom type contributes solely to stacking patterns, whereas atoms located at the perimeter of the molecule (A, B_1, C, D, and E) contribute to both stacking and glide modes.

Summing over all nuclei permitted assignment of the total stacking (S_{st}) and glideforming (S_g) surfaces for any planar aromatic (Table 4.6). From investigation of 32 planar aromatic structures (Figure 4.22), a plot of the ratio of these surfaces versus the total molecular surface (S_m) provided a predictive tool for mapping crystal structures based on molecular structure. As indicated in Figure 4.23, there exists a clear cut division between the herringbone structures and the other three structural patterns. Structures with herringbone arrangements possess linear fused molecules (**18**, **19** and **20**) or polyphenyls (**18**, **22** and **29**). Desiraju and Gavezzotti noted that compounds **21**, **23**, **25**, **27** and **28** deviate slightly from this trend as a result of structural nonlinearity. They also noted that structural features of compounds that lie close to the γ curve exhibit a planar disk-like shape (**35**–**41**, **43**). These compounds lead to molecular stacks promoted by C$\cdots$C interactions (4.7 Å) and dovetailed together by C$\cdots$H interactions. Also indicated in Figure 4.23, molecules that form sandwich patterns (**30**, **31**, **32**, and **34**), straddle the structural gap between herringbone and γ patterns. As seen in γ packing, efficient C$\cdots$C stacking contributes significantly to the sandwich pattern, but the stacked diads are further arranged to form herringbone motifs. These molecules tend to be rectangular in shape

Figure 4.22 Formulae of compounds **18–49**. Crystal structure cutations are in Desiraju and Gavezzotti[77].)

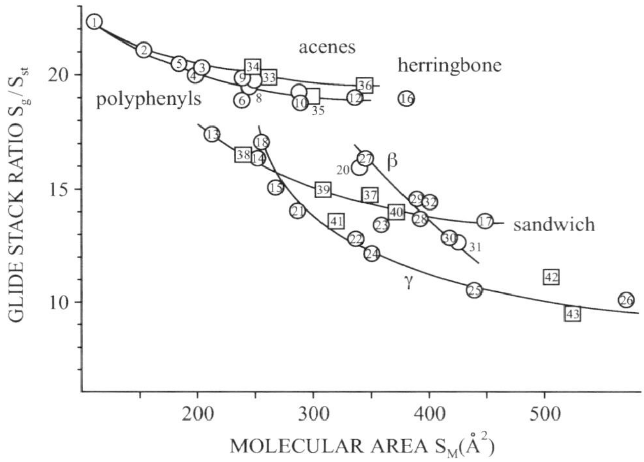

Figure 4.23 A mapping from molecular to crystal structure for polynuclear aromatic hydrocarbons. Glide-stack ratios are plotted against available molecular surface areas. (Reproduced by permission from Desiraju and Gavezzotti[52].)

with both internal C atoms and external H atoms. since both C···C and C···H interactions are important for this structural type, comparable glide and stack promoting interactions exist.

β type structures present yet another motif for the polynuclear aromatic hydrocarbons. These molecules are clustered in a well-defined region in Figure 4.23. The molecules that adopt this pattern are distorted form planarity as a result of E type atoms, thus making stack formation difficult (**44–49**). To avoid excessive nonbonded repulsions, these nonplanar molecules stack with a relatively large perpendicular distance and a small degree of lateral offset to form layered structures.

Based on the mapping of crystal structures in terms of molecular structure (as discussed in the above sections) Desiraju and Gavezzotti made several predictions for polynuclear aromatic hydrocarbons, whose crystal structures were unavailable. Even though the analysis of these compounds resulted in the assignment of the "most probable" crystal structures for these compounds, the true test of the predictive power of this methodology hinges on the analysis of compounds not present in this study, but where the crystal structures are known.

Since the investigation exhausted all known crystal structures of planar aromatic compounds, the authors extended the validity of their analysis by analyzing compounds that were not purely aromatic hydrocarbons. since it was shown that the structural features of each molecule could be condensed into a shape descriptor, it followed that non-polynuclear aromatic hydrocarbons with identical shapes should pack in a similar manner. For example, the elongated

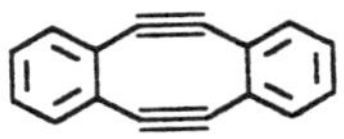

50

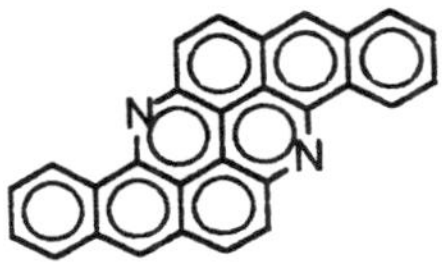

51

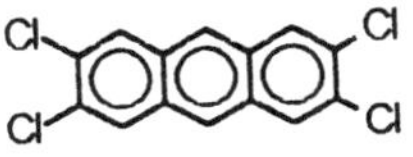

52

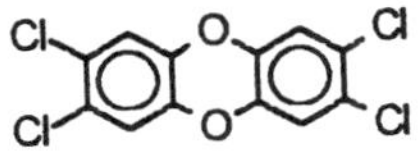

53

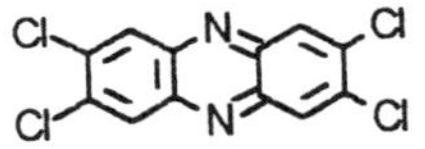

54

Figure 4.24 Formulae of compounds **50–54**. Crystal structure cutations are in Desiraju and Gavezzotti[77]. (Reproduced by permission from Desiraju and Gavezzotti[77].)

molecules dibenzofuran, carbazole, fluorene as well as several dimethylnaphthalenes pack in the expected herringbone pattern. However, diacetylene **50**, adopts the sandwich pattern as a result of the four absent perimeter H atoms. Even though **51** resembles compounds that possess γ- or β-producing shapes, its crystal structure corresponds to β packing patterns, since several H atoms on the fringe of the molecule are absent. Likewise, **52** possesses the expected herringbone motif; however, **53** and **54** possess β structures for the same reason (Figure 4.24). These examples clearly show the utility and the extended application of this approach.

4. REFERENCES

1. Desiraju, G. R. *Prog. Solid St. Chem.* **17**, 295 (1987).
2. Desiraju, G. R. (ed.) *Organic Solid State Chemistry.* Elsevier, Amsterdam (1987).
3. Ref. for information gleaned from crystal structure data.
4. Bürgi, H. B. and Dunitz, J. D. *Acc. Chem. Res.* **16**, 153 (1983).
5. Bürgi, H. B., Dunitz, J. D. and Schefter, E. *J. Am. Chem. Soc.* **95**, 5065 (1973).
6. Birnbaum, G. I. *J. Am. Chem. Soc.* **96**, 6165 (1974).
7. Kaftory, M. and Dunitz, J. D. *Acta Crystallogr B* **31**, 2912 (1975).
8. Kaftory, M. and Dunitz, J. D. *Ibid.* **32**, 1 (1976).
9. Bye, E. and Dunitz, J. D. *Ibid.* **34**, 3245 (1978).

10. Schweizer, W. B., Procter, G., Kaftory, M. and Dunitz, J. D. *Helv. Chim. Acta.* **61**, 2783 (1978).
11. Mighell, A. D., Stalick, J. K. and Himes, V. L. In *Crystallographic Databases*, ed. by F. H. Allen, G. Bergerhoff and Sievers, International Union of Crystallography, Chester (1987), p.134.
12. Rodgers, J. R. and Wood, G. H. *Ibid.* p. 96.
13. Abola, E. E., Berstein , F. C., Bryant, S. H., Koetzle, T. F. and Weng, *J. Ibid.*, p.107.
14. Bergerhoff, G. and Brown, I. D. *Ibid.* p. 77.
15. Allen, F. H., Bellard, S., Brice, M. D., Cartwright, A., Doubleday, A., Higgs, H., Hummelink, T., Hummelink-Peters, B. G., Kennard, O., Motherwell, W. D. S., Rodgers, J. R. and Watson, D. G. *Acta Crystallogr. B* **35**, 2331 (1979).
16. Allen F. H., Kennard, O. and Taylor, R. *Acc. Chem. Res.* **16**, 146 (1983).
17. Orpen, A. G., Brammer, L., Allen, F. H., Kennard, O., Watson, D. G., and Taylor, R. *J. Chem. Soc. Dalton Trans.* S1 (1989).
18. Citation Index reference.
19. CSD manuals reference.
20. Review of articles on how to use CSD.
21. Current ref. of CSD address.
22. Kennard, O., Watson, D. G., Allen, F. H., Isaacs, N. W., Motherwell, W. D. S., Pettersen, R. C. and Town, W. G. (eds), *Molecular Crystals and Dimensions:* International Union of Crystallography (1972).
23. Motherwell, W. D. S. and Clegg, W. *PLUTO*, a program for plotting crystal and molecular structures. Cambridge University, England (1989).
24. D. March list and perils of Cc.
25. Sutton, L. E. (ed.) *Tables of Interatomic Distances and Configuration in Molecules and Ions,* Special Publication 18. The Chemical Society, London (1965).
26. Jones, P. G. and Kirby A. J. *J. Chem. Soc., Chem. Commun.* 288 (1979).
27. Allen, F. H. and Kirby, A. J. *J. Am. Chem. Soc.* **106**, 6197 (1984).
28. Briggs, A. J., Glenn, R., Jones, P. G., Kirby, A. J. and Ramaswany, P. *J. Am. Chem. Soc.* **106**, 6200 (1984).
29. Jones, P. G. and Kirby, A. J. *J. Am. Chem. Soc.* **106**, 6207 (1984).
30. Ether background refs.
31. Blukis, V., Kasai, P. H. and Meyers, R. J. *J. Chem. Phys.* **38**, 2753 (1963).
32. Glidewell, C., Rankin, D. W. H., Robiette, A. G., Sheldrick, G. M., Beagly, B. and Freeman, J. M. *J. Mol. Struct.* **5**, 417 (1970).
33. Blake, A. J., Ebsworth, E. A. V., Henderson, S. G. D. and Dyrbusch, M. *Acta Crystallogr. C* **44**, 1 (1988).
34. Almenningen, A., Bastiansen, O., Hedberg, K. and Traetteberg, M. *Acta Chem. Scand.* **17**, 2455 (1963).
35. Glidewell, C., Rankin, D. W. H., Robiette, A. G., Sheldrick, G. M., Beagly, B. and Freeman, J. M. *J. Mol. Struct.* **5**, 417 (1970).
36. Almenninger, A., Hedberg, K. and Seip, R. *Acta Chem. Scand.* **17**, 2664 (1963).
37. Sauer, J. and Zurawski, B. *Chem. Phys. Lett.* **65**, 587 (1979).
38. Oberhammer, H. and Boggs, J. E. *J. Am. Chem. Soc.* **102**, 7241 (1980).
39. Luke, B. T., Pople, J. A., Krogh-Jesperson, M. B., Apeloig, Y., Chandrasekhar, J. and Schleyer, P. v. R. *J. Am. Chem. Soc.* **108**, 260 (1986).
40. Ernst, C. A., Allred, A. L., Ratner, M. A., Newton, M. D., Gibbs, G. V., Moskovitz, J. W. and Topiol, S. *Chem. Phys. Lett.* **81**, 424 (1981).
41. Grigoras, S. and Lane, T. H. *J. Comput. Chem.* **8**, 84 (1987).
42. Varma, R., MacDiarmid, A. G. and Miller, G. *Inorg. Chem.* **3**, 1754 (1964).
43. Sternbach, B. and MacDiarmid, A. G. *J. Am. Chem. Soc.* **83**, 3384 (1961).

44. Keck, G. E. and Catellino, S. *Tetrahedron Lett.* **28**, 281 (1987).
45. Keck, G. E. and Boden, E. P. *Tetrahedron Lett.* **25**, 265 (1984).
46. West, R., Wilson, L. S. and Powell, D. L. *J. Organomet. Chem.* **178**, 5 (1979).
47. Sauer, J. *Z. Chem.* **22**, 60 (1982).
48. Shambayati, S., Blake, J. F., Wierscke, S. G., Jorgensen, W. L. and Schreiber, S. T. *J. Am. Chem. Soc.* **112**, 697 (1990).
49. Pauling, L. *The Nature of the Chemical Bond*, Ist Edn. Cornell University Press, Ithaca, New York (1939).
50. Hagler and Lifson, p. 18.
51. Hydrogen bond refs.
52. Jeffrey, G. A. and Maluszynska, H. *J. Mol. Struct.* **147**, 127 (1986).
53. Kitaigorodskii, A. I. *Organic Chemical Crystallography.* Consultants Bureau, New York (1961).
54. Whitesell, J. K., Davis, R. E., Saunders, L. L., Wilson, R. J. and Feagins, J. P. *J. Am. Chem. Soc.* **113**, 3267 (1991).
55. Wilson, A. J. C. *Acta Crystallogr. A* **144**, 742 (1990).
56. Filippini, G. and Gavezzotti, A. *Acta Cryst. B* **48**, 230 (1992).
57. Donohue, J. *Acta. Crystallogr A* **41**, 203 (1985).
58. Mighell, A. D., Himes, V. L. and Rodgers, J. L. *Ibid.* **39**, 737 (1983).
59. Scaringe, R. P. In *Electron Crystallography of Organic Molecules,* ed. by J. R. Fryer and D. L. Dorset, pp. 85–113. Kluwer Academic Publishers, Dordrecht (1990).
60. Wilson, A. J. C. *Acta Crystallogr. A* **44**, 715 (1988).
61. Perlstein, J. *J. Am. Chem. Soc.* **116**, 455 (1994).
62. Gavezzotti, A. and Simonetta, M. *Chem. Rev.* **82**, 1 (1982).
63. Wallach, O. *Liebigs Ann. Chem.* **286**, 90 (1895).
64. Brock, C. P., Scheizer, W. B. and Dunitz, J. D. *J. Am. Chem. Soc.* **113**, 9811 (1991).
65. Jacques, J., Collet, A. and Wilen, S. H. *Enantiomers, Racemates and Resolutions*, pp. 23–31. Wiley, New York (1981).
66. Whitesell, J. K., Davis, R. E., Wong, M. -S. and Chang, N. -L. *J. Am. Chem. Soc.* **116**, 523 (1994).
67. Desiraju, G. R., Calabrese, J. C. and Harlow, R. L. *Acta Crystallogr. B* **47**, 77 (1991).
68. Nordman, C. E. *Acta Crystallogr. A* **36**, 747 (1980).
69. Argos, P., Ford, G. C. and Rossman, M. G. *Acta, crystallogr. A* **31**, 499 (1980).
70. Davis, R. E., Chang, N.-L. and Wheeler, K. A. Presented at the *American Crystallographic Association Meeting,* Atlanta, GA (1994) paper UO02.
71. Wheeler, K. A., Chang, N.-L. and Davis, R. E. Presented at the *American Crystallographic Association Meeting,* Albuquerque, NM (1993) paper PH02.
72. Kitaigorodskii, A. I. *Acta Crystallogr.* **18**, 585 (1965).
73. Brock, C. P. and Minton, R. P. *J. Am. Chem. Soc.* **111**, 4586 (1989).
74. Almenningen, A., Bastiansen, O., Fernholt, L., Cyvin, B. N., Cyvin, A. J. and Samdal, S. *J. Mol. Struct.* **128**, 59 (1985).
75. Bastiansen, O. and Samdal, S. *J. Mol. Struct.* **128**, 115 (1985).
76. Eaton, V. J. and Steele, D. J. *J. Chem. Soc., Faraday Trans.* **2**, 1601 (1973).
77. Desiraju. G. R. and Gavezzotti, A. *Acta Crystallogr. B* **45**, 473 (1989).
78. Gavezzotti, A. *J. Am. Chem. Soc.* **105**, 5220 (1983).
79. Gavezzotti, A. *J. Am. Chem. Soc.* **107**, 962 (1985).

Chapter 5

Strategies for the Design and Assembly of Hydrogen-bonded Aggregates in the Solid State

DONOVAN N. CHIN, JONATHAN A. ZERKOWSKI, JOHN C. MACDONALD AND GEORGE M. WHITESIDES
Department of Chemistry, Harvard University, Cambridge, MA, USA

1. INTRODUCTION

Crystalline organic materials are being actively explored for potential application in optoelectronics[1], and as piezoelectrics, photorefractives[2–4], and other types of functional materials. Macroscopic qualities (for example, non-linear optical coefficients, birefringence, toughness, and bioavailability) are important in specific applications[5,6]. To realize the potential of crystalline organic materials in materials science, it would be helpful to be able to correlate molecular structure, crystalline structure, and macroscopic properties to "design" materials: that is, to predict the crystalline packing of an organic compound based on its molecular structure, and to predict its macroscopic materials properties. In practice, chemists usually start their efforts to design materials at the level of molecular components, and build upwards in complexity. Strategies for this sort of molecular engineering, an endeavor we and others include under the broad term "non-covalent synthesis", are in their infancy[7–9].

Organised Molecular Assemblies in the Solid State Edited by James K. Whitesell
© 1999 John Wiley & Sons Ltd.

Predicting the packing of molecules in a crystal is difficult for several reasons. First, there are many possible arrangements of the molecules in space, and no algorithm exists that can sort though these alternatives efficiently. Second, polymorphism is a problem[10–16]. Organic molecules often pack in different crystalline arrangements that are nearly isoenergetic (within ~1–2 kcal/mol)[16], that is, in polymorphs. Third, polar groups contribute to this complexity by introducing long-range intermolecular electrostatic forces that compete with the short-range van der Waals forces[17]. Fourth, the competing kinetics of nucleation and crystal growth are poorly understood[18]. Crystallization may give crystals with structures that are determined primarily by kinetics rather than by thermodynamics.

One promising strategy for minimizing the complexity of the problem of correlating molecular and crystalline structure is to design the molecules so that they link into substructures through non-covalent intermolecular interactions. For several reasons, the hydrogen bond is well suited as an intermolecular connector for controlling the organization of organic molecules in solids. First, its directionality is well characterized[19,20]. Second, since the energy of a single hydrogen bond is relatively small (1–5 kcal/mol)[20–22] and molecules of polar solvent can compete for these donors and acceptors, formation of hydrogen bonds is reversible at room temperature. This reversibility permits practical experimental control over aggregation and growth of the lattice or its substructures. Third, a variety of organic functional groups can be used to introduce hydrogen-bonding capacity into a molecule. Finally, groups that form hydrogen bonds are compatible with many other functional groups[23,24]. These features recommend hydrogen bonds as elements with which to control structure in the organic solid state.

This chapter will examine ways in which hydrogen bonds have been used to rationalize, predict, or control structure in crystals of organic molecules. The first section will present an overview of studies (with emphasis on work of Etter) concerning selectivities of organic functional groups in forming hydrogen bonds, and the most common patterns of hydrogen bonds formed by these functional groups. These patterns can be used to design aggregates. The sections following review examples that have used hydrogen bonds to design crystalline arrays, including our own work with hydrogen-bonded aggregates based on substructures found in the cyanuric acid • melamine (CA • M) lattice. The final section briefly outlines the role of computational simulations in rationalizing and predicting structure in crystalline solids.

The strategies and examples reviewed here comprise one major current thrust in the field of crystal engineering[25,26]. This chapter does not provide a comprehensive overview of crystal engineering: we only consider those cases where the use of hydrogen bonding is the key strategy for design. This chapter also does not cover the fundamentals of hydrogen bonding or the geometrical statistics of hydrogen bonds in crystals, since there are already excellent reviews on those topics[20]. Finally, we make no effort to categorize structures that contain hydrogen-bonded networks that were not the result of conscious efforts at design.

Other important topics, such as the modification of crystal habit by design of face-selective additives[27], and the investigation of reactivity in the solid state[28], are also outside the focus of this chapter.

2. SYSTEMATIC APPROACHES FOR DETERMINING PRINCIPLES AND RULES OF HYDROGEN BONDING

2.1 Patterns, Rules, and Selectivity of Hydrogen Bonds

The research of Etter focused in using hydrogen bonds to control the formation of molecular aggregates made up of small organic molecules containing limited sets of functional groups. Etter's approach is particularly valuable in that it relies on the intermolecular specificity of the hydrogen bonds rather than the size or shape of molecules to direct the formation of aggregates and to control their structure. These studies deliberately excluded competing factors such as preorganized cavities, ionic interactions, and steric interactions. This approach has permitted the evaluation of the independent contributions of hydrogen bonds to the structures of molecular aggregates. Her work can be categorized in three areas:

- characterization of patterns of hydrogen bonds in aggregates of small organic molecules using graph sets
- development of empirical rules that have been reasonably successful in predicting the connectivity of hydrogen bonds in molecular aggregates in the absence of other strong interactions
- determination of the selectivity of common organic functional groups in forming hydrogen bonds

2.1.1 Graph Sets

Etter recognized the need for a general and simple method for recognizing, characterizing, and comparing patterns of hydrogen bonds in molecular aggregates[29]. By applying graph theory to molecular aggregates, she broke down complex networks involving many different types of hydrogen bonds into simpler constituent motifs—that is, into subsets of molecules joined by just one type of hydrogen bond. Surprisingly, all patterns of hydrogen bonds can be broken down into just four simple motifs. Notation describing these motifs is assigned to each different type of hydrogen bond. This notation—referred to as a graph set—provides a complete and accurate description of the patterns of hydrogen bonds in a network at any desired level of complexity[30]. Using graph-set analysis, networks containing many different types of hydrogen bonds have been described, and the similarities or differences in submolecular structure compared easily[31–34].

The process of assigning a graph set beings by identifying the different types of hydrogen bonds contained in a given network; these are determined by the molecular structures of the donors and acceptors[30]. The pattern formed by molecules connected to one another by just one type of hydrogen bond is called a *motif*. A motif is characterized by one of four *designators*. These designators are C(chain), R(ring), and D(dimer) for motifs generated from intermolecular hydrogen bonds. The designator S (self) denotes an intramolecular hydrogen bond. The number of donors and acceptors used in the motif are assigned as subscripts and superscripts respectively (which are dropped when both values are 1). The size of a motif (referred to as the *degree*) is determined by the number of atoms in a ring or the repeat unit of a chain, and is given in parentheses following the designator of the pattern. Figure 5.1 illustrates the procedure for assigning a graph set for the eight-membered ring formed by carboxylic acids. Examples of various motifs with their graph sets are shown in Figure 5.2[30]. The procedure for assigning graph sets has been reviewed recently by Bernstein *et al.*; the authors present many examples that clarify problems encountered frequently when assigning graph sets[34].

Analysis of graph sets can be employed in crystal engineering to identify isographic functional groups. Graph sets have been used systematically to identify patterns of hydrogen bonds that are common to different types of molecules or functional groups[31]. Functional groups that are isographic—that is, functional groups that are different structurally but that have the same graph set—contain the same number of donors and acceptors, and form

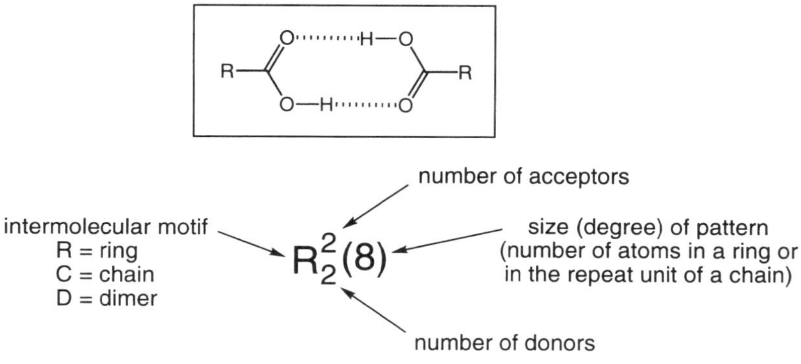

Figure 5.1 Components of a graph set. In this example, the graph set $R_2^2(8)$ is assigned to the ring motif formed by carboxylic acids.

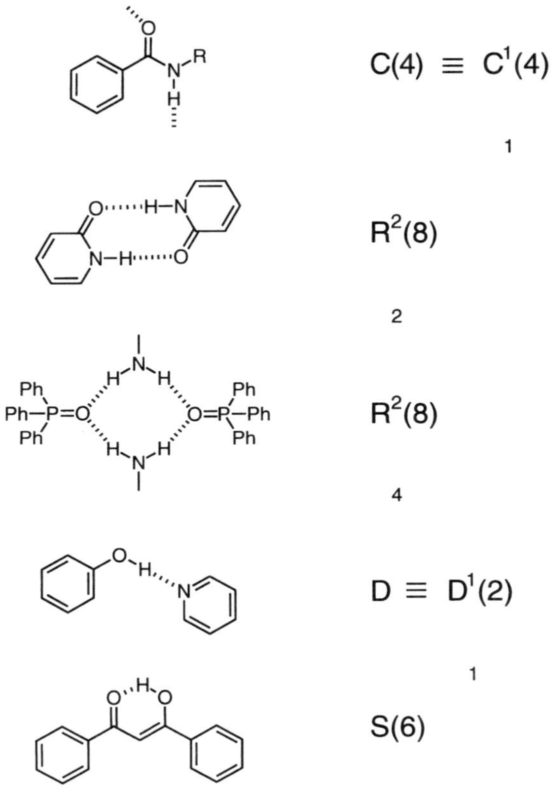

Figure 5.2 Examples of different motifs of hydrogen bonds with their graph sets.

motifs in which molecules are connected by the same number of hydrogen bonds. Examples of several different isographic functional groups that form eight-membered rings (cyclic dimers) having the graph set $R_2^2(8)$ are shown in Figure 5.3.

The concept of isographic relationships provides a convenient method for designing structures in which several different molecular components having the same graph set can be interchanged. This approach has been used extensively by Etter in designing cocrystals. For example, one carboxyl groups of a dimer of carboxylic acids (Figure 5.3a) has been replaced by 2-amino-pyridine [45,46], amide[47,55], imide[50], and urea[51,52,56–58] moieties while preserving the $R_2^2(8)$ motif (Figures 5.3i–5.3l, respectively).

Graph sets also provide a basis for comparing patterns of hydrogn bonds in polymorphs. For example, five motifs occur in different combinations in three polymorphs of iminodiacetic acid ($-O_2CCH_2NH_2+CH_2CO_2H$). Two

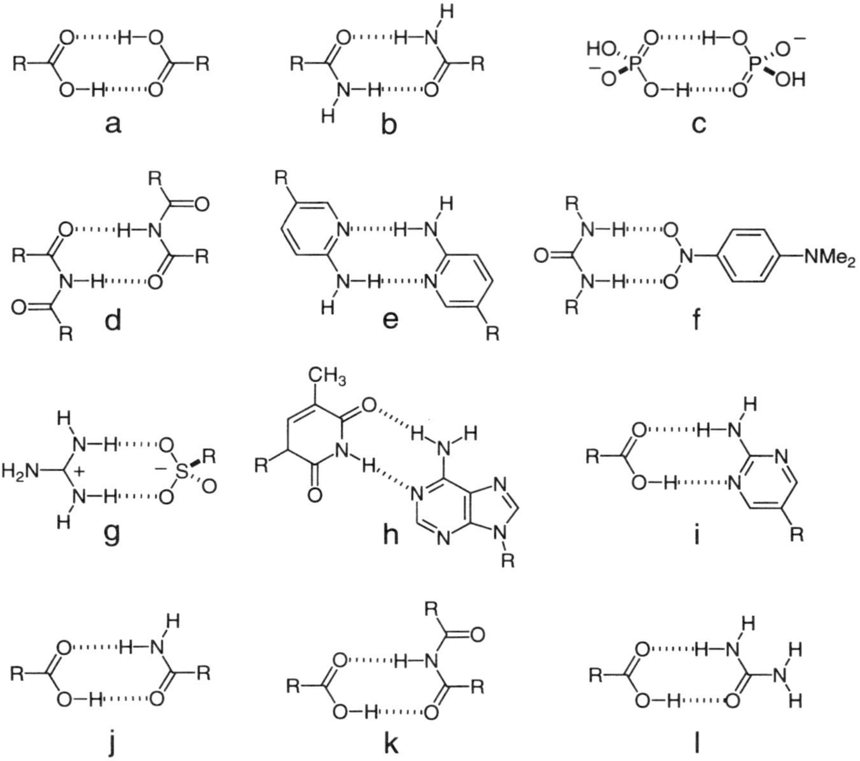

Figure 5.3 Isographic $R_2^2(8)$ motifs formed by (a) carboxylic acids[29,35], (b) primary amides[29,36], (c) anions of dihydrogen phosphate salts[37], (d) acyclic imides[38,39] (e) 2-aminopyridines[40], (f) bis(aryl)urea • N,N-dimethyl-*p*-nitroaniline cocrystal[41,42], (g) guanidinium sulfonate salts[43,44], (h) adenine • thymine base pair[20], (i) carboxylic acid • 2-aminopyrimidine cocrystals[45,46], (j) carboxylic acid • primary amide cocrystals[47–49], (k) carboxylic acid • acyclic imide cocrystals[50], and (l) carboxylic acid • urea cocrystals[51–54].

previous attempts have been made to distinguish these polymorphs based on their patterns of hydrogen bonds[59,60]. These analyses involved comparing geometries of bonds and relationships between the symmetry of neighboring molecules, as well as identifying several motifs consisting of hydrogen-bonded rings. Nevertheless, there was little understanding of the similarities between these structures or about the relative stabilities of the crystals. Structural differences were apparent immediately upon comparing the graph sets for these three structures[32,34]. Two polymorphs have the same graph sets for the first-level network (sum of all motifs), but the graph sets differ in their second-level networks (pattern involving two different types of hydrogen bonds, Figure 5.4).

Polymorph	Motifs	Networks
1	$C(5)R^2_2(10)C(8)$	$R^2_2(14)$
2	$C(5)R^2_2(10)C(8)$	$R^2_4(8)$
3	$C(5)C(5)C(8)$	

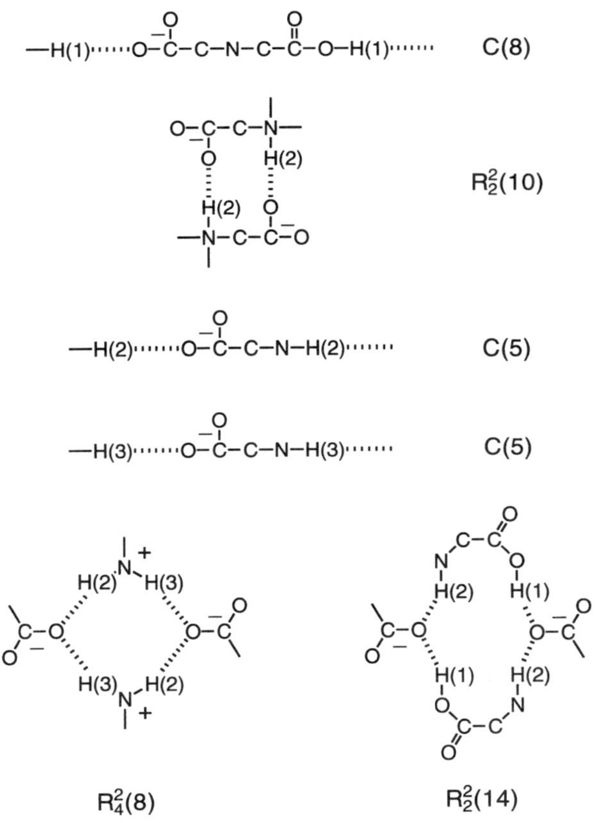

Figure 5.4 Graph set assignments for three polymorphs of iminodiacetic acid. Polymorph 3 differs from polymorphs 1 and 2 by the presence of a second C(5) motif. Polymorphs 1 and 2 are differentiated only when combinations of different kinds of hydrogen bonds (networks) are considered. These networks consist of rings of different sizes.

Computational routines that perform graph set analysis are presently being incorporated into the Cambridge Structural Database (CSD). Graph sets are being used with increasing frequency to design and analyze the structures of aggregates[34]. The CSD presently contains data from $> 10^5$ crystal structures[61]. Using automated algorithms, investigators can in principle, generate, visualize, and compare graph sets for large numbers of crystals structures. Employing graph sets in this manner aids in establishing relationships between the structure of hydrogen-bonded aggregates, and in determining how this structure affects the properties of organic solids.

2.1.2 Rules for Hydrogen Bonds

Etter analyzed the graph sets of a large number of crystal structures to determine distributions of graph sets for classes of molecules containing one particular functional group, and, conversely, distribution of functional groups for structures with a particular graph set[31,33]. Empirical rules of hydrogen bonding are one way to use correlations between functional groups and their patterns of hydrogen bonds, with graph sets being one type of correlation. Graph sets describe the connectivity or configuration of aggregates composed of hydrogen-bonded molecules. Other structural or stereoelectronic factors affecting the primary, secondary, or even tertiary structure of aggregates provide other rules of hydrogen bonding[31].

Three rules with broad but not universal generality for hydrogen bonds that apply to functional groups in neutral organic molecules are: (i) all good proton donors and acceptors are used in hydrogen bonding; (ii) intramolecular hydrogen bonds that form a six-membered ring will occur in preference to intermolecular hydrogen bonds; and (iii) the best proton donors and acceptors remaining after formation of intramolecular hydrogen bonds form intermolecular hydrogen bonds to one another. Additional rules for specific classes of functional groups have been determined, but will not be discussed here[31].

Rules of this sort reflect energetically favorable modes of intermolecular association that occur frequently in the packing patterns of organic crystals. Expected patterns of hydrogen bonds might not occur, however, for a variety of reasons. These reasons include steric effects[62], the presence of multiple sites for hydrogen bonding[63], and competing ionic forces[64]. Examples of such systems are shown in Figure 5.5.

In some cases, different motifs of hydrogen bonds occur by alternative connection of the same set of donors and acceptors; these motifs can give rise to different polymorphic crystalline forms. The patterns of hydrogen bonds found in two polymorphs of diacetimide[65,66] and three polymorphs of 5,5-diethylbarbituric acid[67,68] are shown in Figure 5.6. These examples illustrate that rules of hydrogen bonding cannot always predict which pattern will form

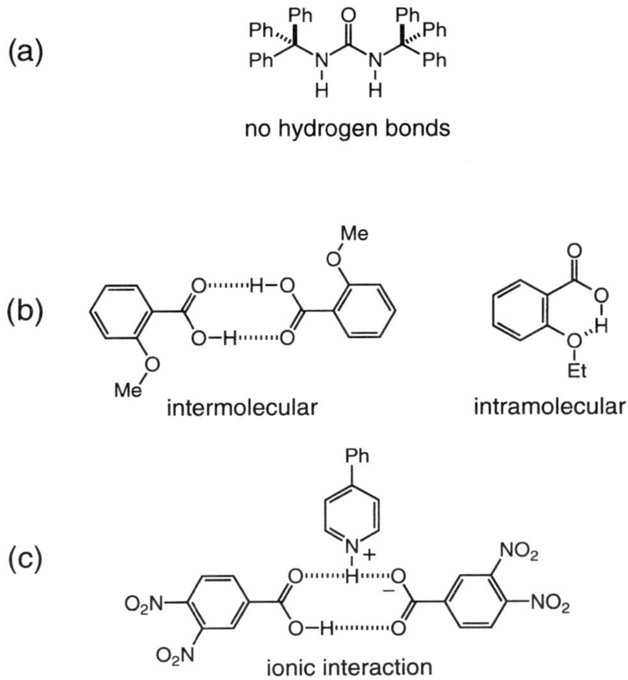

Figure 5.5 Molecules or aggregates that form unexpected patterns of hydrogen bonds. (a) Molecules of ditritylurea do not form intermolecular hydrogen bonds to one another because the urea groups are sterically inaccessible. (b) Molecules of *o*-methoxybenzoic acid form the expected dimer motif of carboxylic acids, while molecules of *o*-ethoxybenzoic acid form an intramolecular hydrogen bond. (c) The cocrystalline complex between 3,4-dinitrobenzoic acid and 4-phenylpyridine forms an unusual ring motif involving one neutral acid and a pyridinium–carboxylate pair.

when molecules can pack in several arrangements that are close in energy ($\sim 1\text{–}2$ kcal/mol)[16].

Studies of hydrogen bonding in organic crystals have shown that certain classes of functional groups almost always form hydrogen bonds when complementary donors or acceptors are available, while other classes of functional groups only occasionally participate in hydrogen bonding. Examples of functional groups that have been classified (loosely) by Etter as good or poor donors and acceptors are shown in Figure 5.7.

The concept of the best donor and acceptor forming a hydrogen bond provides a powerful tool for controlling the composition and structure of molecular aggregates in a crystal. For instance, Etter[48,55,69] and Lechat[70,71] have designed molecular complexes that are supramolecular analogs of nitroanilines by using complementary sets of strong and weak hydrogen bonds

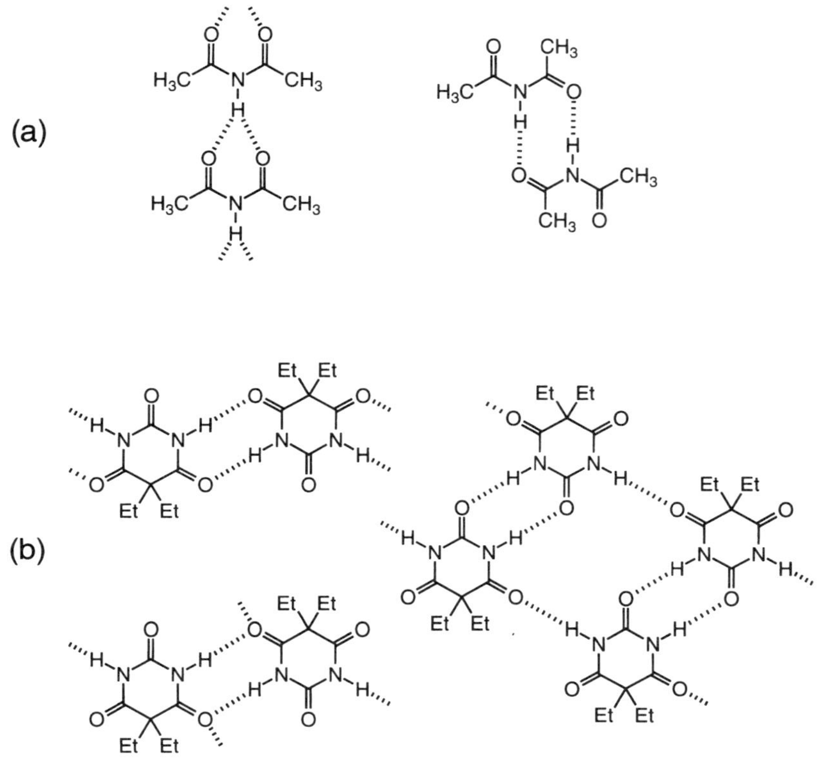

Figure 5.6 The patterns of hydrogen bonds found in (a) two polymorphs of diacetimide and (b) three polymorphs of 5,5-diethylbarbituric acid.

to assemble the molecular components. Examples of cocrystalline supramolecular analogs of *m*-nitroaniline and 3,5-dinitroaniline are illustrated in Figures 5.8 and 5.9, respectively. In each case, the amino and nitro groups were placed on separate molecules and then linked via strong O—H···O interactions to give heterodimers. Formation of a second, weaker set of N—H···O interactions between the amino and nitro groups preserved the patterns of hydrogen bonds (chains or layers) formed by the parent compound. The proposed structure of 3,5-dinitroaniline is believed to be a layer (Figure 5.10) although attempts to solve the crystal structure of this compound have been unsuccessful due to twinning of the crystals.

One of the conditions usually necessary for second-order nonlinear optical properties to occur is that the bulk material has no center of symmetry[72]. The cocrystals in Figures 5.8 and 5.9 demonstrate the use of hydrogen bonds to form one- and two-dimensional structures (chains and layers) that lack a center of

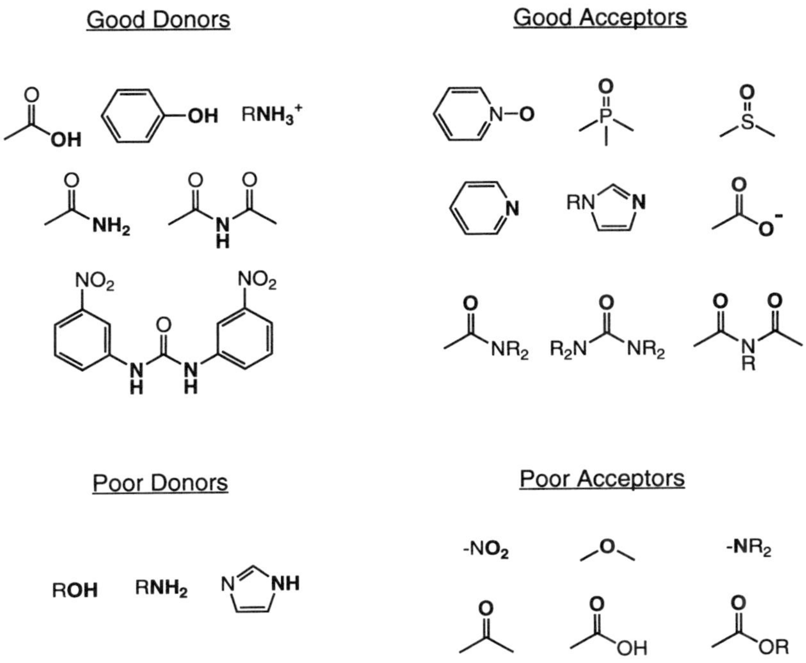

Figure 5.7 Examples of functional groups that form hydrogen bonds in organic crystals. Good donors and acceptors almost always participate in hydrogen bonding, while poor donors and acceptors only occasionally form hydrogen bonds.

symmetry. The layer motif shown in Figure 5.11 is one example in which the final bulk structure of the crystal was found to be acentric[55]. The polar layers of molecules in this structure are parallel and unidirectional, although the nature of the interplanar interactions in this crystal is not clear.

2.1.3 Selectivity of Functional Groups in Forming Hydrogen Bonds

Selectivity of organic functional groups has been determined from studies of competition between a single acceptor (A) for multiple donors (D), and vice versa. Relative abilities of various functional groups in forming hydrogen bonds were determined by examining which donors and acceptors are actually used to form hydrogen bonds, and with what frequency, in crystals. Such studies do not measure the thermodynamic or kinetic contributions of hydrogen bonds to the formation of a particular crystal structure; rather, they determine whether functional groups show consistent patterns of selectivity in the solid state. The extent to which these patterns match those found in solution (or in the gas

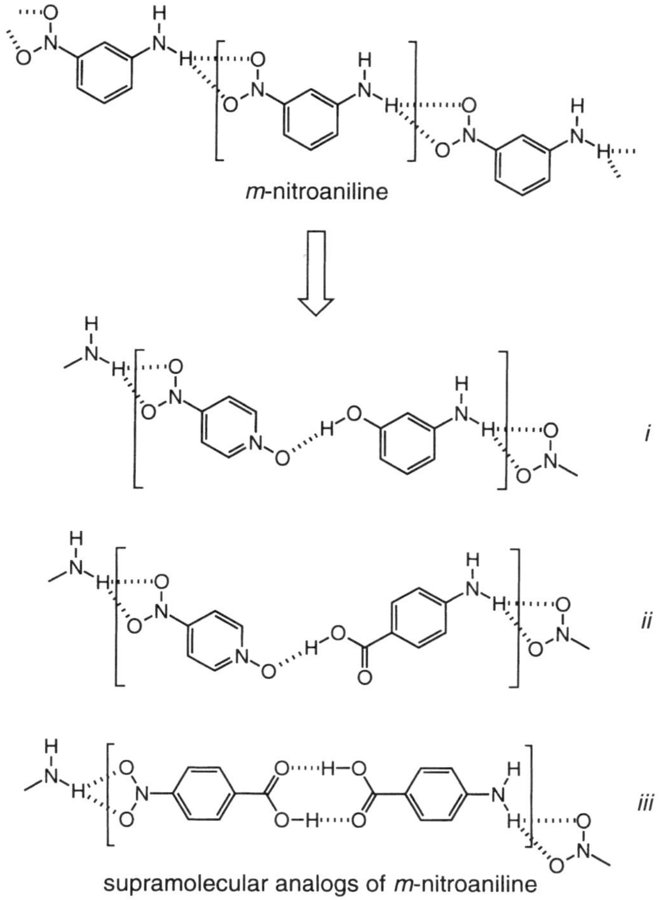

Figure 5.8 Hydroxy, N-oxide, and carboxylic acid functional groups (i–iii) have been used to join aniline and nitrobenzene rings to create extended supramolecular analogs of *m*-nitroaniline.

phase) reflects the degree with which the strengths of hydrogen bonds at equilibrium determine the structures of aggregates in the solid state.

Three general schemes for ranking functional groups are shown in Scheme 5.1. In each case, two different donors compete in forming a hydrogen bond with a single acceptor. Scheme 5.1i represents crystallization of a single-component system where the donors and acceptor reside on the same molecule. Scheme 5.1ii represents cocrystallization between two components, a molecule with two donors and a second molecule with one acceptor. Scheme 5.1iii involves cocrystallization between three components. In this experiment, a

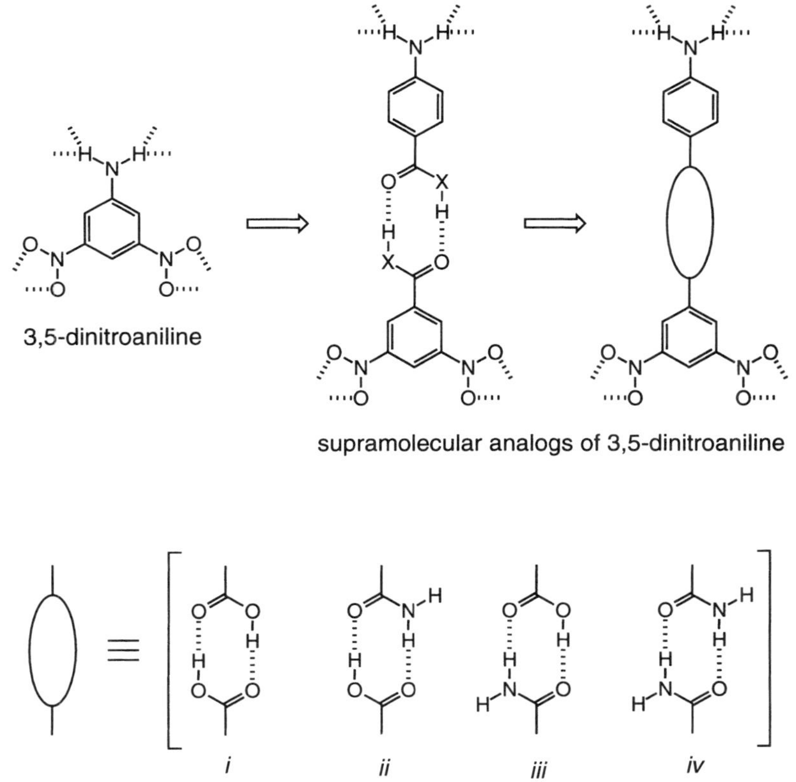

Figure 5.9 Carboxylic acid and amide functional groups (i–iv) have been used to join aniline and dinitrobenzene rings to create extended supramolecular analogs of 3,5-dinitroaniline.

molecule with an acceptor selects one of two different molecules, each containing a single donor, during crystallization. Examples of each of these types of experiment are described below.

Cocrystals of pyridines with benzonic acids. Etter found that experiments involving competitive cocrystallization of a single acceptor with several different, competitive donors (Scheme 5.1iii) provide a convenient method for evaluating the donating (or accepting) ability of functional groups. For example, 4-phenylpyridine and ethyl isonicotinate were both capable of complexing and completely separating the acid of lower pK_a value from a second acid is higher pK_a value by cocrystallization (Figure 5.12)[33]. Even acids with pK_a values that are close to one another, such as 3,4-dinitro- and 3,5-dinitrobenzoic acid ($\Delta pK_{acalc} = 0.04$), were cleanly separated using this technique. These experiments imply that the strength of the hydrogen bond in solution—to the

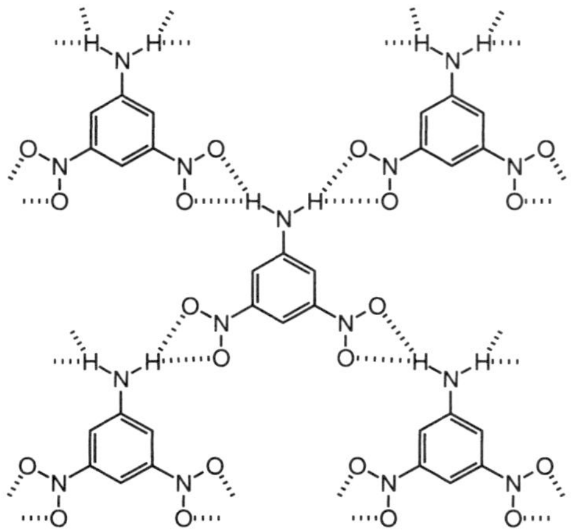

Figure 5.10 The proposed two-dimensional pattern of hydrogen bonds (layer) formed by molecules of 3,5-dinitroaniline.

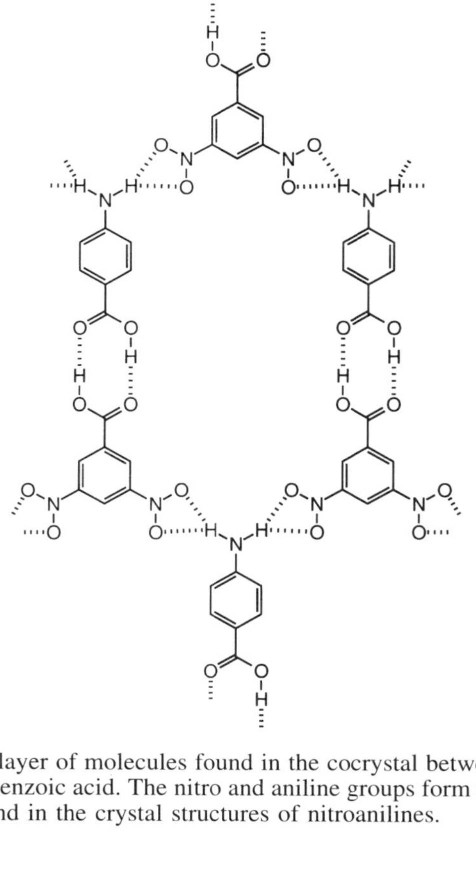

Figure 5.11 Polar layer of molecules found in the cocrystal between 4-aminobenzoic acid and 3,5-dinitrobenzoic acid. The nitro and aniline groups form the same patterns of hydrogen bonds found in the crystal structures of nitroanilines.

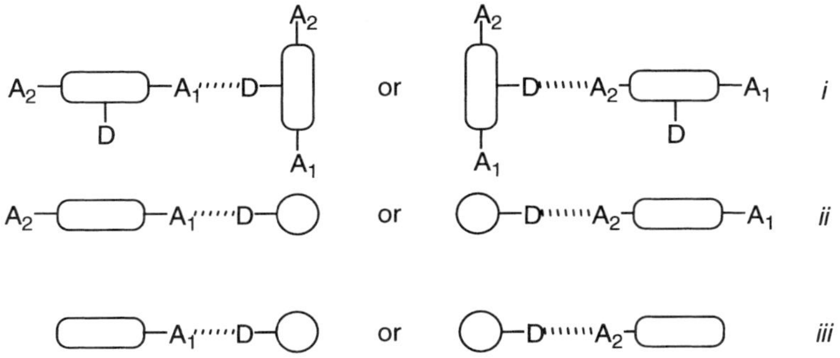

Schemes 5.1 Three general schemes (i–iii) for evaluating the relative abilities of donors (D) and acceptors (A) in forming hydrogen bonds based on competition between functional groups during crystallization.

extent that pK_a values reflect this parameter—determines which aggregate will nucleate the growth of crystals, even when the difference in strength (pK_a value) is small.

Cocrystals with acyclic imides. Acyclic imides formed cocrystals with a variety of guest molecules such as ureas, carboxylic acids, amides, and phenols[50]. In the absence of a guest, acyclic imides crystallized either as chains or dimers (Figure 5.13), depending on the steric bulk of substituents R[38,39]. Guest molecules readily formed complexes with acyclic imides, giving cocrystals with the different motifs (Figure 5.13b). A pattern that recurred frequently when acyclic imides crystallized with phenols was one in which the imide dimer was preserved, and in which the two remaining carbonyl groups were used to bond to the OH groups of phenol. This arrangement is illustrated schematically in Figure 5.14. Etter demonstrated that phenols with two OH groups, such as hydroquinone, could be used to join imide dimers into infinite chains.

Cocrystals of 2-aminopyrimidine with carboxylic acids. Studies of cocrystallization between 2-aminopyrimidines (2AP) and mono- or dicarboxylic acids have shown that selectivity during the formation of cocrystals can be used to monitor subtle changes in the abilities of functional groups to accept hydrogen bonds arising from the effect of substituents. Carboxylic acids, for example, cocrystallized with 2APs to form complexes with ratios (2AP:acid) of 1:2, 1:1, and 2:1, as shown in Figure 5.15[45,46]. The stoichiometry of the cocrystal was controlled, in part, by the ratio of starting materials used. Several carboxylic acids showed a single preferred stoichiometry of 1:1, however, regardless of the ratio of starting materials or of the method of preparation.

Cocrystals with triphenylphosphine oxide (TPPO). Etter has demonstrated that TPPO forms complexes in the solid state with a wide variety of molecules, examples of which are shown in Figure 5.16[48,73]. These complexes were

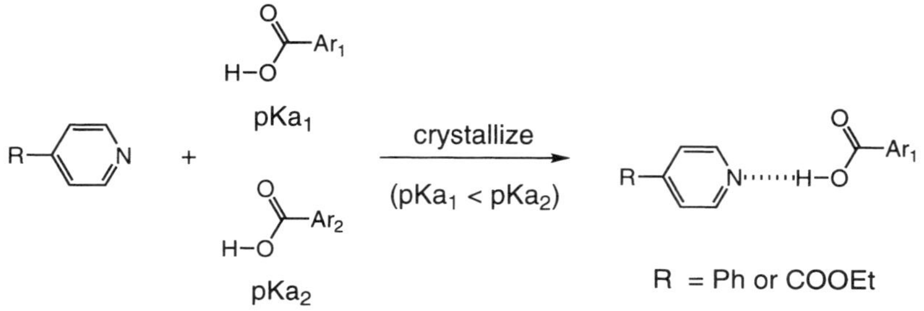

Acid in Cocrystal Ar_1COOH	ΔpKa	Uncomplexed Acid Ar_2COOH
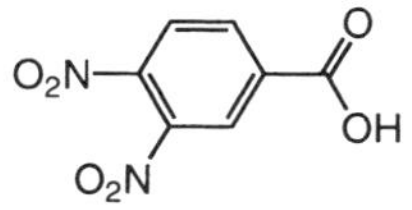	2.35	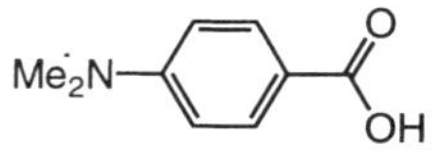
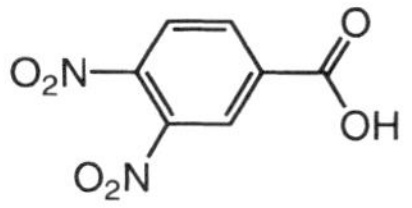	0.74	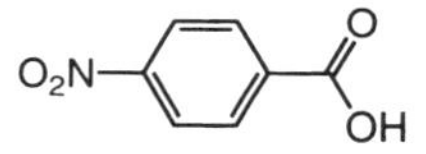
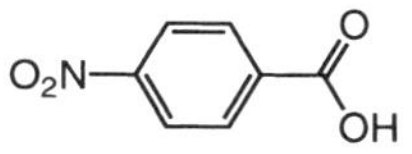	0.41	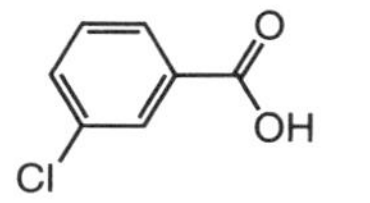
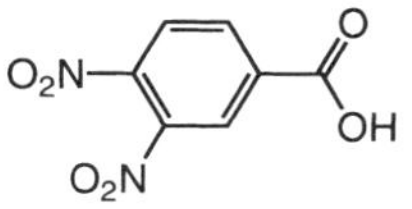	0.04	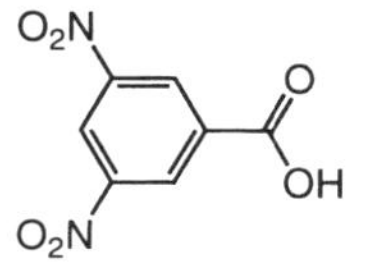

Figure 5.12 Strongly hydrogen-bonded pyridine • benzoic acid dimers were found in cocrystals formed selectively in the presence of a second weaker (higher pK_a) benzoic acid.

stabilized by strong hydrogen bonds between the phosphoryl oxygen and an acidic hydrogen on the guest. This interaction was present in every structure, despite the presence of additional heteroatoms that could compete with TPPO in forming a hydrogen bond. In most cases, the substrates formed chains of hydrogen bonds when crystallized in the absence of TPPO. Complexation of the guest by TPPO usually gave dimers or tetramers (Figure 5.17).

The complexes of TPPO generally formed large, well-developed crystals, while many of the starting materials formed thin needless. Etter suggested that

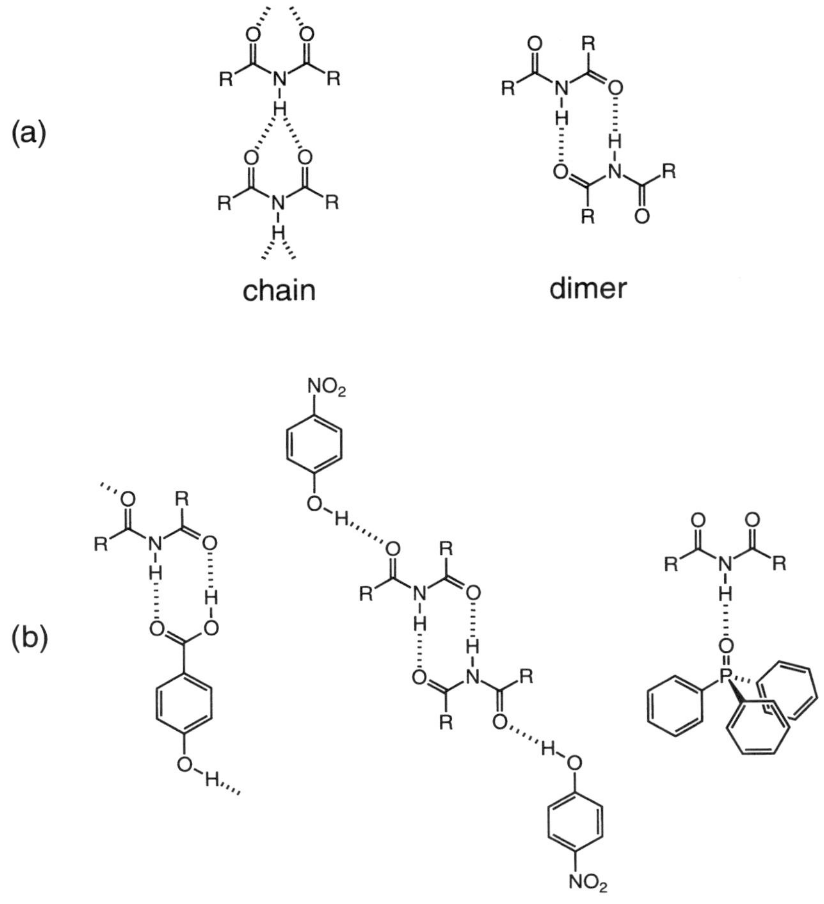

Figure 5.13 (a) Two motifs (chains and dimers) formed by acyclic imides. (b) Cocrystalline complexes formed between acyclic imides and guest molecules having good proton donors and acceptors.

two factors contribute to the effect that TPPO has on the growth of crystals: one is the presence of a strong hydrogen bond that imparts partial ionic character to the crystal; the other is the bulky shape of TPPO, which inhibits the formation of lamellar structures that often induce crystals to grow as thin plates or needles.

Cocrystals with 1,3-bis(m-nitrophenyl)urea (MNPU). Studies of cocrystallization of bis(aryl)ureas with a variety of guest molecules have shown that MNPU, in particular, is capable of forming hydrogen-bonded complexes with molecules containing proton acceptors[41,42]. Examples of several such complexes are shown in Figure 5.18. Guest molecules with strong acceptors

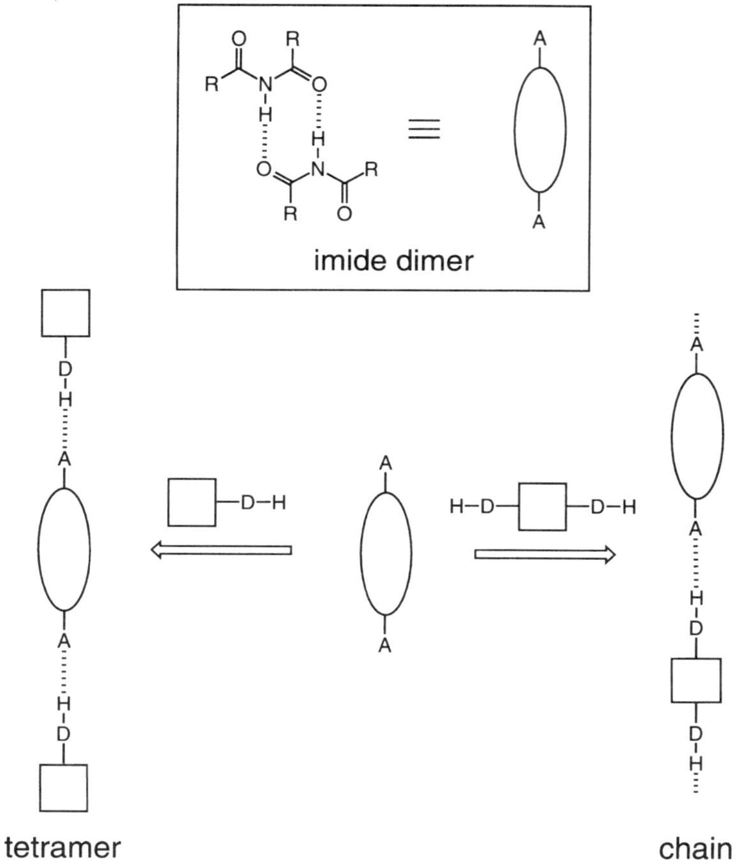

Figure 5.14 Schematic representation of motifs formed by acyclic imides. Acyclic imides that form cyclic dimers (top) have one carbonyl group that is not used in hydrogen bonding. The unbonded carbonyl groups of these dimers serve as sites for hydrogen bonding to molecules that are proton donors (bottom).

such as phosphine oxides and sulfoxides, as well as those with much weaker acceptors such as nitro, ketone, and ether groups, formed complexes in the solid state.

The ability of MNPU to form cocrystals differed from that of other bis(aryl)ureas in several ways. The oxygen of the carbonyl group formed two intramolecular C—H···O interactions with the aryl hydrogens *ortho* to the nitro groups; these weak interactions stabilize the molecule in a planar conformation. In this conformation, the carbonyl group generally did not participate in intermolecular hydrogen bonding. Thus, MNPU is a rare example

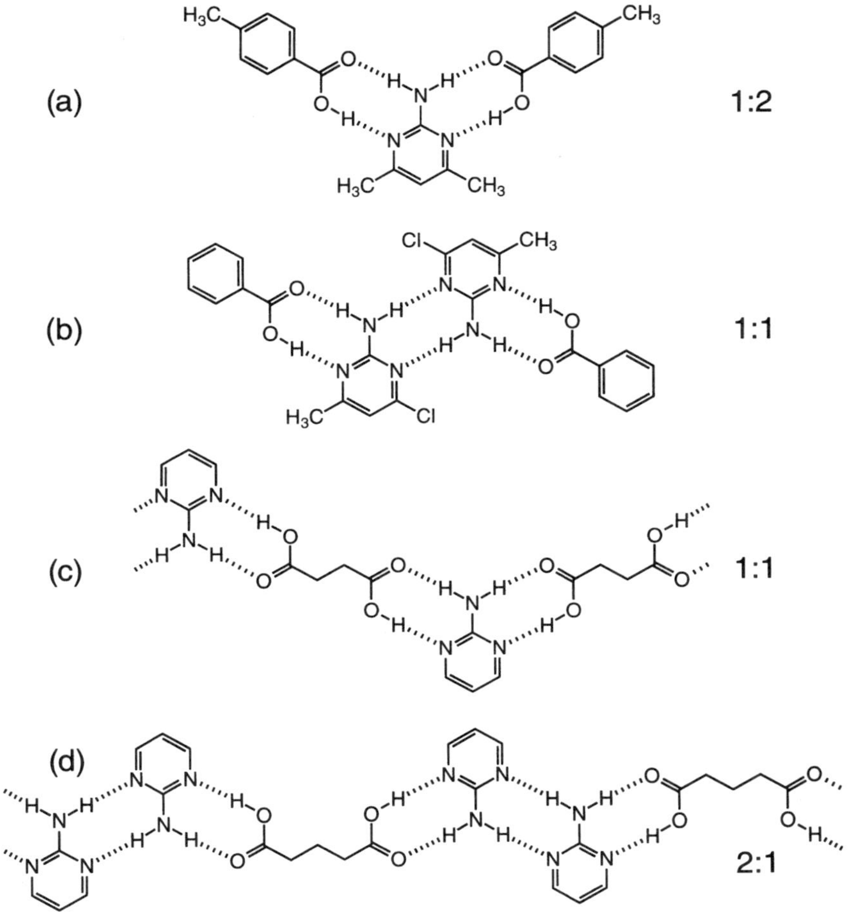

Figure 5.15 Motifs formed in cocrystals with different stoichiometries (pyrimidine:acid) of 2-aminopyrimidines and mono- or dicarboxylic acids.

of a neutral organic molecule that acts solely as a proton donor. Such species are rare because heteroatoms covalently bonded to acidic hydrogen atoms are almost always involved in hydrogen bonding, as stated previously.

Stereoelectronic preferences of functional groups. Etter's work has also demonstrated that stereoelectronic preferences of functional groups can affect the structure of aggregates, especially in crystals containing molecules with a limited number of donors and acceptors and with no other strong intermolecular interactions. The best example of this phenomenon can be seen in the crystal structures of 1,3-cyclohexanediones. These molecules have one donor and one

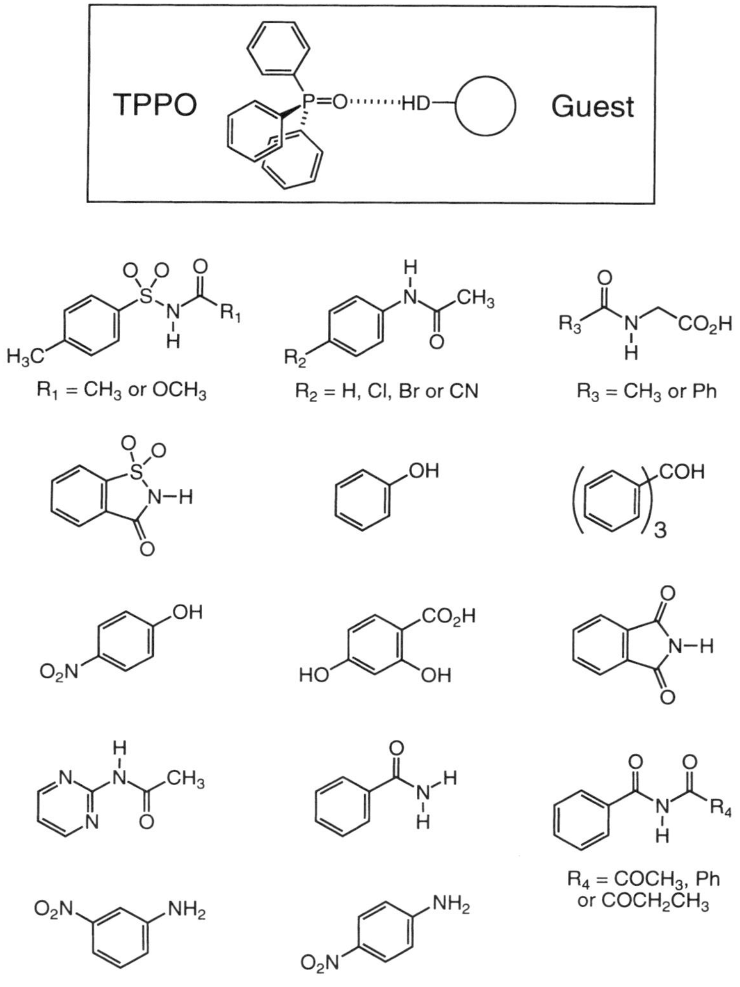

Figure 5.16 Examples of guest molecules that crystallize as hydrogen-bonded host–guest complexes with triphenylphosphine oxide (TPPO).

acceptor. Four stereochemically characteristic modes of association can arise from two different planar conformations of the OH group, and form the choice of *syn* or *anti* lone pairs of electrons on the carbonyl group as the site of the acceptor (Figure 5.19). Energetic preferences for the different modes of association may differentiate cyclic and linear aggregates energetically. The *syn–syn* motif was observed in the crystal structure for 5,5-dimethyl-1,3-cyclohexanedione[74,75], the *anti–anti* motif for 1,3-cyclohexanedione (CHD)[76],

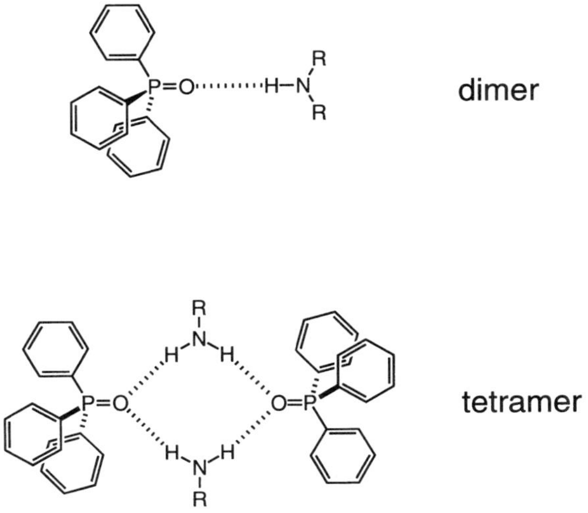

Figure 5.17 Dimer and tetramer motifs found frequently in cocrystals with TPPO.

and the *syn-anti* motif (with a molecule of benzene trapped in the center of the central cavity) for both CHD and 5-methyl-1,3-cyclohexanedione[77]. The cyclic structure—referred to as a *cyclamer*—is similar to that of crown ethers with a hydrogen-bonded rather than a σ-bonded backbone. Etter demonstrated that cyclamers of CHD are able to trap molecules of benzene and thiophene selectively during crystallization.

2.2 Weak Hydrogen Bonds (C—H···O) as Steering Interactions

Desiraju convincingly demonstrated that hydrogen bonds other than those taking part in strong XH···X interactions (X=O and N) can play an important role in determining crystal packing[78–81]. For example, the acidity of the CH donor group again appears to be important in determining the strength of CH···O interactions[82]. Desiraju developed a "crystallographic scale of carbon acidity" in which C···O distances appear to correlate with the acidity of the CH group (as measured independently in DMSO)[83]. Two different correlations emerge; those for CH donors that are sterically hindered, and those for CH donors that are unhindered. The importance of hybridization at the carbon in determining the strength of C—H···O interactions is evident in comparing alkynes with alkenes. For example, Desiraju showed that alkynes form shorter C···O contacts (3.46 Å on average), while alkenes form longer contacts (3.46 Å on average)[84]. While this difference in lengths (0.18 Å) is small, the majority of the shortest contacts (3.1–3.4 Å) involve alkynes.

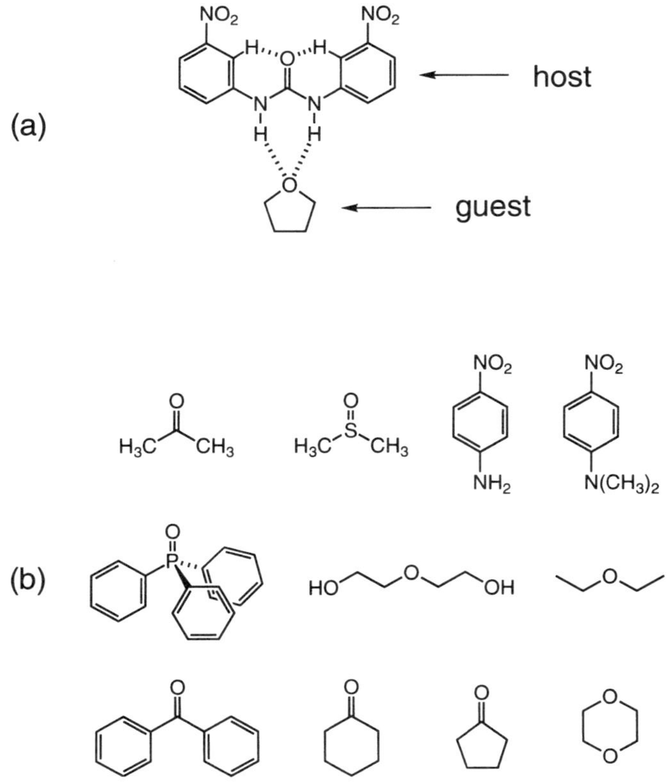

Figure 5.18 (a) Hydrogen-bonded complex between 1,3-bis(*m*-nitrophenyl)urea (host) and tetrahydrofuran (guest). (b) Examples of other guest molecules that form complexes with 1,3-bis(*m*-nitrophenyl)urea.

Desiraju also studied other C—H···O interactions in compounds containing chlorine atoms to determine how electronegative groups affect the acidity of CH hydrogens[82]. The average C···O distances for structures in the series Cl_3CH, Cl_2RCH,ClR_2CH, and R_3CH were 3.32 Å, 3.40 Å, 3.46 Å, and 3.59 Å respectively; thus, in this series, the length of C—H···O interactions varied directly with the number of attached chlorine atoms. Another type of weak interaction studied by Desiraju was XH···C(X=O and N; referred to as a π hydrogen bond) where the acceptor was an alkyne, alkene, or phenyl group[85]. In these studies, the average H···C distances were significantly less than the sum of van der Waals radii for C and H atoms (2.95 Å): observed values ranged from 2.43 Å for OH···C_{phenyl} to 2.82 Å for NH···C_{alkene} contacts.

The work of Desiraju suggests that the more acidic the CH hydrogen involved in a C—H···O interaction is (and therefore the stronger the interaction), the

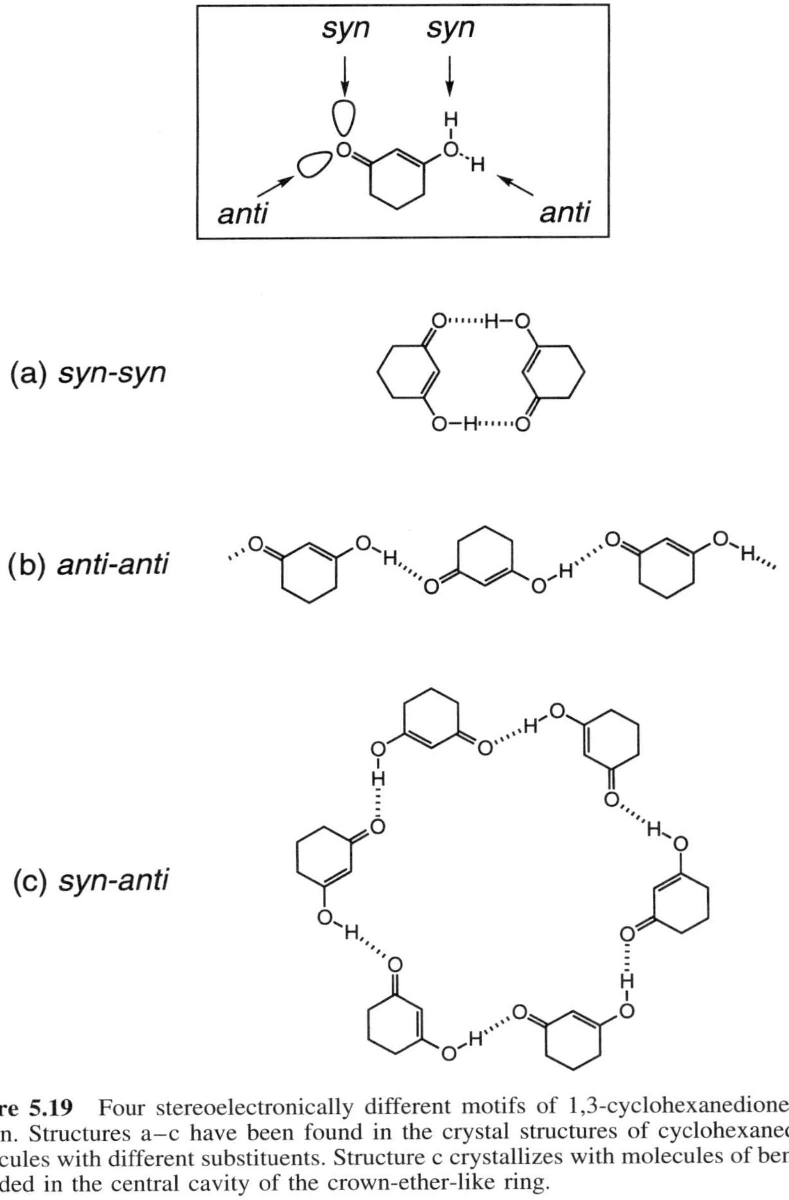

Figure 5.19 Four stereoelectronically different motifs of 1,3-cyclohexanediones are shown. Structures a–c have been found in the crystal structures of cyclohexanedione molecules with different substituents. Structure c crystallizes with molecules of benzene included in the central cavity of the crown-ether-like ring.

(d) *anti-syn*

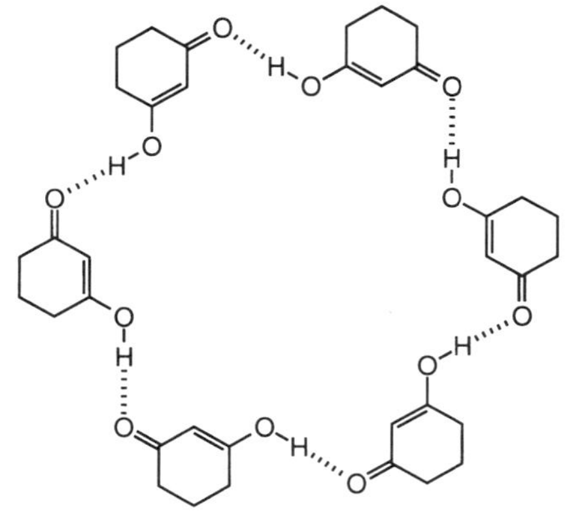

Figure 5.19 (*contd.*)

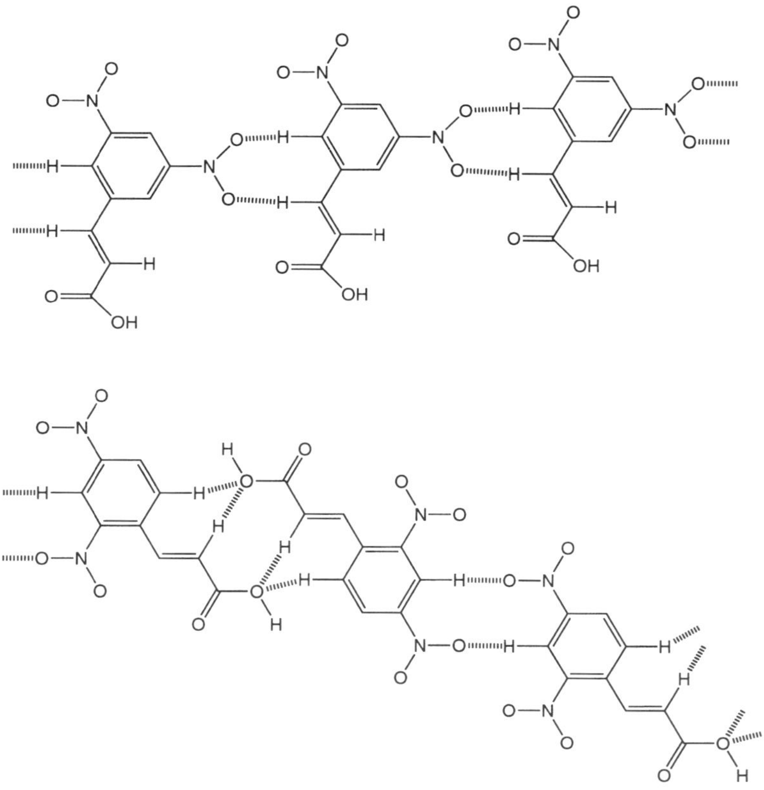

Figure 5.20 Two chain motifs of dinitrocinnamic acids. The top motif is built from one type of CH···O interaction, while that at the bottom uses two different types of CH···O contacts.

more useful thus interaction will be in designing structures. For example, he has used hydrogen bonds involving the alkene protons of cinnamic acids[86], and the aromatic protons of nitroaromatic compounds[87,88] to engineer the structures of crystals. Figure 5.20 shows two recurring patterns formed by these moieties. CH···O interactions are the only type of hydrogen bonds present in both patterns, one of which involves two different hydrogen-bonding motifs to link adjacent molecules in a chain. Another interesting structure involves COOH dimers alternating with an approximately isostructural pattern between dimethylamino and nitro groups to produce a chain of molecules (Figure 5.21).

This last example indicates that these weak CH···O interactions can occur alone in a crystal or in conjunction with stronger hydrogen bonds. In the latter case, Etter's guideline is followed: the structure can be rationalized by assuming that the strongest donors and acceptors pair first, followed by pairing of weaker donors and acceptors. In other words, the geometric demands of stronger hydrogen-bonding moieties usually determine which motifs form (e.g. the uniquitous carboxylic acid dimer), while weaker hydrogen bonds such as CH···O interactions determine how these aggregates pack into higher order aggregates such as stacks or layers. In some cases the weaker hydrogen bonds can influence the type of motif adopted by the strong hydrogen bonds. For example, Desiraju showed that a cocrystal between two carboxylic acids containing no CH···O interactions was dominated by π–π interactions. These π–π forces promote the formation of homodimers between the two carboxylic acids rather than heterodimers, as predicted purely on the strengths of OH···O hydrogen bonds, and the principle that the best donor and acceptor, which reside on different acids, will pair first[89]. This example illustrates one of Desiraju's key points: to control or rationalize crystalline supermolecular architecture, weak interactions—both hydrogen bonds and other kinds—must be taken into consideration[82,90,91].

Desiraju demonstrated the importance of CH···O interactions in controlling molecular packing in a series of experiments. These experiments showed that aromatic hydrocarbons pack almost exclusively in different structural categories depending on the ratio of carbon to hydrogen (C : H): herringbone patterns

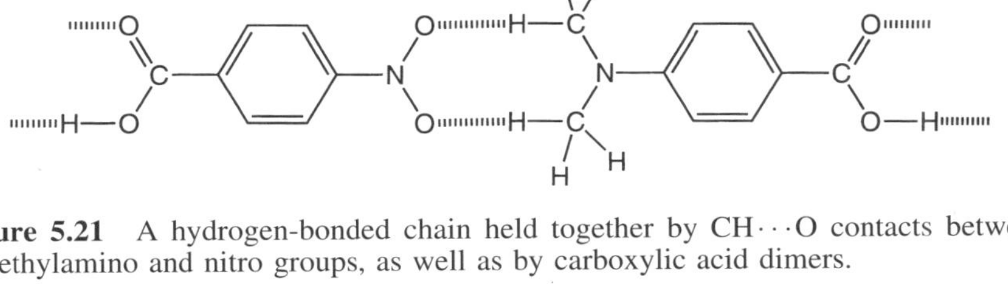

Figure 5.21 A hydrogen-bonded chain held together by CH···O contacts between dimethylamino and nitro groups, as well as by carboxylic acid dimers.

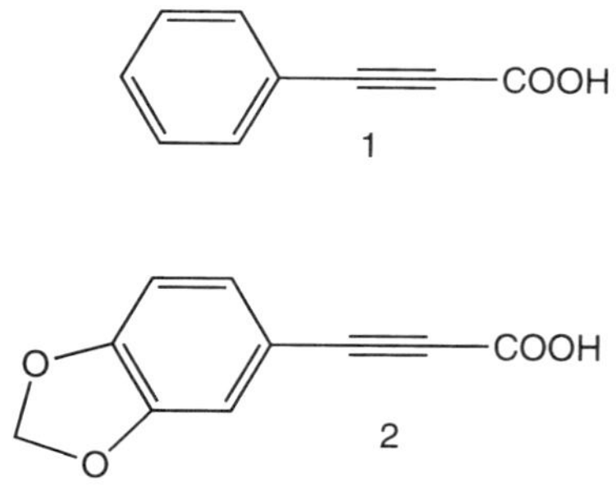

Figure 5.22 The added oxygen atoms in **2** allow the formation of intermolecular CH···O contacts that favor adoption of a stacked, sheet-like packing arrangement, whereas **1** packs as dimers that adopt a herringbone motif.

(called β-structures) at low C : H ratios; columns and sheets at high C:H ratios[78,82,92]. Desiraju reasoned that adding oxygen atoms to the molecules would promote the formation of CH···O bonds—that is, the hydrogen atoms would be less likely to engage in the H···C contacts necessary for herringbone packing—thus increasing the effective C : H ratio, and driving the molecules to pack in the β-structure (Figure 5.22). Desiraju also reasoned that intermolecular CH···O contacts should favor formation of layered structures. Experimentally, a higher content of oxygen led to packing in stacks and layers. This work has practical ramifications for reactivity in the solid state. For example, stacked acetylene groups exhibit thermal reactivity as diene/dienophile pairs for cycloaddition reactions in the solid state[82,92,93].

2.3 Resonance-Assited Hydrogen Bonding (RAHB)

Gilli's group has shown, using surveys of the crystallographic database as well as their own experimental results, that the strengths of hydrogen bonds are intimately related to electronic features of the molecules involved[94–97]. Their Resonance-Assited Hydrogen Bonding (RAHB) model states that there is "synergistic coupling between the increased π- resonance and hydrogen-bond strengthening"[94]. While most of their examples are based on intramolecular hydrogen bonding (which is outside the scope of this review) particularly in β-diketone enols or related structures, they have also shown that the phenomenon is relevant to intermolecular hydrogen bonding. The key feature of their model is that a change-separated resonance form enhances hydrogen bonding when they hydrogen-bonded groups are conjugated (see **3** and **4**). The extent of resonance or delocalization can be quantified by a parameter derived from bond lengths along the conjugted backone. This parameter does correlate with the strength of the hydrogen bond, as measured by the intermolecular oxygen–oxygen distance and the length of the covalent O—H bond. That electronic effects are important is shown by structures of ketoesters, which form longer hydrogen bonds than diketones: electron donation from the extra ester oxygen disfavors the charge-separated resonance form (**3**) that would otherwise enhance hydrogen bonding[95]. This opposition of inductive and resonance effects also arises in intramolecular cases, notbaly in *para*-substituted diketo-arylhydrazones, where O···O distances correlate with Hammett σ-values of the aryl substituents[94]. Other conjugated functional groups, such as those involving nitrogen (enaminones), should also exhibit this phenomenon. The authors suggest that the RAHB effect could have ramifications for the design of non-linear optical materials, since the hydrogen-bonded "resonant chains" have extensive electron delocalization and ground-state dipole moments[95].

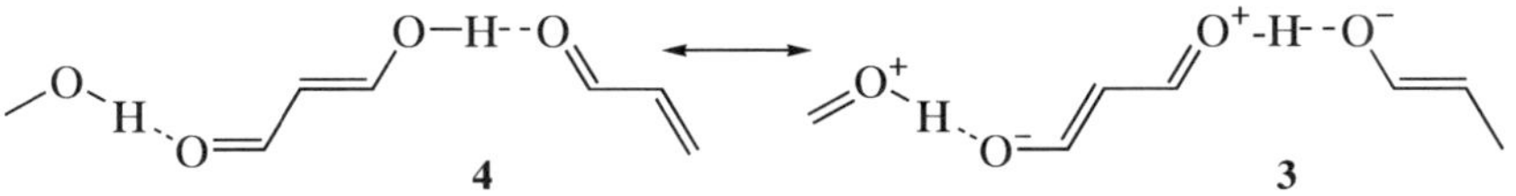

4 **3**

2.4 Packing Motifs of Carboxylic Acids and Primary Amides

Leiserowitz has extensively catalogued the motifs that carboxylic acids and amides are likely to adopt in crystals, and used symmetry, steric, and electrostatic arguments to rationalize these motifs. The broad outlines of his findings are as follows. Carboxylic acids "almost invariably" form cyclic hydrogen-bonded dimers[98]. While in principle a catermeric motif could form from either the *anti*planar (**5**) or the *syn*planar (**6**) COOH conformers, Leiserowitz found no examples of the former. Several examples of the latter are however known. The scarcity of the catermeric motif can be ascribed to either O···O repulsion, or weak distored hydrogen bonds resulting from the steric

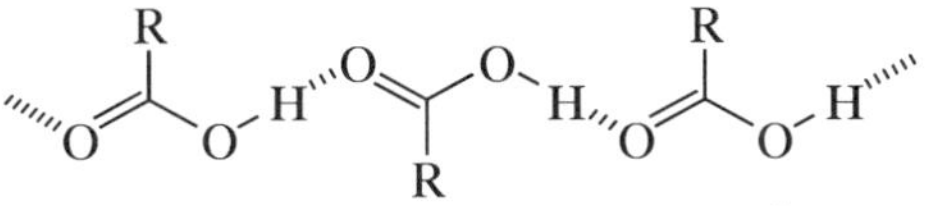

5

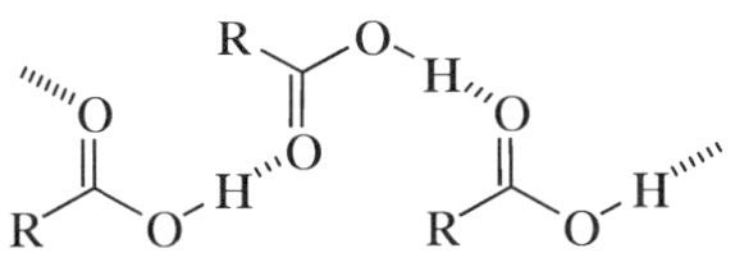

6

repulsion of the R groups. The only way in which carboxylic acids are observed to form hydorgen-bonded chains is if the molecule is a dicarboxylic acid. If one of the carboxylic acid groups is replaced with a primary amide, then there are two options for chain formation: one with acid–acid and amide–amide rings (homodimers, **7**) and one with acid–amide rings (heterodimers, **8**). In either chain, the second proton of the NH_2 group is unused in forming the chain and can link adjacent chains together. Consideration of the geometry of these layers of chains shows why the homodimer chains are not observed: optimal interchain hydrogen bonds would lead to electrostatic repulsion between oxygen atoms on adjacent chains of this class.

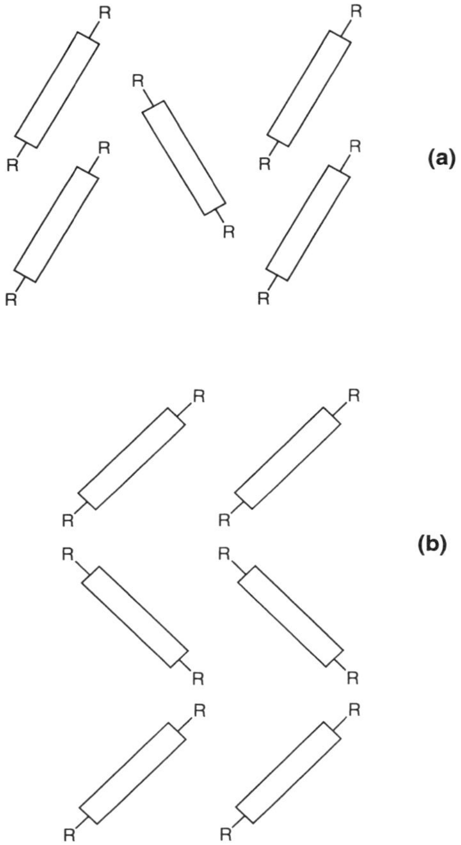

Figure 5.23 End-on views of hydrogen-bonded rods of primary amides in two possible packing arrangements. The top one, a herringbone motif, is observed, while the corrugated motif at the bottom is not.

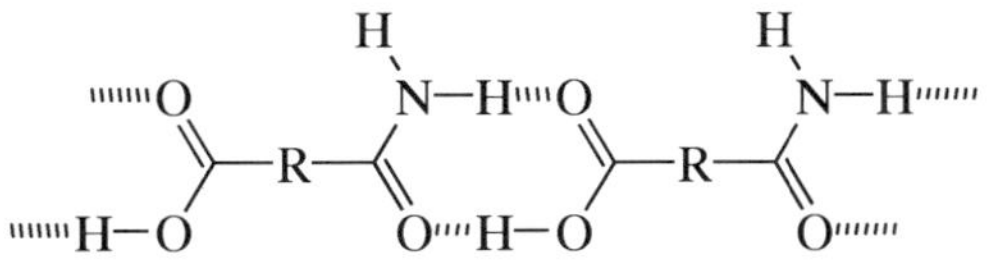

7

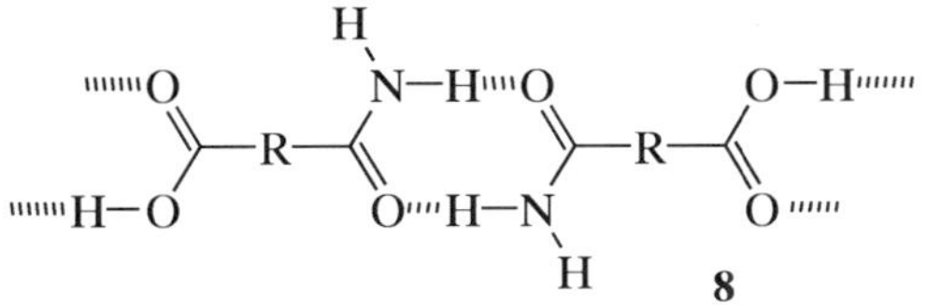

8

Similar issues are relevant to a consideration of the packing of primary amides[99]. Again, a ring structure similar to the carboxylic dimer, is overwhelmingly formed. Here, however, the *anti* amide protons that are external to the ring can be conveniently used to link rings into chains. Leiserowitz and Hagler apply symmetry operations to these chains to first generate two-dimensional layers, and then three-dimensional lattices. Only some of these possibilities are likely to be observed, because not all symmetry operations are consistent with the close packing of R groups and favorable interlayer electrostatic contacts. These arguments are able to rationalize, for example, why herringbone and not corrugated packing is observed (Figure 5.23). The authors suggest that a symmetry analysis of this sort could be used to construct a small number of possible packing arrangements that could then be evaluated by calculating the lattice energy (calculation of the lattice energy is discussed in Section 5).

3. SURVEY OF DESIGNS FOR CRYSTAL ENGINEERING

Recently, a number of innovative strategies that use hydrogen bonds to control crystalline structure have been explored[100]. Studies in this area focused deliberately on designing structures with specific shapes and sizes through non-covalent synthesis, and consequently led to the discovery of new structural principles and interesting topologies of packing. More importantly, these systems all used various elements of design to control the spacing and arrangement of structural units in the crystal. These elements of design provide chemists with a variety of choices for linking molecules together in an effort to design new crystalline systems. We will discuss each of these elements—charged and uncharged hydrogen bonds, rigid and flexible components, aliphatic and aromatic units—with respect to a given molecular system in the sections that follow.

3.1 Tubulands from Alicyclic Diols

Bishop and Dance discovered that alicyclic diols of varying ring size will self-assemble to give isostructural crystalline lattices[101–106]. The key structural features of these crystals are trigonal hydrogen-bonded "spines" interspersed with parallel channels (Figure 5.24). As a result of these channels, inclusion of molecules of solvent or other guests usually occurs. By controlling the shapes of the components of the host lattice, a range of topologies and diameters can be attained within the channel. More interestingly, the structure of the host lattice can be controlled by the geometrical and chemical structure of the included guest. In some cases, empty host lattices are also observed[106] (a characteristic that is unusual for organic hosts)[107]. One molecule of diol can be replaced in

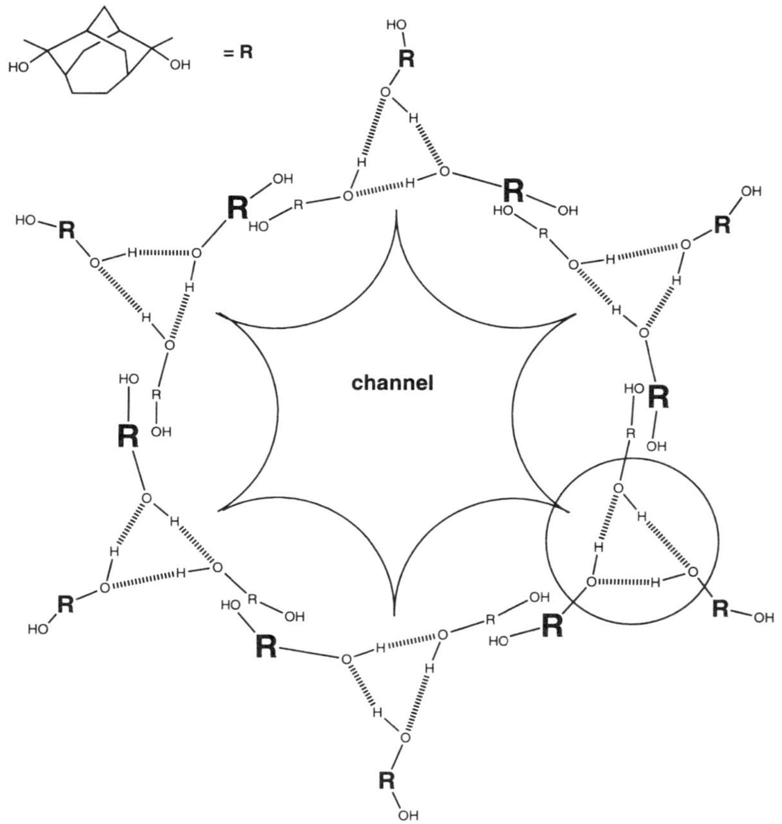

Figure 5.24 Schematic view of the packing of an alicyclic diol into the tubuland lattice, looking down one of the channels for guest enclathration. A hydrogen-bonded trigonal spine is circled.

the lattice of the host by another molecule capable of hydrogen bonding (e.g. a phenol) to give cocrystals[104]. In this way, other sorts of intermolecular interactions, such as Cl···Cl contacts or different kinds of hydrogen bonds, can be introduced into the crystal. Bishop and Dance have proposed guidelines describing the molecular features that must be present to form trigonal spines[102].

3.2 Ionic Hydrogen-Bonded Layers

Aakeröy and Seddon used hydrogen-bonded ionic salts to create robust crystalline architecture in one and two dimensions[108]. They generated entire families of crystal structures using various organic cations complexed with either a phosphate[37], malate[109], tartrate[110,111], or carboxylate anion[112]. The members of these families were compared for structural regularities; graph-set notation was indispensable in classifying the hydrogen-bonded networks[110–112]. A common motif consisted of chains of anions held together by strong $OH \cdots O^-$ interactions (Figure 5.25). These chains were further linked into layers or three-dimensional networks by other hydrogen bonds, including weaker $NH \cdots O$ interactions involving ammonium counterions. Using different cations generally did not change the motifs of the anions (one-dimensional chains, two-dimensional layers, and three-dimensional networks) except in crystals built from phosphates. One of the goals of these authors was to use the

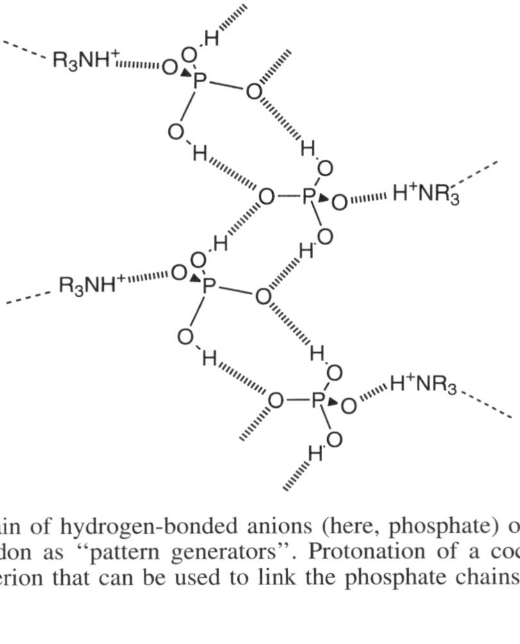

Figure 5.25 Chain of hydrogen-bonded anions (here, phosphate) of the sort used by Aakeröy and Seddon as "pattern generators". Protonation of a cocrystallized amine generates a counterion that can be used to link the phosphate chains into layers.

anions as "pattern generators"—that is, as reliable biulding blocks for crystal engineering—while using the cations to impart useful properties to the crystal[112]. For example, they hoped to use chiral components, either cations or anions, in conjunction with cations possessing a high degree of electron-delocalization to obtain materials that exhibit second-harmonic generation.

3.3 Conformational Control of Architecture: Chains and Helices

Hamilton found that bis(acylamino)pyridines cocrystallize with dicarboxylic acids to give 1:1 complexes that form extended chains (Figure 5.26)[113,114]. The crucial strategy for crystal engineering that these workers employed was to use components of complementary or mismatched sizes. When the two units were roughly equal in lengths, as in the case of 1,8-octanedicarboxylic acid and N,N′-bis[2-(6-methyl)pyridyl]-2,7-napthyldicarboxamide[114], both molecules were oriented approximately perpendicular to the ditrection of propagation of the hydrogen-bonded chain (Figure 5.27). When one component was substantially longer than the other (e.g. when 1,12-dodecanedicarboxylic acid is used)[114] a "slippage" of the molecules occurred that gave more acute tilt angles between the two components within the chain (Figure 5.27). By simply changing the size of either component, the width of the chains and the spacing of groups along the chains (e.g. the methyl groups or some other substituent on the pyridyl components) could be controlled reliably.

Control over the structure of the chains depended significantly on the conformtion of the acylamino moieties: the two acylamino groups had to adopt an *anti* orientation with respect to one another in order to form extended chains

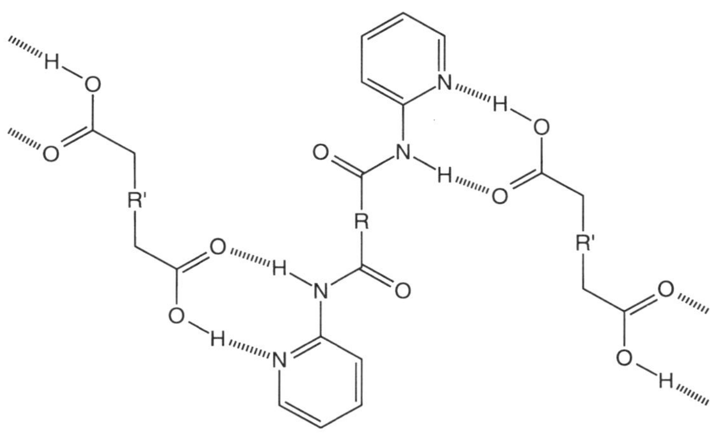

Figure 5.26 Hydrogen-bonded chains of bis(acylamino)pyridines and dicarboxylic acids.

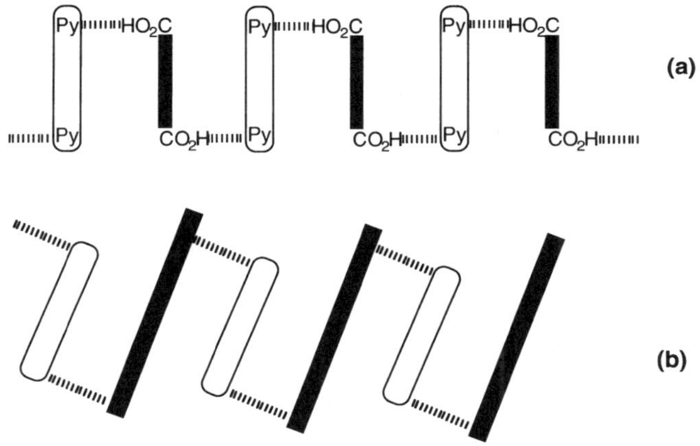

Figure 5.27 Schematic diagram showing the importance of matching the sizes of the two hydrogen-bonded components. (a) When they are approximately the same length, the molecular units are roughly perpendicular to the direction of the chain. (b) When one component is significantly longer than the other, both molecules tilt with respect to the chain axis.

(Figure 5.26). Adopting a *syn* orientation presented a convergent binding pocket that gave a closed dimeric complex[113,115]. When the acylamino groups were substituted *meta* with respect to one another on the aromatic ring, however, the *syn* conformation formed a binding pocket that was selective for only small

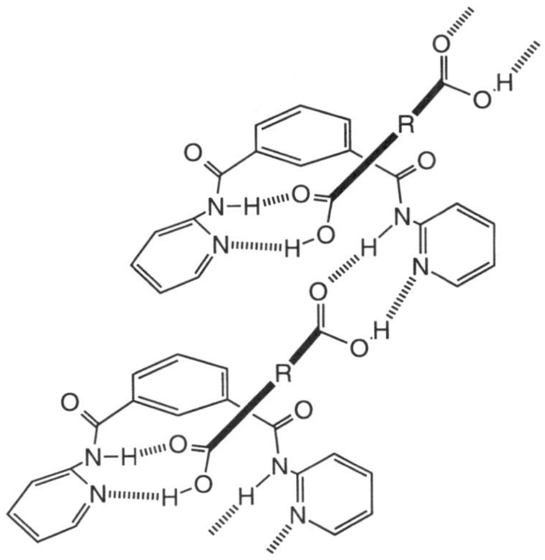

Figure 5.28 Hydrogen-bonded helix constructed from aminopyridine units linked to an isophthalate spacer and a dicarboxylic acid.

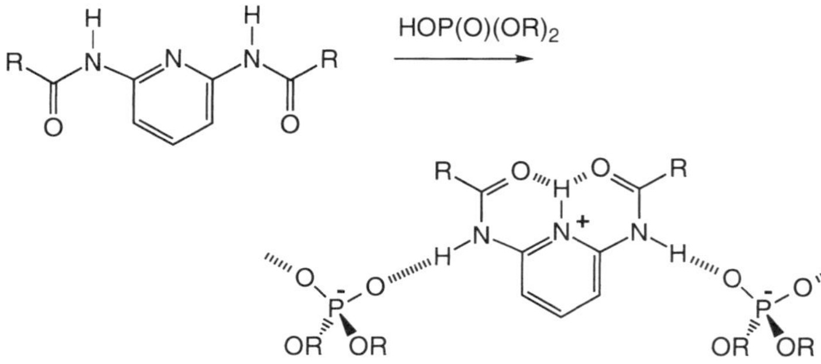

Figure 5.29 A conformational "switch". Protonation of the pyridine nitrogen swings the acylamino groups around to form intramolecular hydrogen bonds, and positions the NH groups to form intermolecular hydrogen bonds that form chains.

dicarboxylic acids. A mismatch in size between larger dicarboxylic acids and this pocket forced the acid to twist out of the plane of the bisacylaminopyridine to form a helical structure (Figure 5.28)[115]. Closest packing was achieved when helices of opposite handedness intercalated with one another. Aromatic stacking interactions appeared to be important in favoring this mode of packing. One drawback in controlling the formation of helical substructures was the occurrence of polymorphism. In one case, both a helix and an extended chain were observed[113].

Hamilton also reported preliminary work in controlling the *syn/anti* conformation by changing the state of protonation at selected sites[116]. When the pyridine ring of a 2,6-diamidopyridine was protonated, the two carbonyl oxygens pointed inward and formed a hydrogen bond with the pyridinium NH group (Figure 5.29). The formation of this complex forced the two NH group into an exo conformation, where they formed hydrogen-bonded chains with counterions such as a diarylphosphate.

3.4 One- and Two-Dimensional Aggregates Predicted Using Analyses of Symmetry

Fowler and Lauher used analyses of symmetry to predict and control the ways in which one-dimensional hydrogen-bonded assemblies aggreated into two-dimensional layers[117,118]. Their guiding principle is that "all intermolecular interactions between molecules of different asymmetric units will correspond to a specific symmetry operation:" a rule set by the symmetry of the lattice[118]. Hence the symmetry of a one-dimensional aggregate (which the authors call a rod group) can be combined with another symmetry operation to predict the symmetry of the resulting two-dimensional aggregate (which the authors call a layer group). These authors demonstrate that a maximum of 15 layer groups can

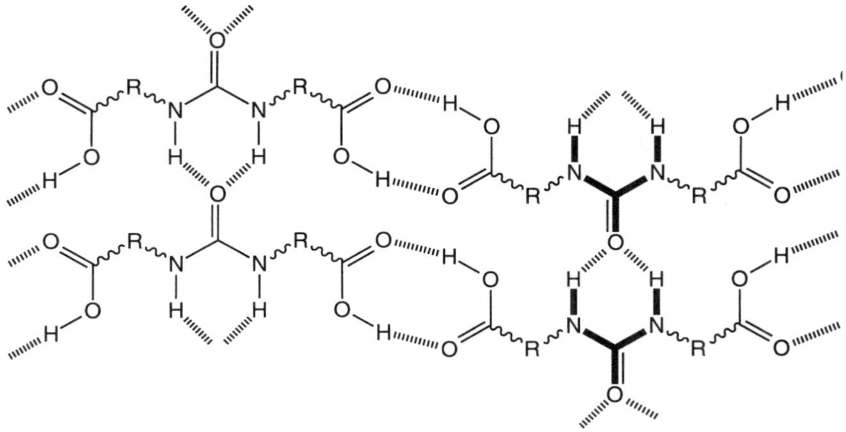

Figure 5.30 Layered motif of ureylene-dicarboxylic acids, with urea columns (bold) held together by carboxylic acid dimers.

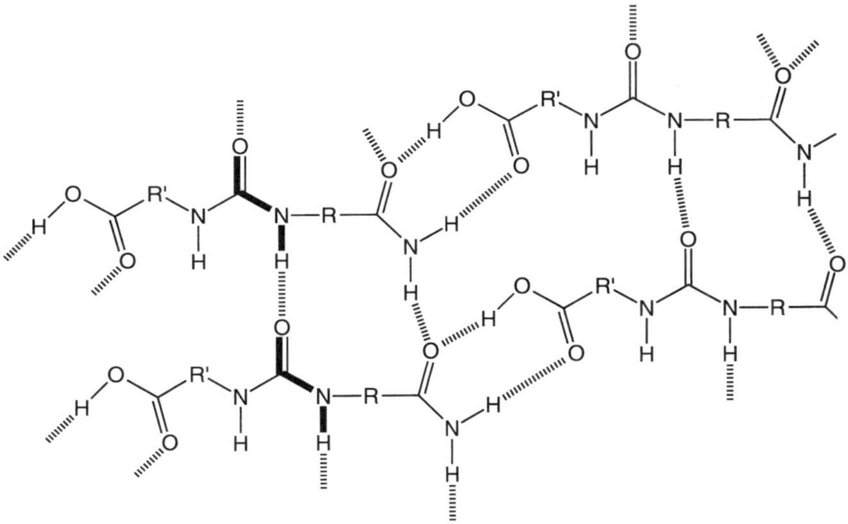

Figure 5.31 Non-equivalent terminal groups (here an acid and an amide) can lead to disruption of the familiar urea columns to give a different kind of urea chain (bold), again held together by acid–amide heterodimers.

form because some symmetry operations are inconsistent with the demands of closest packing in the crystal.

From a chemical standpoint, this scheme requires one motif of hydrogen bonds to generate the rod group, and another to tie the rods together into a layer. Fowler and Lauther's conceptual starting point was the chain motif formed by ureas. Using N,N′-distributed ureas with carboxylic acids or primary amides as the substituents, they usually obtained rods crosslinked by chains of carboxylic acids (Figure 5.30)[118,119]. When the trwo substituent groups were not equivalent, however, unexpected patterns of hydrogen bonds occurred (for example, Figure 5.31). One strategy used by the authors to simplify the analysis of symmetries was to incorporate a chiral substituent into the urea, since there were fewer symmetry operations that could be used to pack chiral molecules[118].

Lauher and Fowler also expanded the skeleton of the urea by replacing it with an aminopyridone[120]. This replacement gave a rod motif similar to that of ureas, but increased the spcing between the rods from 4.6 Å to 6.7 Å. As with ureas, introducing additional hydrogen-bonding group to the ring of the pyridone promoted linking of the rods into layers. The most elegant of these designs used a nitrile to form a hydrogen bond to the *syn*-NH_2 proton, the one that is not used in the motif to build rods (Figure 5.32).

3.5 Open Lattices Constructed from "Tectons"

Wuest *et al.* used pyridones attached to rigid, linear, or tetrahedral cores (tectones, **9**) to assemble a range of lattice sizes and shapes.They have used these tectons as part of a supramolecular "building kit" to design zeolite-like

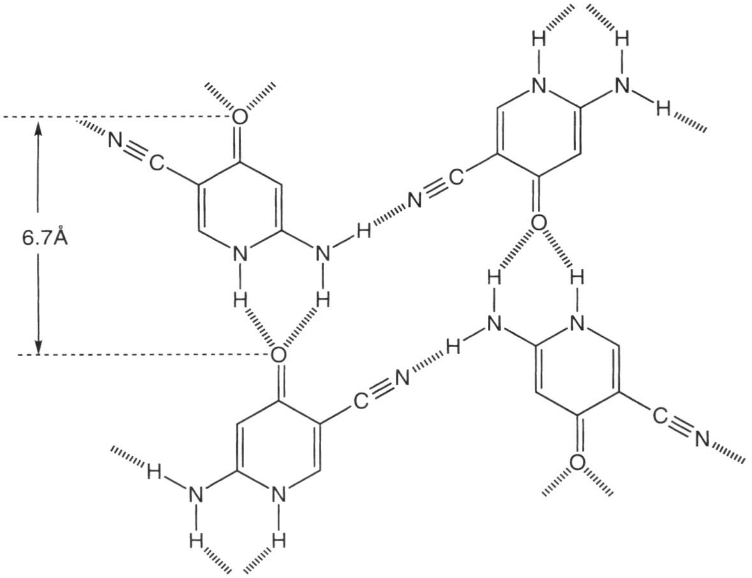

Figure 5.32 Incorporation of a vinylogous urea into a heterocycle leads to one-dimensional hydrogen-bonded rods than can again be joined into layers with an extra hydrogen-bonding moiety, here a nitrile.

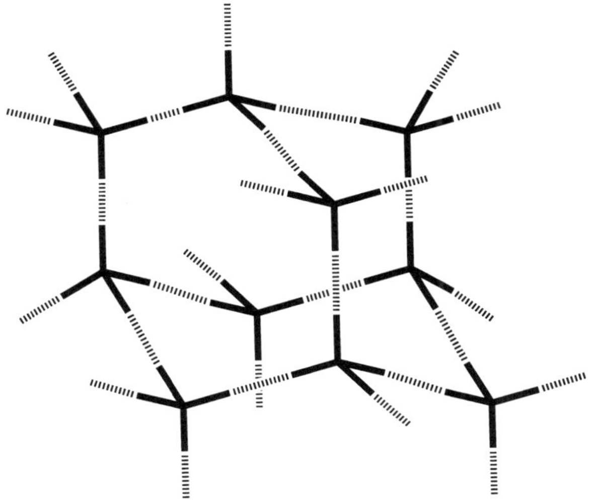

Figure 5.33 Schematic view of a diamondoid lattice obtained by self-assembly of a tetrahedral hydrogen-bonding molecule.

cages and diamondoid lattices in organic crystals (Figure 5.33). By changing the tetrahedral core from $C(C_6H_4C_2)_4$ to the analogous compounds of Si or Sn, channels with different volumes, or microporosities, could be engineered[121,122]. Different guests (usually carboxylic acids) were enclathrated with stoichiometries ranging from one molecule of valeric acid to four molecules of acetic acid per tecton. These cages did not form unless a guest of the proper size and shape to match the incipient channels was present. Moreover, when the size of the guest did not fit the channel, the tectons formed hydrogen bonds with the carboxylic acids[121,122].

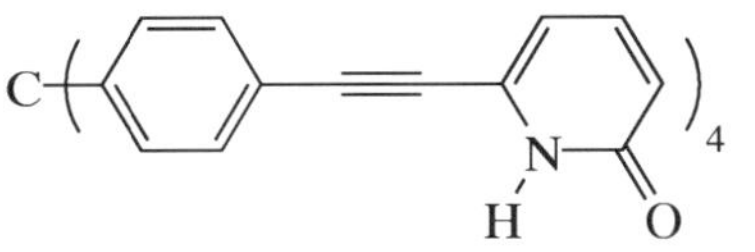

9

These authors also used rigid acetylenic units to link dipyridones together[123]. Depending on the site at which the C2 unit was joined to the pyridone ring, self-complementary dimers or open-ended chains resulted (Figure 5.34). An attempt at designing a cyclic trimer using this strategy was, however, unsuccessful, as competition with DMSO (as the solvent) during crystallization yielded an open-chain form[122]. Attempts to design structures based on dipyridones with flexible linkers were unsuccessful due to competition between intra- and intermolecular hydrogen bonding between the pyridone units[124].

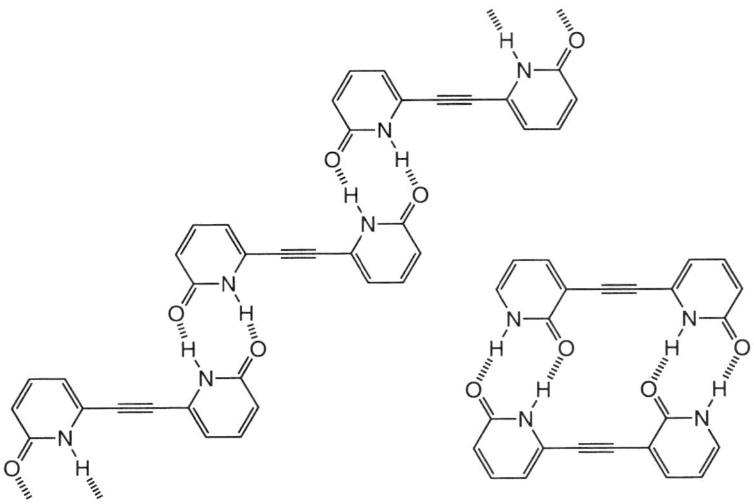

Figure 5.34 Two isomers of pyridones linked by an acetylene spacer can hydrogen bond to form a self-complementary dimer or an open-ended chain.

3.6 Layered Materials from Guanidinium Sulfonates

Etter and Ward have investigated salts formed from alkane- or arenesulfonates and guanidine for structural regularities that can be used in crystal design[125]. They predicted that the complementarity in shape and size of the three-fold symmetric ions—planar $C(NH_2)_{3^+}$ and tetrahedral RSO_{3^-}—would lead them to pair up into hydrogen-bonded dimers (Figure 5.35). In all 15 cases examined, dimers do form and are in turn linked into ribbons. They predicted that these ribbons would also aggregate into layers, which was in fact observed in 11 of these structures. These layers are not perfectly flat, but have variablility in the value of the inter ribbon angle. Three of the cases, for example, have inter ribbon angles less than 90°; the layers are formally present but are highly corrugated. In one structure (*para*-carboxyphenyl) layers are not formed at all. The authors rationalize these differences in planarity of the layers on the basis of the hydrogen-bonding ability of the alkene or arene substituents. Groups that hydrogen bond well, such as NO_2, OH, and COOH, compete with the guanidinium and sulfonate groups, and distort the inter-ribbon hydrogn bonds.

Another structural rationalization relates the size of the substituents to the kind of layer motif adopted. A bilayer structure, where two guanidinium sulfonate layers are stacked directly on top of each other and the substituents interdigitate with each other in a hydrophobic region, is formed if the substitutents are small (Figure 5.36). When the substituents are large, close packing of the hydrophobic groups cannot be achieved by this same kind of

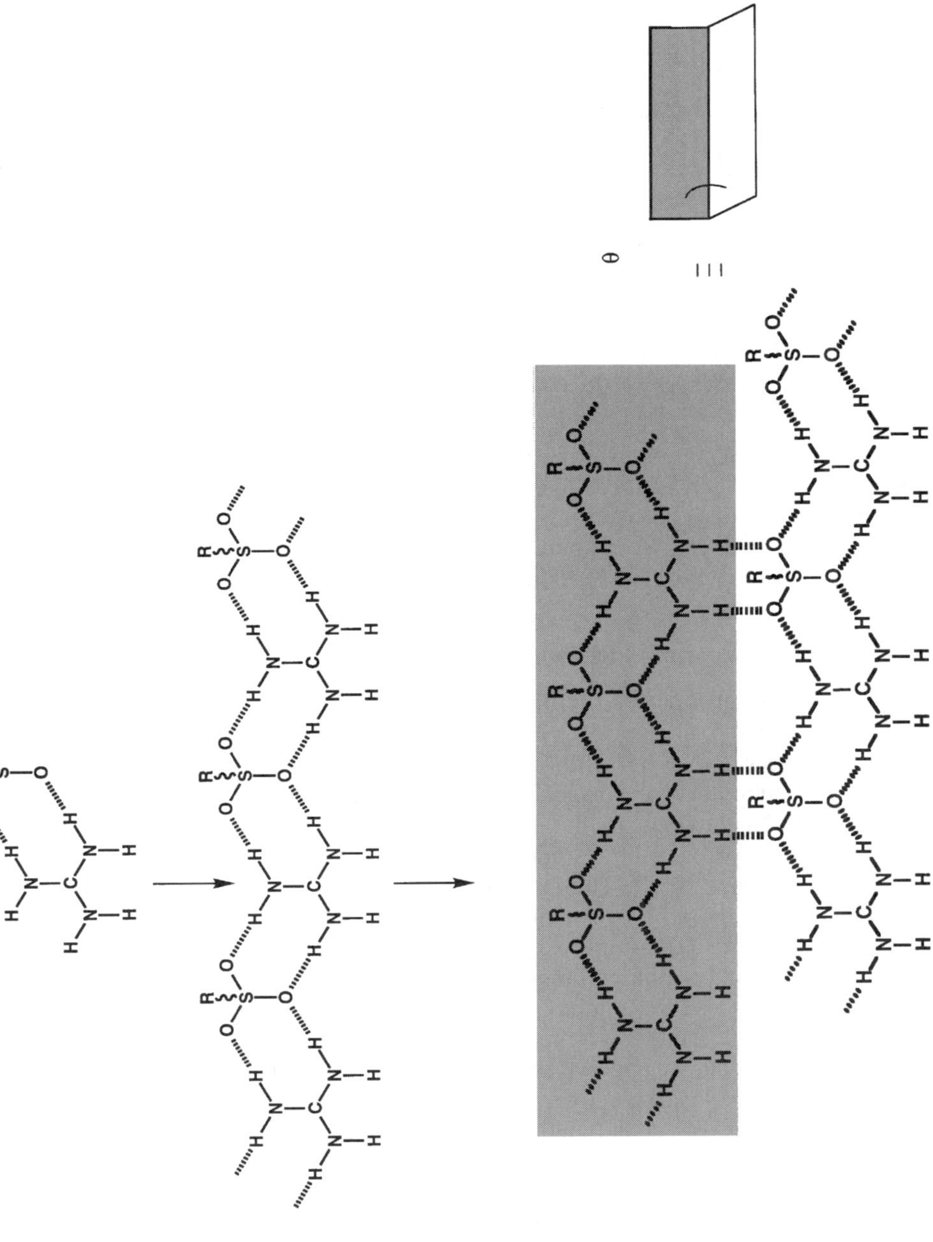

Figure 5.35 Etter and Ward's design for layered materials: guanidinium cations and sulfonate anions can hydrogen bond to form dimers, which in turn can link up into ribbons and sheets. The ribbons that form sheets do not need to be coplanar, so there is an inter-ribbon angle θ that depends on the steric and hydrogen-bonding properties of the sulfonate R group.

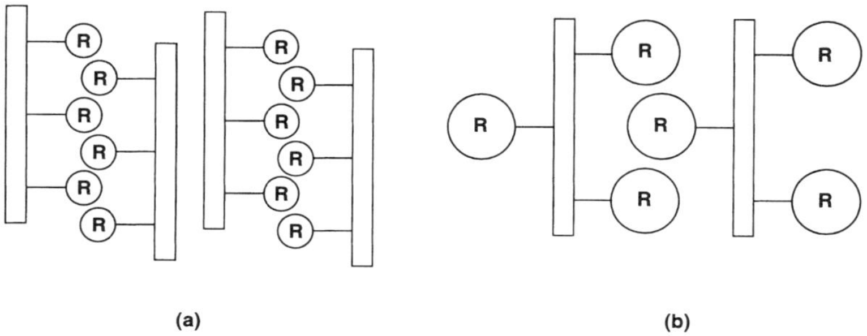

Figure 5.36 Two packing motifs for guanidinium sulfonate layers: (a) small substituents can interdigitate to give a bilayer arrangement; (b) larger substituents can only pack as single layers if close packing of the alkyl or aryl groups is to be achieved. The rectangles represent end-on views of idealized, perfectly planar hydrogen-bonded guanidinium sulfonate layers.

interdigitation, so a single-layer motif, with substituents alternating on opposite sides of the hydrogen-bonded layer, is formed. The authors hope to expand upon thse initial designs to build acentric materials: they speculate that the hydorgen-bonded layers could screen dipole–dipole interactions that lead to centrosymmetric packing by Coulomb interactions.

3.7 Layered Cocrystals Incorporating Urea Columns

Hollingsworth's group has applied spectroscopic methods to inclusion compounds of α,ω-disubstituted alkane guests in urea host lattices to obtain fundamental thermodynamic information about weak intermolecular interactions such as methyl/nitrile contacts[126,127]. While most of their work is outside the scope of this review, in the course of their studies on dinitriles they discovered that hydrogen bonds can form between the nitrile groups and the urea *syn*-NH protons[128–130]. The interesting structural feature is that there is a sudden change in the type of packing adopted depending on the length of the methylene chain. For 1:1 complexes of urea with $NC(CH_2)_nCN$ where $n = 3–5$, the NH···NC contacts link columns of urea molecules with rows of extended dinitriles to form layers (Figure 5.37)[128]. Longer guests, however, pack as inclusion complexes, either commensurate or incommensurate. The authors have performed lattice energy calculations and considered close-packing arguments to show that the switch in packing with length is due to the inherently poor packing of the methylene chains in the 1:1 layered complexes. As the number of CH_2 groups increases, the gain in van der Waals energy due to better chain packing in the inclusion compounds outweighs the cost of losing NH···NC hydrogen bonds (which appear to be less extensive in the channel-type inclusion compounds) from the layered motif.

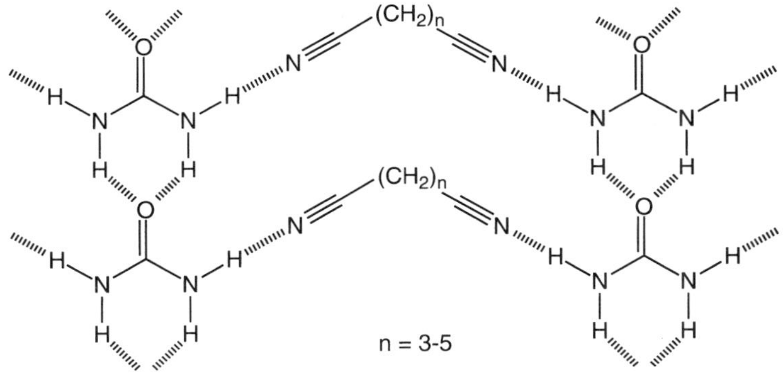

Figure 5.37 Layered packing of 1:1 complexes of short dinitriles with urea. The methylene chains are not in van der Waals contact with each other.

In terms of crystal engineering, Hollingsworth's group developed a clever way to circumvent this switch in packing motif, which occurs with longer guests: Their solution was to add hydrogen-bonding capacity at the middle of the methylene chain, so that 1:1 layer packing would not be energetically disfavored relative to inclusion-type packing[128]. The moiety that they incorporated was another urea group (Figure 5.38). The methylene chains are still not closely packed, but the added hydrogen-bonding interaction makes up for this situation. The resulting structure is thus reminiscent of Fowler and Lauher's layered structures[120], except that while one urea column is covalently a part of the crystal, the other column is held in place by intermolecular NH· · ·NC interactions.

In addition, Hollingsworth *et al.* have expanded upon a statistical study of packing efficiency by Gavezzotti, who had demonstrated a correlation between

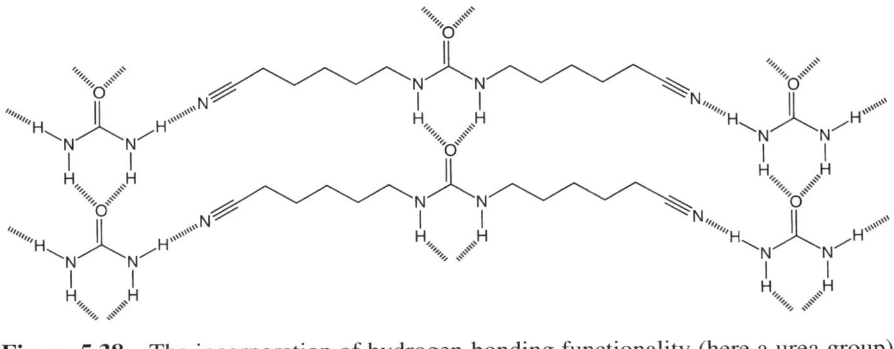

Figure 5.38 The incorporation of hydrogen-bonding functionality (here a urea group) in the middle of a long-chain dinitrile allows formation of a 1:1 layered complex with urea instead of a switch to a channel-type inclusion compound.

the average electron density of a crystal (in units of e/Å^3) and average atomic number of the unit cell[131–133]. Gavezzotti's study did not include data from hydrogen-bonded crystals, and Hollingsworth has now explicitly done so[128]. The result shows that there is no intrinsic need for hydrogen-bonded crystals to pack with decreased efficiency—that is, with open lattices that scarifice close packing for a greater number of hydrogen bonds. This correlation also explains the switch from 1:1 layered cocrystals to inclusion compounds in the dinitrile system under investigation: the hypothetical electron density value for the unobserved layer-type complex with $NC(CH_2)_6CN$ is very low, while that for the observed inclusion compound is in good agreement with the statistical trend. Calculations of this sort could also be useful for rationalizing an abrupt change in packing motif in other families of structural homologues.

3.8 Diamondoid Networks

Zaworotko's group succeeded in designing microporous materials using metal carbonyl clusters of $[M(CO)_3(m_3\text{-}OH)]_4$, (where M = Mn or Re) as building blocks (**10**)[134–136]. In these clusters, the bridging OH groups were oriented tetrahedrally; therefore cocrystallization with hydrogen-bond acceptors as spacer units gave diamondoid lattices that interpenetrated to different degrees, usually 2-fold or 4-fold (Figure 5.39). The tetrahedral geometry of the OH groups gave structures that were two-dimensional: they gave a "doubly interwoven carpet-like structure". Spacers used by Zaworotko included diamines, hexamethylenetetramine, and 4,4′-dipyridine, which formed OH···N hydrogen bonds; and toluene or naphthalene, which formed OH···arene hydrogen bonds[134]. Three-component networks, where hydrogen bonds with alcohols link the metal clusters and 4,4′-dipyridine, have also been formed[135]. These structures, however, were two-dimensional, also giving a "doubly interwoven carpet-like structure". Zaworotko controlled the size of the micropores in these lattices by changing the length of the spacer to enclathrate different molecules of solvent selectively.

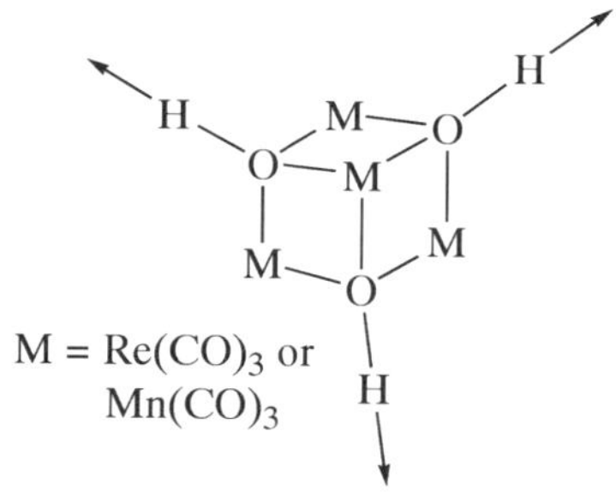

10

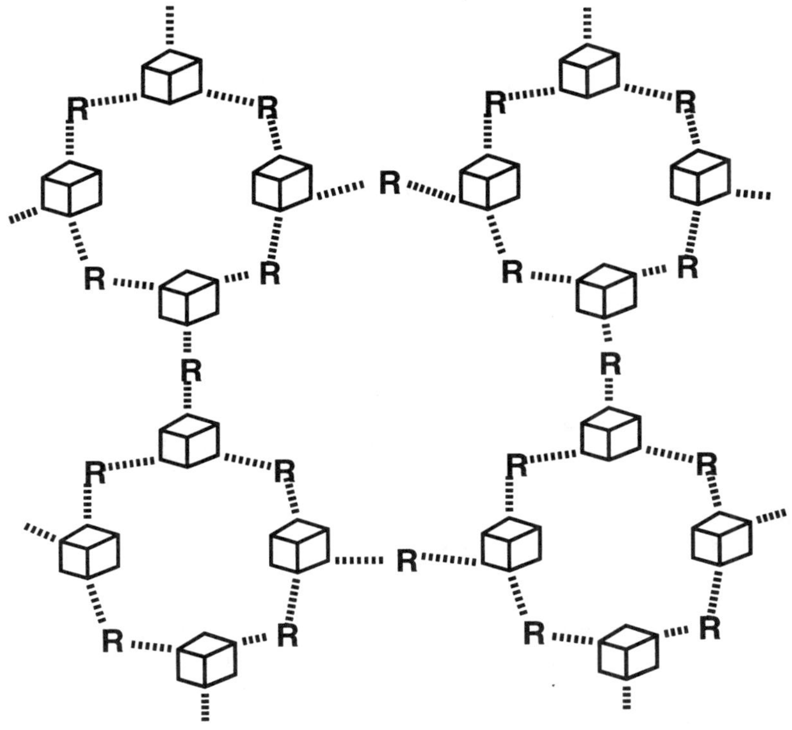

Figure 5.39 Schematic view of a two-dimensional fragment of a diamondoid lattice, where the cubes represent $M(CO)_3(OH)_4$ clusters and the spacers R could be diamines, bis-phosphine oxides, or aromatic hydrocarbons. Interpenetration of two-dimensional layers through the central pore can lead to construction of the three-dimensional lattices.

In another study, Zaworotko replaced the terahedral metal cluster with a functionally-equivalent tetraprotonated cyclam as a proton donor to form the network[137,138]. The counterions ranged from chloride and trifluoroacetate to ferrocenedicaboxylate, all of which bridged neighboring cyclams using charged hydrogen bonds. Zaworotko has also obtained two-dimensional networks with microchannels for solvent with these components. These results demonstrate the versatility of Zaworotko's approach to controlling microporosity by interchanging tectons with different geometries.

4. DESIGNS BASED ON CYANURIC ACIDS AND MELAMINES (THE CA • M LATTICE)

Our strategy has focused on controlling the various motifs present in the 1:1 complex between cyanuric acid and melamine (CA • M) shown in Scheme 5.2.

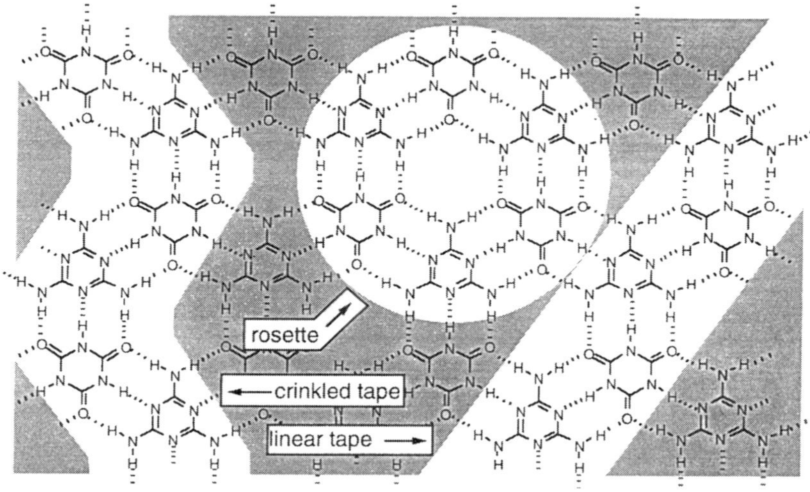

Scheme 5.2 Schematic view of the cyanuric acid and melamine (CA • M) substructures: linear tape; crinkled tape; and rosette.

We and others believe the structure of CA • M consists of infinite hydrogen-bonded sheets[139]. A report has also appeared in the literature describing the crystal structure of a 1:1 complex of cyanuric acid with melamine, although it consists of solvated, protonated dimers and not a layer structure, since it was crystallized from 3 M HCl[140]. Although no crystal structure is available for CA, data from X-ray powder diffraction (XPD) are consistent with this structure: one of the strongest reflections gave a *d*-spacing of 3.4 Å, the expected value for interlayer separation[141].

We have used the group of three hydrogen bonds formed between the components of CA • M to construct linear, crinkled, and cyclic motifs in organic crystals (Scheme 5.2 and Figure 5.40)[142–146]. Linear and crinkled motifs are substructures that we refer to as "tapes"; cyclic motifs are referred to as "rosettes". Systems based on CA • M are attractive because the derivatives of the components are easy to synthesize, and thus a range of substituents or functional groups can be easily introduced. We used derivatives of cyanuric acid rather than barbiturates for their ease of synthesis and more convenient solubilities. Also from a paragmatic view, we have never been able to obtain diffraction-quality crystals of complexes incorporating derivatives of cyanuric acid. This failure is most likely due to the greater acidity of cyanurates[142], which leads to the formation of aggregates that are more insoluble than those based on barbituric acids. In addition to these solid-state structures, we have constructed a range of soluble aggregates basd on the cyclic motif[147].

Our goal was to determine what structural features of the molecules would allow us to control the formation of the linear, crinkled, and rosette motifs. We

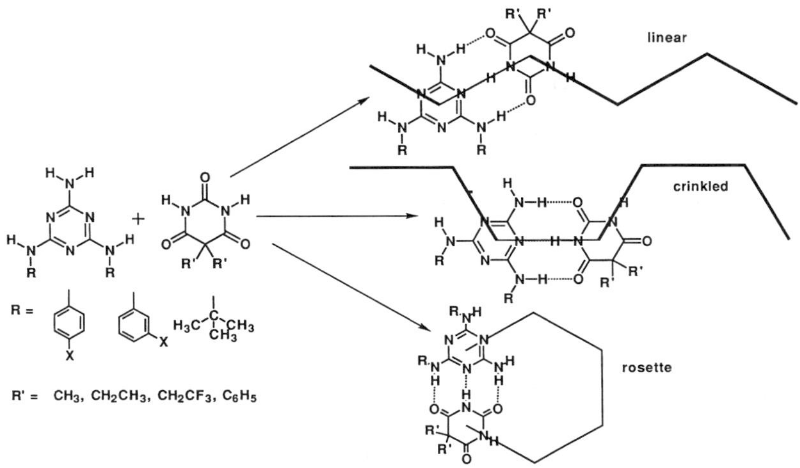

Figure 5.40 Schematic view of the proposed two-dimensional sheet of the complex between cyanuric acid and melamine (CA • M). Substructures are outlined: rosette; crinkled tape; and linear tape.

reasoned that blocking hydrogen bonding in some directions was a logical tactic that would promote the formation of a small number of crystalline substructures: these substructures are geometrically and conceptually simple; this simplicity aids in their classification.

Our experiments focused on structures of complxes between barbituric acids and three different classes of melamines:

- N,N′-bis(*para*-X-phenyl)melamine, where X = H, F, Cl, Br, I, CH_3, CF_3, CO_2Me, and $C(CH_3)_3$, cocrystallized with 5,5-diethylbarbituric acid (**11**)
- N,N′-bis(*tert*-butyl)melamine cocrystallized with 5,5-di-X-substituted barbituric acids, where X = CH_3, CH_2CH_3, CH_2CF_3, and C_6H_5 (**12**)
- N,N′-bis(*meta*-X-phenyl)melamine, where X = H, F, Cl, Br, I, CH_3 (**13**) CF_3, CO_2Me, and $C(CH_3)_3$, cocrystallized with 5,5-diethylbarbituric acid

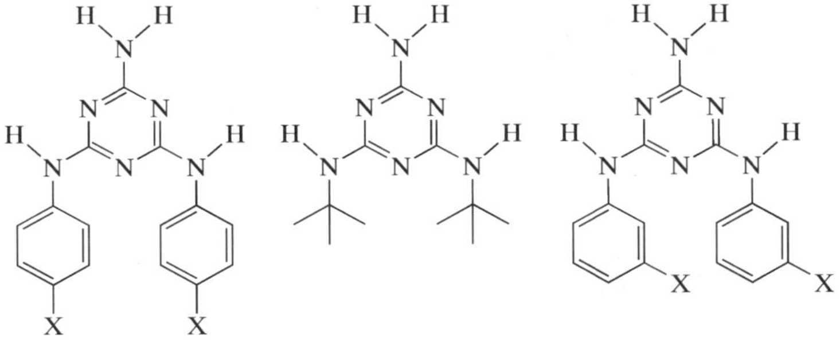

11 **12** **13**

The appproach we followed in designing these families is a familiar one in physical–organic chmistry, namely incremental variation in the size of substituents (X). Considering entire families of complexes is necessary if we are to discern trends in packing patterns and relate them to molecular structure. This strategy, which relates molecular structure to crystalline structure, is central to crystal engineering.

To facilitate analyses, we classified the substructures of the crystal into several hierarchical levels. These levels constitute a scheme that is intended to provide points of attack for purposes of design at these different levels: we hope to determine the molecular features responsible for these substructures, and then demonstrate that these features can be manipulated rationally to give the desired crystalline structure or property. The various levels of substructures are shown in Figure 5.41. The primary level of substructure is simply the identity, stoichiometry, sequence, and conformation of the constituent molecules. The secondary level consists of the motif produced by aggregation of the individual molecules—in this case, tapes. These tapes can be linear, crinkled, or perhaps a more complicated hybrid of both. The tertiary level is constructed in turn from these secondary levels. For example, we found that tapes can aggregate as dimers or as sheets. We expect that tapes should pack with their long axes parallel to each other in the crystal to maximize van der Waals contacts: as a

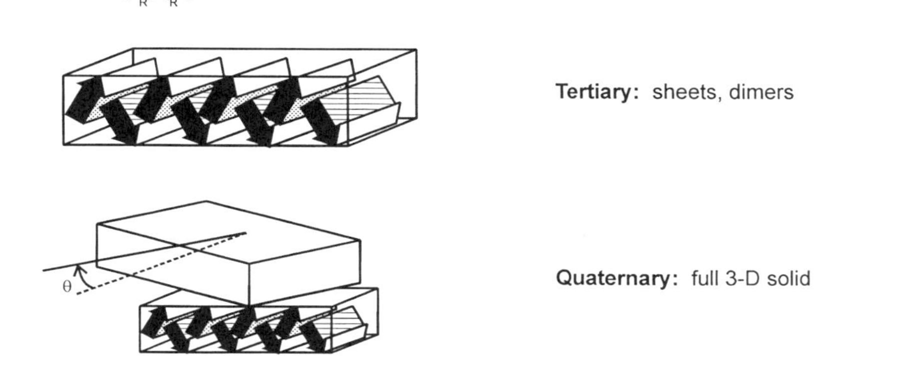

Figure 5.41 Hierarchical levels of crystalline architecture.

broad generalization, the ridges and grooves that tapes present along the surface of a sheet or dimer should pack snugly against each other to minimize empty space in the crystal. We also hypothesized that steric repulsion between adjacent melamines would be unavoidable at some point if we increased the size of the substituents on the phenyl rings of melamine: large substituents should not be able to pack in the linear motif.

The most important points we will stress in the discussion of our work are that:

- the CA • M lattice provides several robust structural motifs for the design of organic solids
- some motifs can be controlled by a choice of substituents
- the possibility for variations in molecular conformation complicates the interpretation of packing because of polymorphism

4.1 The *para*-substituted Diphenylmelamine Family

4.1.1 Linear Tapes

Cocrystals built from N,′-bis(*para*-X-phenyl)melamine, where X = H, F, Cl, Br, I, CH_3, CF_3, formed linear tapes (Figure 5.42)[142]. Figure 5.42 shows the structural features of these linear tapes from two different viewpoints. One important feature is that the substituents on melamine are separated from the ethyl groups of barbital on opposite sides of the tape. The backbone of these tapes is held together by three hydrogen bonds as in the CA • M lattice. Increasing the size of substituents on melamine from H to CF_3 did not interfere with the formation of these linear tapes. Reliability of this sort is essential for studies attempting to compare a range of substituents in the solid state: if a system adopts a fundamentally new motif with every change of substituent, then it would be useless for purposes of design unless each new motif could be predicted beforehand.

While the packing of each memeber of the *para*-family has interesting structural features, the *trends* in packing for this family are more instructive in developing guidelines for crystal engineering. One of the strengths of this cocrystalline system is that different types of substructures are identified easily.

4.1.2 Crinkled Tapes

We further reasoned that *para*-substituents larger than those that gave linear tapes would force the tape from linear to crinkled. The substituent that we used to test this hypothesis was the CO_2Me group. Figure 5.43 shows the crinkled tape obtained from a complex of N,N′-bis(4-methoxycarbonyl-phenyl)melamine with barbital[143]. In this structure, the CO_2Me groups are oriented

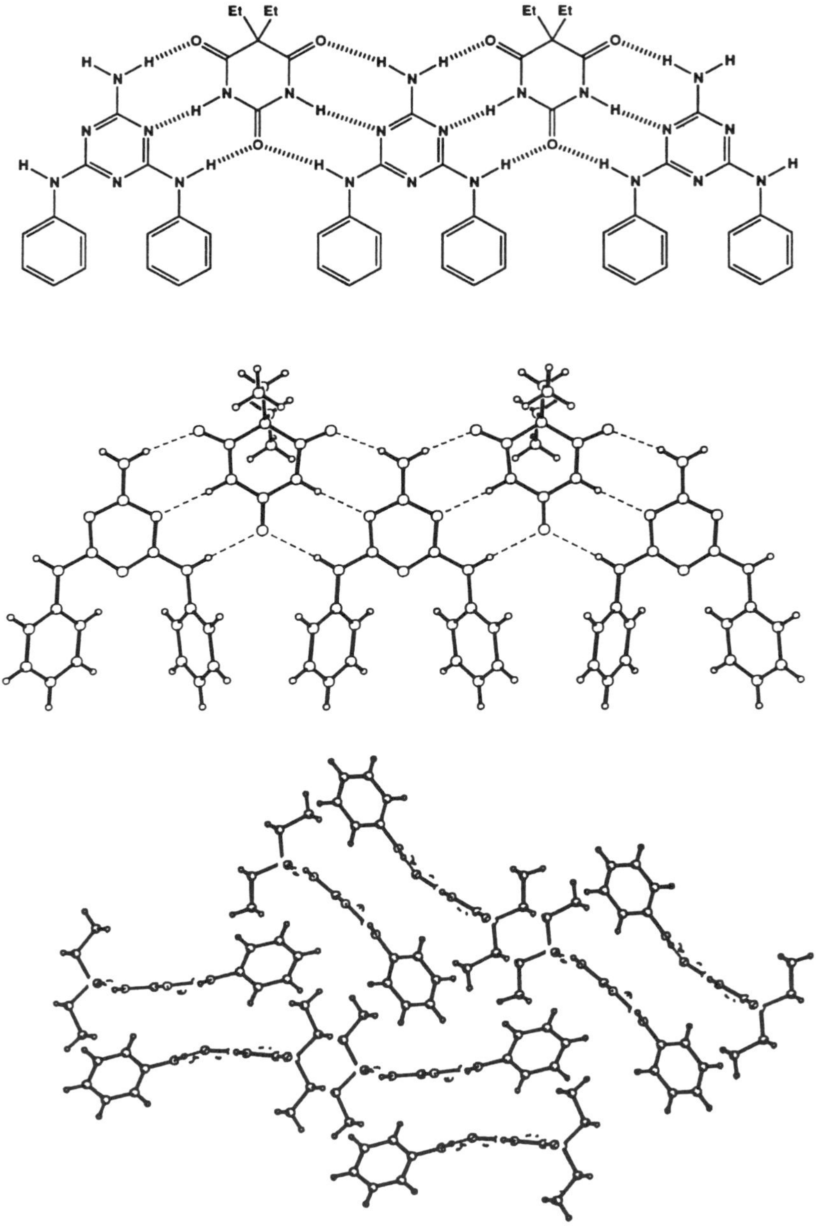

Figure 5.42 Two views of the crystal structure of the complex between N,N-diphenylmelamine and 5,5-diethylbarbituric acid. Middle: a view of the hydrogen-bonded linear tape from the top. Bottom: end-on views of the packing of the linear tapes, looking down their long axes.

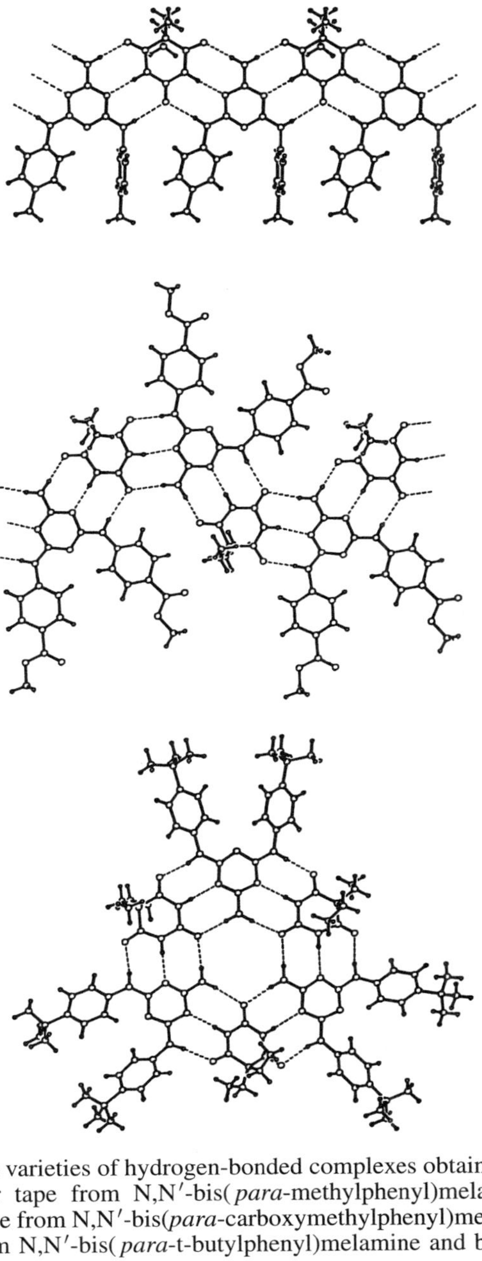

Figure 5.43 Three varieties of hydrogen-bonded complexes obtained from the CA • M system. Top: linear tape from N,N′-bis(*para*-methylphenyl)melamine and barbital. Middle: crinkled tape from N,N′-bis(*para*-carboxymethylphenyl)melamine and barbital. Bottom: rosette from N,N′-bis(*para*-t-butylphenyl)melamine and barbital.

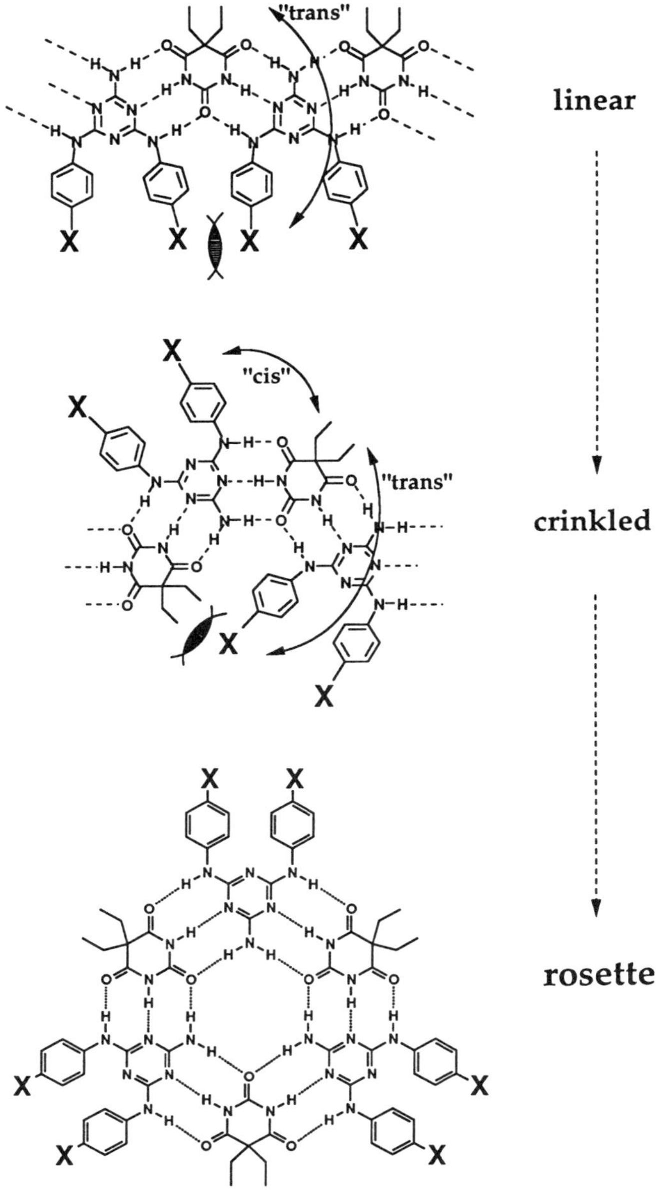

Figure 5.44 Steric repulsion between adjacent melamines leads to changes in the tape motif.

approximately coplanar with the hydrogen-bonded backbone of the tape, rather than pependicular to the backbone. The CO_2Me groups most likely adopted this coplanar orientation to pack efficiently at the tertiary level. This inference highlights an important aspect of crystalline structure, namely that there can be an interplay between the various forces of packing at different levels of substructure. In other words, the motif that is adopted at the secondary level—linear or crinkled—can depend on the packing at higher levels.

4.1.3 Rosettes

We sought to circumvent problems in predicting the packing of tapes with flexible substituents that have anisoptropic shapes (the CO_2Me group) by using those that are isotropic, and also to ensure that linear tapes could not be formed by using a substitutent of sufficient size. A *tert*-butyl group satisfies these requirements. The complex of N,N′-bis(*tert*-butylphenyl)melamine with barbital did not, however, adopt any of the tape motifs, but instead formed a cyclic hexamer, or rosette (Figure 5.43)[143]. We believe that this complex forms and packs as a rosette to permit maximal separation of the substituents from each other (Figure 5.44).

4.2 The Bis(t-butyl)melamine Family

4.2.1 Crinkled Tapes

The results from the *para*-substituted diphenylmelamine series suggest an upper limit in size for the groups attached to the melamines (that is, both the phenyl ring and its substituents) that can be accommodated and still maintain the crinkled motif; groups larger than this limit produce the rosette motif. With this observation in mind, we found that the crinkled motif could be maintained by replacing the phenyl rings of melamine with *tert*-butyl groups[144]. Four complexes built from N,N′-bis(*tert*-butyl)melamine with different barbiturates all crystallized in the crinkle motif (Figure 5.45).

These example illustrate the value of steric effects in controlling *secondary* substructure in the compounds of these crystals.

4.3 The Family of *meta*-Substituted Diphenylmelamines

We examined the crystal packing for N,N′-bis(*meta*-X-phenyl)melamines where X = H, F, Cl, Br, I, CH_3, and CF_3, cocrystallized with barbital (Figure 5.46) [145]. We wished to determine of moving substituents from the *para* position to the *meta* position would significantly alter the type of motif that formed during crystallization. We found that derivatives of melamine from this

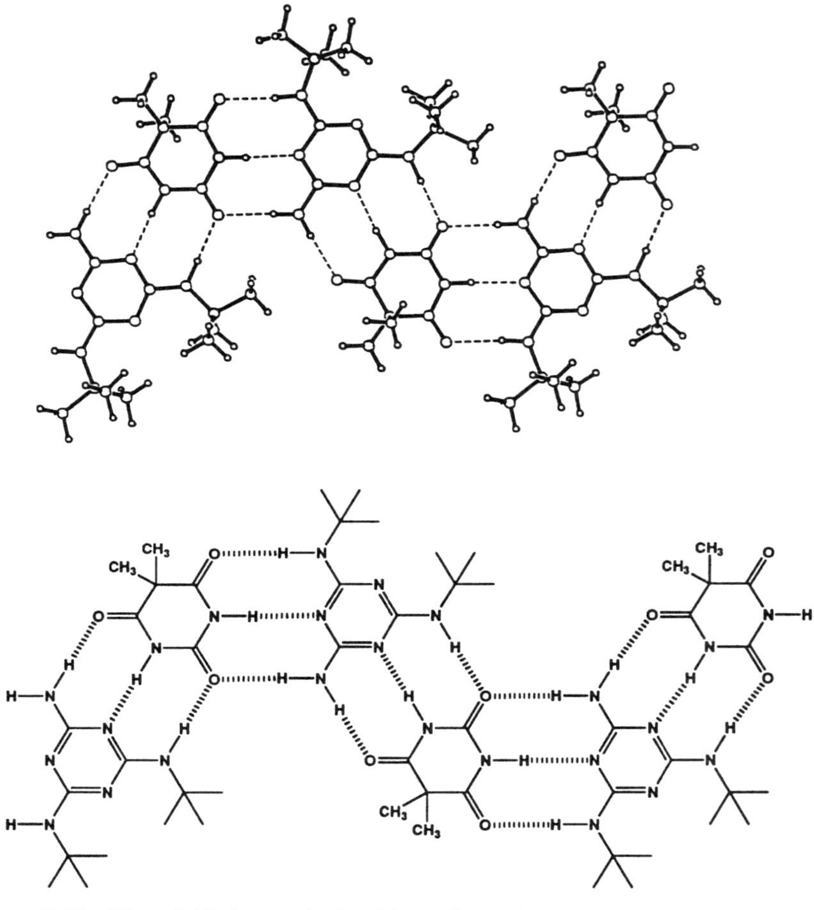

Figure 5.45 The crinkled tape obtained from di-*tert*-butylmelamine and 5,5-dimethylbarbituric acid.

meta-substituted series showed no correlation between the size of the substituents and the resulting packing motif. The complexes with the larger substituents CH_3, I, and CF_3 packed as linear tapes, while those with Br and Cl adopted the crinkled motif (Figure 5.46; none of these structures was checked for polymorphism). In the latter two cases, there was a contact between a phenyl hydrogen and a barbital oxygen across the bay region of the crinkled tape, which appeared to stabilize the crinkled motif.

A serious problem with the *meta*-family that hindered us from controlling structure was the high frequency with which molecules of solvent were trapped in the crystalline lattice. We believe that the propensity for molecules of solvent

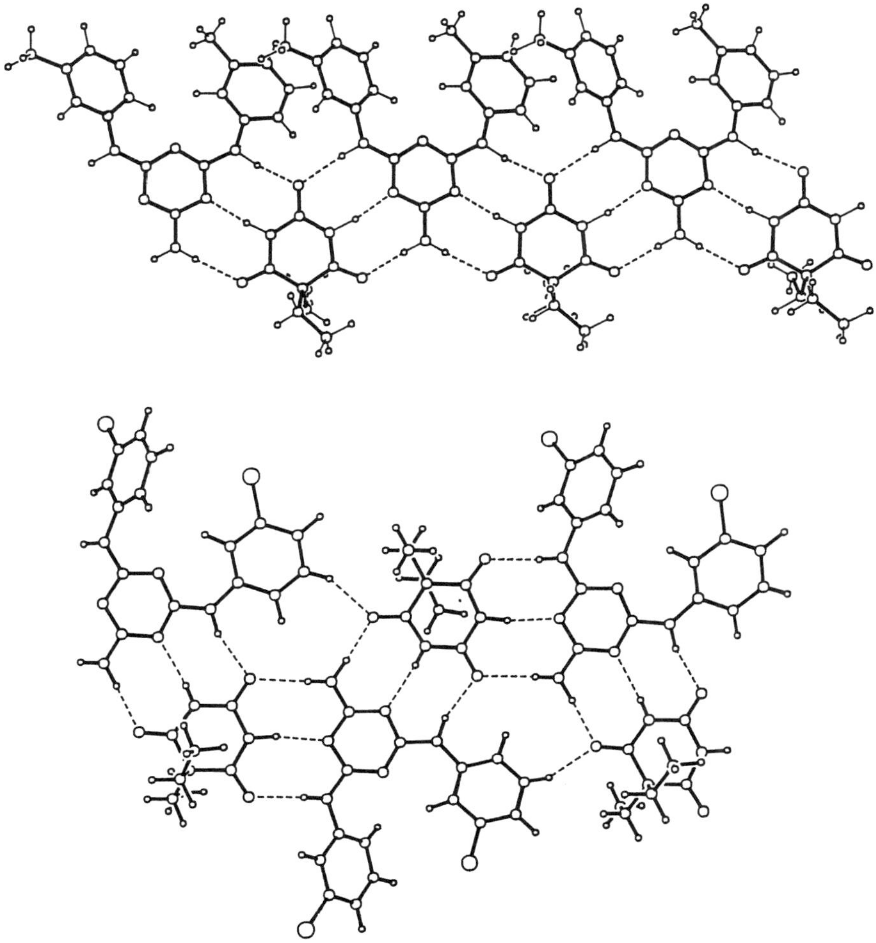

Figure 5.46 Two tapes from the *meta*-family. Top: linear tape from N,N′-bis(*meta*-methylphenyl)melamine and barbital. Bottom: crinkled tape from N,N′-bis(*meta*-chlorophenyl)melamine and barbital, showing intra-tape CH···O contacts.

to be incorporated into the lattice of these *meta*-substituted diphenylmelamines reflect the fact that they do not pack efficiently.

4.4 Polymorphism

Studies in crystal engineering are incomplete without searching for the occurrence of polymorphism, because organic molecules often can pack in isoenergetic, or nearly isoenergetic arrangements[10–15]. Moreover, anticipating whether a given series of molecule will be prone to polymorphism is difficult, because so many factors can affect packing. One of our goals has been to find robust systems that limit the possibility of polymorphism. There are several experimental methods that can be conveniently used to search for and characterize polymorphs: X-ray powder diffraction[148,149], and solid-state NMR spectroscopy are especially useful[150–152]; thermochemistry is also very useful, but substantially more difficult experimentally.

We tried to induce polymorphism by crystallizing members of the *para*-series from a variety of different solvents[153]. All but two complexes gave just one crystalline form. It is important to stress, however, that we crystallized these complexes under a limited range of conditions: we did not vary temperature or control the rate of evaporation.

Attempts to obtain single crystals of polymorphs were successful for N,N′-bis(*p*-Br-phenyl)melamine•barbital, and unsuccessful for N,N′-bis(*p*-CF_3-phenyl)melamine•barbital[153]. A comparison of the packing of the two Br polymorphs (based on single-crystal diffraction) show that one packs isomorphously to the Cl complex, while the other packs similarly to the I and CH_3 complexes (Figure 5.47). Since bromine is intermediate in size for the substituents in the *para* series (H to CF_3), melamines with this substituent appear to have access to the packing arrangement associated with either smaller or larger substituents.

4.5 Strengths and Weaknesses of the CA • M System

The CA • M lattice represents a useful starting point for crystal engineering because it predictably forms hydrogen-bonded substructures: tapes or rosettes. These motifs are generated reliably: the triplet of hydrogen bonds that holds the components together are strong enough so that they are not easily disrupted. We have not, however, succeeded in introducing polar peripheral substituents, where strong polar interactions may disrupt these motifs. We have developed a model for crystal engineering based on the hierarchical packing of substructures in this system that permits limited studies of packing in crystals. We were also able to rationalize each level of packing by identifying the molecular features responsible for promoting each substructural motif. Our clearest success has come in rationalizing general trends in secondary architecture in complexes with *para*-substituted diphenylmelamines[143]. By altering the size of the substituents, we were able to control whether the molecules adopted a linear tape, a crinkled tape, or a rosette motif[143,154].

Despite these virtues, the number of families of complexes based on the CA • M lattice that comply with this scheme of analysis and design, and the potential of this system in crystal engineering, may be limited. Several features of this system contribute to these limitations. First, growing diffraction-quality

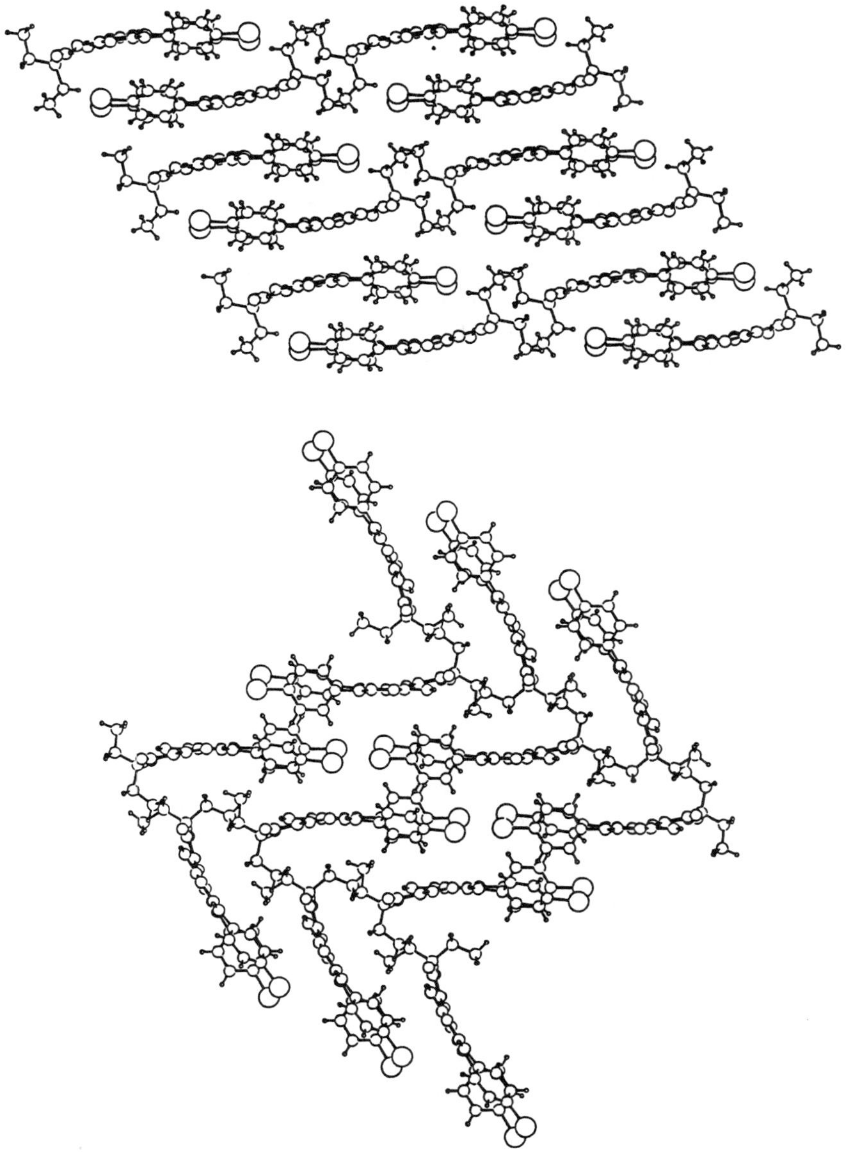

Figure 5.47 End-on packing views of linear tapes for two polymorphs of the complex between N,N′-bis(*para*-bromophenyl)melamine and barbital. The top polymorph packs isomorphously to the *para*-chloro complex, while the bottom polymorph packs similarly to the *para*-iodo and *para*-methyl complexes.

crystals in this series remains slow and difficult; for an ideal system, crystallization would be straightforward. Second, the system does not tolerate polar substituents well. Third, the presence of polymorphism decreases the level of control that can be exercised in the system: complexities in packing due to conformational isomerism may be one of the biggest hurdles hindering the rational design of molecular crystals. In principle, each accessible conformation can pack as a separate crystalline phase. Polymorphism of this sort occurs both occasionally in the *para*-family (where it is associated with size of substituents) and in the *meta*-family. In both series, the ethyl groups of the diethylbarbital can also contribute to conformational polymorphism. Presently, we cannot predict the packing of these tapes at the tertiary level of substructures.

We are currently studying other model systems containing rigid molecules that form tapes[155]. The use of rigid molecules should limit the conformational isomerism we encountered in the CA • M system. While the CA • M system has many attractive characteristics, it is probably too complex to serve as a basis for extensive studies at the current state of development of the field.

5. COMPUTATIONAL APPROACHES TO PREDICTING THE STRUCTURE OF ORGANIC CRYSTALS

A number of new computational strategies have emerged that promise to be useful in crystal engineering. In this section, we will highlight these recent computational efforts that ultimately intend to *predict* the crystal structure of organic molecules using only information about the geometric structure of a single molecule. Other methods for predicting the crystalline structure of organic molecules existed before those reviewed here[156,157], but these earlier methods typically required either additional experimental information or the presence of a plausible hypothetical crystal structure[158], before an accurate prediction of the final crystal structure could be made. In other words, the methods excluded from this section are only appropriate if given much information about the crystalline environment.

The methods reviewed here rely on two important assumptions about the experimental crystal structures: that molecules of solvent do not affect the crystal structure, and that the arrangement of the molecules in the crystal represents the thermodynamic minimum in free energy. These assumptions are only valid for robust crystal systems—that is, for systems with a well-defined global minimum in free energy that do not form polymorphs. These methods also neglect entropy in their calculations. In the future, methods such as normal mode analyses[159,160], and molecular dynamics[161], which can estimate the vibrational entropy, may be useful in conjunction with these methods.

At the heart of the computational strategies to predict the packing of organic molecules we consider here is the calculation of the lattice energy (U_{latt} using atom–atom potential functions*:

$$U_{\text{latt}} = \sum_{i,j} (E_{\text{vdw}} + E_{\text{elec}}) \tag{1}$$

In equation 1, the van der Waals (E_{vdw}) and electrostatic (E_{elec}) energies are summed between atoms i in a single molecule and atoms j in the other molecules included in the computation. The components in Equation 1 can vary, but they are empirical descriptions of the energy of the crystal[162]. The parameters in the components of Equation 1 are derived from experimental and quantum mechanical data; therefore, calculating U_{latt}, rather than relying on methods that use statistical correlations, represents the most physically meaningful way of evaluating crystal structures.

Currently, the most promising methods for predicting the crystalline state of organic molecules, based solely on the geometric structure of a single molecule, involve computing U_{latt} for the possible arrangements of rigid and semi-rigid molecules in the crystal[25,163]. There are, however, two significant limitations in computing U_{latt} of these arrangements. First, the very large number of spatial arrangements—based on translational and rotational changes—intrinsic to small organic molecules can render the calculations intractable†. A computer evaluates U_{latt} for each of these arrangements to find the lowest value, which is then assumed to represent the arrangement observed in reality. In addition, the possibility of conformational isomerism in most organic molecules substantially complicates the problem of rationalizing crystal structures by broadening the range of structures that has to be examined. Second, the accuracy of the force field used in the computations can limit the ability of the computations to differentiate between polymorphs (often different packing arrangements differing in energy by less than ~ 1 kcal/mol). (The force field is described by the potential energy functions and its parameters.).

The methods that we will discuss use three strategies to reduce the computational burden resulting from the numerous possible arrangements of molecules in crystals: first, they use efficient search algorithms that tackle the complexity associated with the large number of arrangements of molecules. Second, they exploit experimental generalities that most organic molecules seem to crystallize in only a few symmetry groups, or that a crystal can be

* The use of empirical potential energy functions is a practical one. Calculations involving first principles, or quantum mechanics, are only appropriate for systems comprised of a few atoms.

† For example, if we allow a *single* rigid molecule to translate in all three directions within a 5 Å window (0.1 Å increments), and to rotate about all three axes within a 360° window (5° increments), then this molecule will have $50^3 \times 72^3$, or 4.7×10^{10} possible spatial configurations. This number dramatically increases if we consider more molecules, or if the molecules are conformationally flexible. In reality, this number is slightly less since it is known that organic molecules will tend to adopt only a few symmetries and will therefore eliminate some of these configurations[10].

decomposed into substructures, each of which is close to its global minimum in free energy. Third, they use supramolecular structures as the "molecular" unit.

The principle search algorithm used[164] involves Monte Carlo sampling of crystalline arrangements of molecules to find those that are plausible—that is, arrangements with low (favorable) values of U_{latt}. Monte Carlo algorithms, in the context of crystal engineering, generate crystalline structures by randomly changing the values of the crystalline lattice (a b, c, α, β, γ), and the geometry (primarily through torsions) or spatial orientation of the molecules. The value of U_{latt} of each newly generated crystalline structure is compared to that of the previous crystalline structure: If this new value of U_{latt} is lower than that of the previous structure, then this new crystalline structure is accepted. If this value of U_{latt} is higher than that of the previous structure, then the difference in these two values is weighted by a Boltzmann probability, and a random decision is made to keep or reject the new crystalline structure; the greater the difference in energies, the less probable that the new crystalline structure will be accepted. This procedure of generating and evaluating new crystalline structures is typically repeated many times (for example, on the order of 500 000 iterations) until the crystalline structure with the lowest value of U_{latt} is revealed.

Experimental generalities can help to guide strategies for predicting crystal structures. For example, most crystals of organic molecules contain a hierarchy of only a few structural motifs—one-dimensional stacks, two-dimensional layers, etc.[165]. Using these generalizations in a scheme for prediction helps to reduce the large number of crystalline arrangements of molecules by limiting the search to only a few symmetry types. Analogously, the use of supramolecular structures—for example, systems based on CA • M—as the basic structural unit that packs in a crystal reduces the degrees of freedom that are open to unconstrained molecules, and therefore significantly reduces the amount of computation involved[166].

The empirical force fields used in the computations reviewed here seemed adequate, but these force fields were only tested on a limited set of molecules: weaknesses appeared when polymorphism*, or extensive hydrogen bonding†, occurred. It is clear, therefore, that much work is still needed in improving the quality of these force fields to a level that makes them generally useful for crystal engineering.

Any procedure for predicting the crystalline state of organic molecules that must search the large number of spatial arrangements of molecules—even for crystals comprised of tapes based on the CA • M lattice—to find the

---

* This may or may not be a serious problem: some of these methods were able to correctly suggest that polymorphism might be present; although they could not distinguish which polymorph was the thermodynamically stable one.

† The hydrogen-bonding networks in the crystal structure of these compounds did not form the well-defined motifs of the types described in this review.

arrangement with the lowest U_{latt} can be computationally time consuming for even the simplest case. Fortunately, the methods discussed here demonstrate that predicting the crystal structures of some moderately complex organic molecules, based solely on their molecular structure, is becoming practical.

5.1 Generation of Clusters

Gavezotti developed a strategy for predicting the structure of organic crystals that relies on building a small cluster of two to four molecules using certain statistically favored space groups*. His method involves a systematic search to find the most energetically favorable arrangements of the molecules in this cluster[167]. Trial crystalline structures are then built by translating in space the most energetically favorable clusters, and determining which of these crystalline structures have favorable values of U_{latt}. Gavezzotti's method successfully predicted the crystalline structures of a variety of rigid hydrocarbons (the data were taken from the CSD)[61]. One shortcoming of his method is that in generating the cluster, no procedure is used that accounts for the effect of an infinite crystalline system (these procedures, which model the presence of surrounding unit cells, are referred to as "periodic boundary conditions"). The lack of periodic boundaries may limit the use of this method with compounds containing heteroatoms because the distance over which electrostatic interactions are important are typically greater than those explored in the small clusters examined in this method, and their omission may (or may not) compromise the predictive ability of this method.

Gavezzotti also discussed the difficulty in comparing crystal structures determined from different methods—that is, those from computation with those from experiments. For example, ambiguities can rise if two structures are compared based only on the parameters of the lattice (a, b, c, α, β, γ) which can be assigned arbitrarily. He suggests two more impartial metrics for comparing crystal structures: first, using the value of the distance between the centers-of-mass of molecules, and using the value of the difference in energies between a reference molecule and its neighbors; second, categorizing the "orientation of distinguishable molecular features in a crystal"—for example, the geometries of aromatic planes.

5.2 Generating Crystal Structures Using Simulated Annealing Monte Carlo in the Presence of Periodic Boundaries

Karfunkel and Gdanitz co-authored the first paper that demonstrated the successful prediction of crystal structures of organic molecules containing

* These spaces groups are comprised of translation, inverse, screw, and glide operations, and account for over 80% of all observable crystal structures in the Cambridge Crystallographic Database.

heteroatoms[168]. They used a simulated annealing Monte Carlo (SAMC) algorithm to optimize the structure of a crystal using a unit cell of arbitrary dimensions that contained a small number of molecules (1 to 4) under periodic boundary conditions. This algorithm uses high temperatures (usually values well above the temperature of sublimation for the crystal) in the initial stages of a MC procedure to increase the probability of accepting new crystalline arrangements of molecules that have values of U_{latt} that are more unfavorable relative to the previous arrangement. In this way, molecular arrangements in the crystal can escape a local minimum, and explore other minima. Karfunkel and Gdanitz eliminate "evaporation" of the molecules within the system caused by these high temperatures by constraining the molecules so that they cannot separate beyond a certain distance. The system is then "cooled" slowly over many cycles of computation to allow ample time to search for local minima. Variables in their method include the translational and rotational degrees of freedom for a molecule, torsions, and the appropriate lattice constants (a, b, c, α, β, γ). When analyzing the results from their computations, they eliminate ambiguities in comparing lattice vectors (as discussed by Gavezzotti) by converting the lattice vectors of each crystal structure to a common frame of reference (Niggli-reduced cell)[169]. Their implementation of SAMC is useful for rigid and semi-rigid molecules, but not yet general enough for conformationally flexible organic compounds. Their method has also proved useful for predicting crystalline structures containing just hydrocarbons[170].

Karfunkel and Gdanitz found that the choice of partial charges assigned to molecules containing heteroatoms was crucial for obtaining a predicted crystal structure that agreed with the experimental crystal structure: the electrostatic component of the potential energy was very sensitive to its choice of parameters. Also, Karfunkel *et al.* reported that the task of identifying similar crystal structures was not trivial, and therefore suggested that comparing the calculated X-ray powder diffraction patterns for crystal structures obtained from different computational methods may be the best means for comparing them[171].

5.3 Simulated Annealing Monte Carlo on Aggregates of Molecules in Conjunction with Principles of Crystals

Perlstein predicted the crystal structures of a number of organic molecules using a single molecule as the asymmetric unit[172,173]. He used a SAMC method similar to that of Karfunkel and Gdanitz; his approach to constructing the crystal, however, differed in the treatment of the crystalline environment, in that it closely follows the principles outlined by Kitaigorodsky[165]. These principles state that molecules initially assemble into one-dimensional aggregates; these aggregates then assemble into two-dimensional layers;

and finally these layers assemble to form the crystal structure in three dimensions. Perlstein titled these principles the "Kitaigorodsky Aufbau Principle" or KAP. Perlstein also bases his strategy on other experimental generalizations: only four types of symmetry in one-dimensional aggregates occur for 92% of all crystals (translation, inversion, screw, and glide)[174]; and only seven types of layers are formed form these aggregates[175]. Thus, Perlstein restricts his search to one-dimensional aggregates using only these four types of symmetry operators. When computing U_{latt}, Perlstein does not use periodic boundary conditions, but instead relies on molecular interactions between a central molecule and its nearest and next-nearest neighbors. More specifically, his approach uses SAMC to find one-dimensional stacks of molecules with the lowest U_{latt}, and then compares these molecular aggregates to those found in the crystal. This method works surprisingly well. Perlstein found that, for translational aggregates, the observed structure was less than 2 kcal/mol above the global minimum predicted for the aggregate; and for inversion, screw, and glide aggregates, the observed structure was less than 6 kcal/mol above the global minimum predicted for the aggregate.

5.4 Simulated Annealing Monte Carlo and Molecular Dynamics in Conjunction with Molecular Tapes

We have used computations to predict and rationalize the crystalline structure of molecular tapes comprised of hydrogen-bonded organic molecules. We believe that the restricted configurational freedom of molecular tapes, used in concert with existing computational methodologies, is a strategy that will give a good opportunity to predict the structures of organic molecular crystals. Our method for predicting crystal structures is similar to the SAMC method of Perlstein except that it allows the molecular tapes in the system to translate and rotate in three dimensions. Preliminary results from this work have focused on 4,5-disubstituted 2-benzimidazolones. We chose this class of cyclic ureas because they are conformationally more rigid, and therefore less prone to polymorphism, than many other organic molecules capable of forming hydrogen-bonded tapes[176]. As a class, these cyclic ureas often, but not invariably, form tapes[155]: when the substituents are large (iodine) or electronegative (fluorine) they instead adopt a three-dimensional motif[176]. If we *assume* that the cyclic ureas form tapes, we have successfully predicted the packing of tapes having methyl substituents based solely on the knowledge of the geometry of a single tape (Figure 5.48)[177]. Other computational methodologies—especially simulated annealing molecular dynamics (SAMD)[177,178]—also being explored for use in predicting crystal structures of these cyclic ureas: initial results are promising.

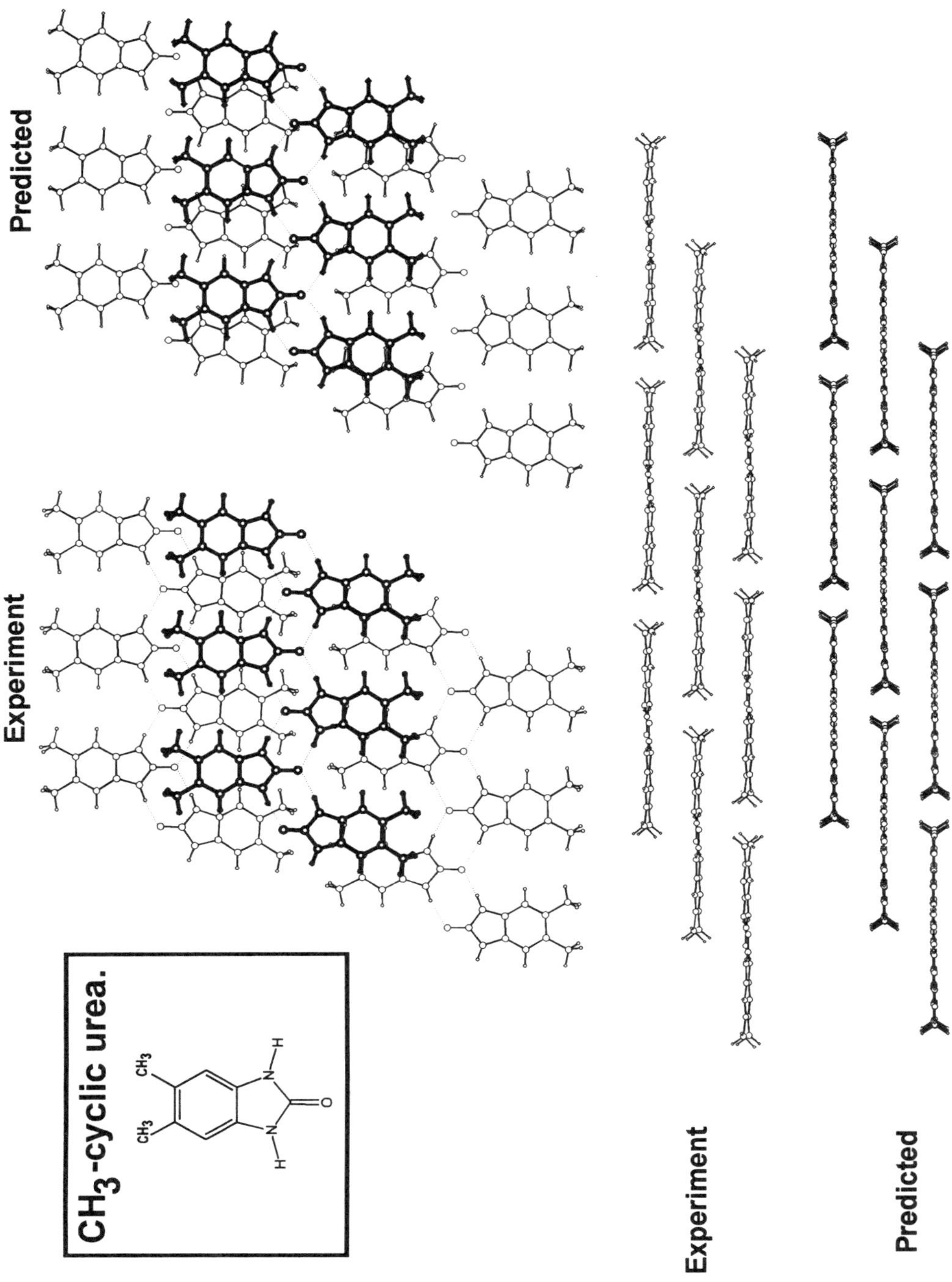

Figure 5.48 Computations successfully predicted the crystal structure of 4,5-dimethyl 2-benzimidazolone

6. CONCLUSIONS

What is the outlook for the approaches to engineering the organic solid state suggested by the work reviewed in this chapter? The answer is, "Still uncertain". There is no doubt that the primary objective—designing and controlling the structures of crystals—is important: it is a broadly relevant problem in molecular recognition; it is a central intellectual challenge in solid-state chemistry; it offers possibilities for applications in areas from optoelectronics to drug-delivery systems. Despite the excellent work in the field, however, there has been no new dramatic advance in technique—such as that provided by the development of self-assembled monolayers in organic surface science, or the introduction of molecular beam epitaxy in semiconductor materials—that has rendered previously very difficult problems much easier.

The lack of revolutionary ideas notwithstanding, some areas of solid chemistry have seen very substantial advances in the last 20 years. X-ray crystallography of small organic molecules is now almost routine, and other techniques—such as solid state NMR spectroscopy and electron diffraction—have also contributed to the solution of structures in the solid state. These techniques have resulted in several crystallographic databases that provide a wealth of information about structure with which to formulate hypotheses. Based on these types of data, the work by Etter, Kitaigorodsky, Desiraju, Kennard, Dunitz and others has provided some very useful approaches to generalizations about the solid state. In our own work, the concepts of space filling and avoiding unfavorable steric interactions have proved to be especially useful approaches to simplifying design. the use of directional intermolecular interactions—such as hydrogen bonds—to limit the range of orientations open to molecules clearly simplifies the problem, and helps to make the goals of crystal engineering more approachable with present techniques. Very substantial difficulties still remain, however. Among them are the following:

- *Crystallization.* It remains difficult to grow diffraction-quality crystals of arbitrary organic compounds. As a result, crystallization is now the rate-limiting step in many investigations in this field, and until there are rapid ways of obtaining the structures of organic solids, the rate of progress in correlating molecular and crystalline structure will be slow.
- *Ranking crystalline arrangements.* The underlying problem is that organic molecules seem to be able to pack in a number of arrangements and conformations with similar energies. Every accessible conformation of a flexible organic molecule is, in principle, the basis of a new solid state structure. The overall energy of a crystal is the sum of many individual small interactions, and for this type of problem—as for the problem of molecular recognition of ligands by proteins—there are currently no search

routines that can sort the possible structures. Also, the problem of estimating the energies of different packing arrangements is too complex to be carried out by a system of simple rules, and computation will be necessary. Although the appropriateness of the commonly used potential functions is now well-established, computation is only just beginning to be useful in rationalizing known structures. Computation has, unfortunately, not yet reached the point of offering useful generalizations about organic molecules of even modest complexity.

- *Polymorphism.* The potential for polymorphism continues to cloud fundamental approaches to the energetics of crystals. There is no way of determining that there is not a lower energy polymorph than the one observed; measuring the thermochemical properties of crystals is difficult; and there is very little understanding of the processes governing the rates of crystal nucleation and growth.
- *Macroscopic properties.* The final objective of the field is to be able to predict the properties of organic crystals, and to design crystals having known properties. It is very difficult, however, to calculate properties accurately, even when the structure of the crystalline solid is known. To reliably predict properties reliably is still some years away.

7. ACKNOWLEDGMENTS

This project, and D.N.C., were supported by The National Science Foundation through Grant CHE-91-22331 to G.M.W., and by the Merck Corporation through a fellowship award to J.C.M.

8. REFERENCES

1. Lehn, J. M. *From Molecular to Supramolecular Nonlinear Optical Properties*, Vol. 455. American Chemical Society, Washington, DC (1991).
2. Ducharme, S., Scott, J. C., Twieg, R. J. and Moerner, W. E. *Phys. Rev. Lett.* **66**, 1846 (1991).
3. Günter, P. and Huignard, J. P. (eds) *Photorefractive Materials and Their Applications*, Vol. 1. Springer-Verlag, Berlin (1988).
4. Günter, P. and Huignard, J. P. (eds) *Photorefractive Materials and Their Applications*, Vol. 2. Springer-Verlag, Berlin (1989).
5. Byrn, S. R., Pfeiffer, R. R., Stephenson, G., Grant, D. J. W. and Gleason, W. B. *Chem. Mater.* **6**, 1148–1158 (1994).
6. Silverman, R. B. *The Organic Chemistry of Drug Design and Drug Action*. Academic Press, New York (1992).
7. Whitesides, G. M., Simanek, E. E., Mathais, J. P., Seto, C. T., Chin, D. N., Mammen, M. and Gordon, D. M. *Acc. Chem. Res.* **28**, 37–44 (1995).
8. Vogtle, F. *Supramolecular Chemistry*. John Wiley & Sons, New York (1991).

9. Ball, P. *Designing the Molecular World.* Princeton University Press, New Jersey (1994).
10. Bernstein, J. (ed.) *Polymorphism in Drug Design and Delivery*, pp. 203–215. Liss, New York (1988).
11. Bernstein, J. (ed.) *Polymorphism and the Investigation of Structure–Property Relations in Organic Solids*, pp. 6–26. Oxford University Press, Oxford (1991).
12. Bernstein, J. *J. Phys. D: Appl. Phys.* **26**, B66–B76. (1993).
13. Byrn, S. R. *Solid-State Chemistry of Drugs.* Academic Press, New York (1983).
14. Haleblian, J. K. *J. Pharmaceut. Sci.* **64**, 1269–1288 (1975).
15. McCrone, W. C. (ed.) *Polymorphism*, Vol. 2, pp. 725–767. John Wiley & Sons, New York (1965).
16. Bernstein, J. (ed.) *Conformational Polymorphism*, Vol. 32, pp. 471–518 . Elsevier, New York (1987).
17. Israelachvili, J. N. *Intermolecular and Surface Forces.* Academic Press, Orlando (1985).
18. Dunitz, J. D. and Bernstein, J. *Acc. Chem. Res.* **28**, 193 (1995).
19. Taylor, R. and Kennard, O. *Acc. Chem, Res.* **17**, 320 (1984).
20. Jeffrey, G. A. and Saenger, W. *Hydrogen Bonding in Biological Structures.* Springer-Verlag, Berlin (1991).
21. Vinogradov, S. N. and Linnel, R. H. *Hydrogen Bonding.* van Nostrand Reinhold, New York (1971).
22. Joesten, M. D. and Schaad, L. J. *Hydrogen Bonding.* M. Dekker, New York (1974).
23. Braga, D. and Grepioni, F. *Acc. Chem. Res.* **27**, 51 (1994).
24. Braga, D., Grepioni, F., Biradha, K., Pedireddi, V. R. and Desiraju, G. R. *J. Am. Chem. Soc.* **117**, 3156 (1995).
25. Desiraju, G. R. *Crystal Engineering: The Design of Organic Solids*, Vol. 54. Elsevier, New York (1989).
26. Desiraju, G. R. (ed.) *Organic Solid State Chemistry.* Elseveir, Amsterdam (1987).
27. Addadi. L., Berkovitch-Yellin, Z., Weissbuch, I., Mil, J. v., Shimon, L. J. W., Lahav, M. and Leiserowitz, L. *Angew. Chem., Int. Ed. Engl.* **24**, 466 (1985).
28. Wheeler, K. A. and Foxman, B. M. *Chem. Mater.* **6**, 1330 (1994).
29. Etter, M. C. *Isr. J. Chem.* **25**, 312–319 (1985).
30. Etter, M. C., MacDonald, J. C. and Bernstein, J. *Acta Crystallogr.* **B46**, 256–262 (1990).
31. Etter, M. C. *Acc. Chem. Res.* **23**, 120–126 (1990).
32. Bernstein, J., Etter, M. C. and MacDonald, J. C. *J. Chem. Soc., Perkin Trans.* **2**, 695–698 (1990).
33. Etter, M. C. *J. Phys. Chem.* **95**, 4601–4610 (1991).
34. Bernstein, J., Davis, R. E., Shimoni, L. and Chang, N.-L. *Angew. Chem. Int. Eng. Ed.* **34**, 1555–1573.
35. Leiserowitz, L. *Acta Crystallogr.* **B32**, 775–802 (1976).
36. Leiserowitz, L. and Schmidt, G. M. J. *J. Chem. Soc. A* 2372–2382 (1969).
37. Aakeröy, C. B., Hitchcock, P. B., Moyle, B. D. and Seddon, K. R. *J. Chem. Soc., Chem. Commun.* 1856 (1989).
38. Reutzel, S. M. and Etter, M. C. *J. Phys. Org. Chem.* **5**, 44–54 (1992).
39. Etter, M. C., Britton, D. and Reutzel, S. M. *Acta Crystallogr.* **C47**, 556–561 (1991).
40. Schwalbe, C. H., Williams, G. J. B. and Koetzle, T. F. *ACA Trans.* **23**, 101 (1987).
41. Etter, M. C., Urbanczyk-Lipkowska, Z., Zia-Ebrahimi, M. and Panunto, T. W. *J. Am. Chem. Soc* **112**, 8415–8426 (1990).
42. Etter, M. C. and Panunto, T. W. *J. Am. Chem. Soc.* **110**, 5896–5897 (1988).

43. Russell, V. A., Etter, M. C. and Ward, M. D. *J. Am. Chem. Soc.* **116**, 1941–1952 (1994).
44. Russell, V. A., Etter, M. C. and Ward, M. D. *Chem. Mater.* **6**, 1206–1217 (1994).
45. Etter, M. C. and Adsmond, D. A. *J. Chem. Soc., Chem. Commun.* 589–591 (1990).
46. Etter, M. C., Adsmond, D. A. and Britton, D. *Acta Crystallogr.* **C46**, 933–934 (1990).
47. Leiserowitz, L. *Acta Crystallogr.* **B33**, 2719–2733 (1977).
48. Etter, M. C., Huang, K. S., Frankenbach, G. M. and Adsmond, D. A. (eds) *Control of Symmetry and Asymmetry in Hydrogen-Bonded Nitroaniline Materials*, Vol. 455, pp. 446–456. Americal Chemical Society, Washington, DC (1991).
49. Etter, M. C., Frankenbach, G. M. and Adsmond, D. A. *Mol. Cryst. Liq. Cryst.* **187**, 25–39 (1990).
50. Etter, M. C. and Reutzel, S. M. *J. Am. Chem. Soc.* **113**, 2586–2598 (1991).
51. Wiedenfeld, H. and Knoch, F. *Acta Crystallogr.* **C46**, 1038 (1990).
52. Emsley, J., Reza, N. M. and Kuroda, R. *J. Cryst. Spectrosc.* **16**, 57 (1986).
53. Harkema, S. and Brake, J. H. M. t. *Acta Crystallogr.* **B35**, 1011 (1979).
54. Harkema, S., Bats, J. W., Weyenberg, A. M. and Feil, D. *Acta Crystallogr.* **B28**, 1646 (1972).
55. Etter, M. C. and Frankenbach, G. M. *Chem. Mater.* **1**, 10–12 (1989).
56. Hummel, G. J. v. and Helmholdt, R. B. *Acta Crystallogr.* **C47**, 213 (1991).
57. Benetollo, F., Bombieri, G. and Truter, M. R. *J. Heterocycl. Chem.* **26**, 981 (1989).
58. Harkema, S., Brake, J. H. M. t. and Helmholdt, R. B. *Acta Crystallogr.* **C40**, 1733 (1984).
59. Bernstein, J. *Acta Crystallogr.* **B35**, 360–366 (1979).
60. Boman, C.-E., Herbertsson, H. and Oskarsson, A. *Acta Crystallogr.* **B30**, 378–382 (1974).
61. Allen, F. H., Kennard, O. and Taylor, R. *Acc. Chem. Res.* **16**, 146–156 (1983).
62. Hart, H., Lin, L.-T. W. and Goldberg, I. *Mol. Cryst. Liq. Cryst.* **137**, 277–286 (1986).
63. Goldberg, I., Lin, L.-T. W. and Hart, H. J. *Incl. Phenom.* **2**, 377–389 (1984).
64. MacDonald, J. C. and Etter, M. C., Unpublished results.
65. Kuroda, Y., Taira, Z., Uno, T. and Osaki, K. *Cryst. Struct. Commun.* **4**, 321–324 (1975).
66. Matias, P. M., Jeffrey, G. A. and Ruble, J. R. *Acta Crystallogr.* **B44**, 516–522 (1988).
67. Craven, B. M., Vizzini, E. A. and Rodrigues, M. M. *Acta Crystallogr.* **B25**, 1978–1993 (1969).
68. Craven, B. M. and Vizzini, E. A. *Acta Crystallogr.* **B27**, 1917–1924 (1971).
69. Etter, M. C. and Huang, K.-S. *Chem. Mater.* **4**, 824–827 (1992).
70. Lechat, J. R., Santos, R. H. d. A. and Bueno, W. A. *Acta Crystallogr.* **B37**, 1468 (1981).
71. Lechat, J. R., Personal communication.
72. Prasad, P. N. and Williams, D. J. *Introduction to Nonlinear Optical Effects in Molecules and Polymers*. John Wiley & Sons, Inc., New York (1991).
73. Etter, M. C. and Baures, P. W. *J. Am. Chem. Soc.* **110**, 639–640 (1988).
74. Semmingsen, D. *Acta Chem, Scan.* **B28**, 169–174 (1974).
75. Singh, I. and Calvo, C. *Can. J. Chem.* **53**, 1046–1050 (1975).
76. Etter, M. C., Urbanczyk-Lipkowska, Z., Jahn, D. A. and Frye, J. S. *J. Am. Chem. Soc.* **108**, 5871–5876 (1986).
77. Etter, M. C., Parker, D. L., Ruberu, S. R., Panunto, T. W. and Britton, D. *J. Inclusion Phenom. Mol. Recogn. Chem.* **8**, 395–407 (1990).

78. Sarma, J. A. R. P. and Desiraju, G. R. *Acc. Chem. Res.* **19**, 222 (1986).
79. Ramamurthy, V. (ed.) *Photochemistry in Organized and Constrained Media.* VCH, New York (1991).
80. Goud, B. S., Panneerselvam, K., Zacharias, D. E. and Desiraju, G. R. *J. Chem. Soc. Perkin Trans.* II 325 (1995).
81. Dhurjati, M. S. K., Sarma, J. A. R. P. and Desiraju, G. R. *J. Chem. Soc., Chem. Commun.* 1702 (1991).
82. Desiraju, G. R. *Acc. Chem. Res.* **24**, 290 (1991).
83. Pedireddi, V. R. and Desiraju, G. R. *J. Chem. Soc., Chem. Commun.* 988 (1992).
84. Desiraju, G. R. *J. Chem. Soc., Chem. Commun.* 454 (1990).
85. Viswamitra, M. A., Radhakrishnan, R., Bandekar, J. and Desiraju. G. R. *J. Am. Chem. Soc.* **115**, 4868 (1993).
86. Desiraju, G. R. and Sharma, C. V. K. M. *J. Chem. Solc., Chem. Commun.* 1239 (1991).
87. Biradha, K., Sharma, C. V. K., Panneerselvam, K., Shimoni, L. C., Carrell, H. L., Zacharias, D. E. and Desiraju, G. R. *J. Chem. Soc., Chem. Commun.* 1473 (1993).
88. Sharma, C. V. K., Panneerselvam, K., Pilati, T. and Desiraju, G. R. *J. Chem. Soc., Chem. Commun.* 832 (1992).
89. Sharma, C. V. K., Panneerselvam, K., Pilati, T. and Desiraju, G. R. *J. Chem. Soc. Perkin Trans. II* 2209 (1993).
90. Desiraju, G. R., Murty, B. N. and Kishan, K. V. R. *Chem. Mater.* **2**, 447 (1990).
91. Thalladi, V. R., Panneerselvam, K., Carrell, C. J., Carrell, H. L. and Desiraju, G. R. *J. Chem. Soc., Chem. Commun.* 341 (1995).
92. Desiraju, G. R. and Kishan, K. V. R. *J. Am. Chem. Soc.* **111**, 4838 (1989).
93. Kishan, K. V. R. and Desiraju, G. R. *J. Org. Chem.* **52**, 4640 (1987).
94. Bertolasi, V., Nanni, L., Gilli, P., Ferretti, V., Gilli, G., Issa, Y. M. and Sherif, O. E. *New J. Chem.* **18**, 251 (1994).
95. Gilli, G., Bertolasi, V., Ferretti, V. and Gilli, P. *Acta Crystallogr.* **B49**, 564 (1993).
96. Bertolasi, V., Gilli, P., Ferretti, V. and Gilli, G. *J. Am. Chem. Soc.* **113**, 4917 (1991).
97. Gilli, G., Bellucci, F., Ferretti, V. and Bertolasi, V. J. *J. Am. Chem. Soc.* **111**, 1023 (1989).
98. Leiserowitz, L. *Acta Crystallogr.* **B32**, 775 (1976).
99. Leiserowitz, L. and Hagler, A. T. *Proc. R. Soc. London A* **388**, 133 (1983).
100. Subramanian, S., and Zaworotko, M. J. *Coord. Chem. Rev.* **137**, 357 (1994).
101. Bishop, R. and Dance, I. G. *Inclusion Compounds*, Vol. 4. Oxford University Press, Oxford (1991).
102. Ung, A. T., Bishop, R., Craig, D. C., Dance, I. G. and Scudder, M. L. *Chem. Mater.* **6**, 1269 (1994).
103. Bishop, R., Craig, D. C., Dance, I. G., Scudder, M. L., Marchand, A. P. and Wang, Y. *J. Chem. Soc. Perkin Trans.* II 937 (1993).
104. Ung, A. T., Bishop, R., Craig, D. C., Dance, I. G. and Scudder, M. L. *Tetrahedron* **49**, 639 (1993).
105. Ung, A. T., Bishop, R., Craig, D. C., Dance, I. G. and Scudder, M. L. *J. Chem. Soc. Perkin Trans.* II 861 (1992).
106. Ung, A. T., Bishop, R., Craig, D. C., Dance, I. G. and Scudder, M. L. *J. Chem. Soc., Chem, Commun.* 1012 (1991).
107. Atwood, J. L., Davies, J. E. D. and MacNicol, D. D. (eds). *Inclusion Compounds*, Vol 1–3. Academic Press, London (1984).
108. Aakeröy, C. B. and Seddon, K. R. *Chem. Soc. Rev.* **22**, 397 (1993).
109. Aakeröy, C. B. and Nieuwenhuyzen, M. *J. Am. Chem. Soc.* **116**, 10983 (1994).
110. Aakeröy, C. B. and Hitchcock, P. B. *Acta Crystallogr.* **C50**, 759 (1994).

111. Aakeröy, C. B., Hitchcock, P. B. and Seddon, K. R. *J. Chem. Soc., Chem. Commun.* **553**, (1992).
112. Aakeröy, C. B., Bahra, G. S., Hitchcock, P. B., Patell, Y. and Seddon, K. R. *J. Chem. Soc., Chem, Commun.* **152**, (1993).
113. Fan, E., Vicent, C., Geib, S. and Hamilton, A. D. *Chem. Mater.* **6**, 1113 (1994).
114. Garcia-Tellado, F., Geib, S. J., Goswami, S. and Hamilton, A. D. *J. Chem. Soc.* **113**, 9265 (1991).
115. Geib, S. J., Vicent, C., Fan, E. and Hamilton, A. D. *Angew. Chem. Int. Ed. Engl.* **32**, 119, (1993).
116. Geib, S. J., Hirst, S. C., Vicent, C. and Hamilton, A. D. *J. Chem. Soc., Chem. Commun.* 1283 (1991).
117. Lauher, J. W., Chang, Y.-L. and Fowler, F. W. *Mol. Cryst. Liq. Cryst.* **211**, 99 (1992).
118. Chang, Y.-L., West M.-A., Fowler, F. W. and Lauher, J. W. *J. Am. Chem. Soc.* **115**, 5991 (1993).
119. Zhao, X., Chang, Y.-L., Fowler, F. W. and Lauher, J. W. *J. Am. Chem. Soc.* **112**, 6627 (1990).
120. Toledo, L. M., Lauher, J. W. and Fowler, F. W. *Chem. Mater.* **6**, 1222 (1994).
121. Simard, M., Su, D. and Wuest, J. D. *J. Am. Chem. Soc.* **113**, 4696 (1991).
122. Wang, X., Simard, M. and Wuest, J. D. *J. Am. Chem. Soc.* **116**, 12119 (1994).
123. Ducharme, Y. and Wuest, J. D. *J. Org. Chem.* **53**, 5787 (1988).
124. Gallant, M., Viet, M. T. P. and Wuest, J. D. *J. Org. Chem.* **56**, 2284 (1991).
125. Russell, V. A., Etter, M. C. and Ward, M. D. *J. Am. Chem. Soc.* **116**, 1941 (1994).
126. Hollingsworth, M. D. and Palmer, A. R. *J. Am. Chem. Soc.* **115**, 5881 (1993).
127. Hollingsworth, M. D. and Cyr. N. *Mol. Cryst. Liq. Cryst.* **187**, 135 (1990).
128. Hollingsworth, M. D., Brown, M. E., Santarsiero, B. D., Huffman, J. C. and Goss, C. R. *Chem. Mater.* **6**, 1227 (1994).
129. Hollingsworth, M. D., Santarsiero, B. D. and Harris, K. D. M. *Angew. Chem. Int. Ed. Engl.* **33**, 649 (1994).
130. Hollingsworth, M. D., Santarsiero, B. D., Oumar-Mahamat, H. and Nichols, C. J. *Chem. Mater.* **3**, 23 (1991).
131. Gavezzotti, A. *Acta Crystallogr.* **B46**, 275 (1990).
132. Gavezzotti, A. *J. Am. Chem. Soc.* **105**, 5220 (1983).
133. Gavezzotti, A. New J. Chem. **6**, 443 (1982).
134. Zaworotko, M. J. *Chem. Soc. Rev.* **23**, 283 (1994).
135. Copp, S. B., Subramanian, S. and Zaworotko, M. J. *Angew. Chem. Int. Ed. Engl.* **32**, 706 (1993).
136. Copp, S. B., Subramanian, S. and Zaworotko, M. J. *J. Am. Chem. Soc.* **114**, 8719 (1992).
137. Christie, S. D., Subramanian, S., Thompson, L. K. and Zaworotko, M. J. *J. Chem, Soc., Chem. Commun.* 2563 (1994).
138. Subramanian, S. and Zaworotko, M. J. *Can. J. Chem.* **71**, 433 (1993).
139. Zerkowski, J. A., Seto, C. T., Wierda, D. A. and Whitesides, G. M. *J. Am. Chem. Soc.* **112**, 9025 (1990).
140. Wang, Y., Wei, B. and Wang, Q. *J. Cryst. Res.* **20**, 79 (1990).
141. Zerkowski, J. A. and Graham, R. Unpublished results.
142. Zerkowski, J. A., MacDonald, J. C., Seto, C. T., Wierda, D. A. and Whitesides, G. M. *J. Am. Chem. Soc.* **116**, 2382 (1994).
143. Zerkowski, J. A., Seto, C. T. and Whitesides, G. M. *J. Am. Chem. Soc.* **114**, 5473 (1992).
144. Zerkowski, J. A. and Whitesides, G. M. *J. Am. Chem. Soc.* **116**, 4298 (1994).

145. Zerkowski, J. A., Mathias, J. P. and Whitesides, G. M. *J. Am. Chem. Soc.* **116**, 4305 (1994).
146. Zerkowski, J. A., MacDonald, J. C. and Whitesides, G. M. *Chem. Mater.* **6**, 1250 (1994).
147. Simanek, E. E., Mammen, M., Gordon, D. M., Chin, D., Mathais, J. P., Seto, C. T. and Whitesides, G. M. *Tetrahedron* **51**, 607 (1995).
148. Bish, D. L. and Post, J. E. (eds) *Modern Powder Diffraction*. Mineralogical Society of America, Washington, DC (1989).
149. Klug, H. P. and Alexander, L. E. *X-Ray Diffraction Procedures*. John Wiley & Sons, New York (1954).
150. Etter, M. C. and Hoye, R. C. *Trans. Am. Crystallogr. Assoc.* **22**, 31 (1986).
151. Etter, M. C., Reutzel, S. M. and Vojta, G. M. *J. Mol. Struct.* **237**, 165 (1990).
152. Etter, M. C. and Vojta, G. M. *J. Mol. Graphics* **7**, 3–11 (1989).
153. Zerkowski, J. A., MacDonald, J. C. and Whitesides, G. M. *Chem. Mater.* **9**, 1933–1941 (1997).
154. Whitesides, G. M., Zerkowski, J. A., Macdonald, J. C. and Chin, D. N. *Mater. Res. Soc. Symp. Proc.* **328**, 3 (1994).
155. MacDonald, J. C. and Whitesides, G. M. *Chem. Rev.* **94**, 383–420 (1994).
156. Timofeeva, T. V., Chernikova, N. Yu. and Zorkii, P. M. *Russ. Chem. Rev.* **49**, 966 (1980).
157. Dauber, P. and Hagler, A. T. *Acc. Chem. Res.* **13**, 105–112 (1980).
158. Guo, Y., Karasawa, N. and Goddard, W. A. *Nature* **351**, 464–467 (1991).
159. Warshel, A. and Lifson, S. *J. Chem. Phys.* **53**, 582–589 (1970).
160. Lacks, D. J. and Rutledge, G. C. *J. Phys. Chem.* **98**, 1222–1231 (1994).
161. Karplus, M. and Kushick, J. N. *Macromolecules* **14**, 325–332 (1981).
162. Lipkowitz, F. and Boyd, D. B. (eds) *Reviews in Computational Chemistry.* VCH Publishers, New York (1990).
163. Kitaigorodsky, A. I. *Molecular Crystals and Molecules*, Vol. 29. Academic Press, New York (1973).
164. Allen, M. P. and Tildesley, D. J. *Computer Simulations of Liquids.* Oxford University Press, Oxford (1987).
165. Kitaigorodsky, A. I. *Organic Chemical Crystallography.* Consultants Bureau, New York (1961).
166. This idea is analogous to that for predicting the packing of rigid polymer chains: Segerman, E. *Acta Crystallogr.* **19**, 789 (1965).
167. Gavezzotti, A. *J. Am. Chem. Soc.* **113**, 4622–4629 (1991).
168. Karfunkel, H. R. and Gdanitz, R. J. *J. Comp. Chem.* **13**, 1171–1183 (1992).
169. Hahn, T. (ed.) *International Tables for Crystallography*, Vol. A. D. Reidel, Dordrecht (1983).
170. Gdanitz, R. J. *Chem. Phys. Lett.* **190**, 391–396 (1992).
171. Karfunkel, H. R., Rohde, B., Leusen, F. J. J., Gdanitz, R. J. and Rihs, G. *J. Comp. Chem.* **14**, 1125–1135 (1993).
172. Perlstein, J. *J. Am. Chem. Soc.* **114**, 1955–1963 (1992).
173. Perlstein, J. *J. Am. Chem. Soc.* **116**, 455–470 (1994).
174. Scaringe, R. P. and Perez, S. J. *J. Phys. Chem.* **91**, 2394–2403 (1987).
175. Scaringe, R. P. *Electron Crystallography of Organic Molecules.* Kluwer, Dordrecht (1990).
176. Schwiebert, K. E., Chin, D. N., MacDonald, J. C. and Whitesides, G. M. *J. Am. Chem. Soc.* **118**, 4018–4029 (1995).
177. Chin, D. N. MRS Transactions, Washington, 1998, in press.
178. Bruccoleri, R. E. and Karplus, M. *Biopolymers* **29**, 1847–1862 (1990).

Index

Index compiled by Geoffrey Jones, Information Index